T0320790

CATIA v5

CATIA v5 is the world's leading 3D CAD engineering and design software, used in a variety of industries to design, innovate, simulate, analyse and manufacture products. *CATIA* is taught at thousands of academic institutions around the globe to teach today the great engineers of tomorrow.

This book is more than an introduction to *CATIA v5* Finite Element Analysis, providing a practical approach to the subject. The basic concepts of finite element analysis (FEA) in *CATIA v5* are explained and augmented with examples and figures for a thorough understanding of the subjects.

This book is intended to be used by students from programs with a mechanical or industrial engineering background, but also by design and control engineers from various industries (automotive, aerospace, military, heavy machinery, medical technology, etc.). These users need to work and verify their 3D parts and assemblies by applying various methods. Among them, the finite element method (FEM) is a very important tool because it provides information on how the stresses are distributed in the component parts, how the loads are applied and the values and orientations of the resulting displacements.

All the content is organized in a logical manner, with chapters that cover both theoretical concepts and practical issues addressed through the use of modelling, assembly and FEA. The presented applications are clearly written and easy to understand, with step-by-step instructions and ample explanations, illustrations and figures. Many of the tutorials start from the beginning, including the parametric modelling of the part and the interpretation of FEM analysis results.

From students to engineers, all are advised to open and follow the pages of this book with interest and perseverance, to patiently go through all the explanations of the presented tutorials, to explore the proposed FEM problems and then to successfully apply the knowledge acquired in their professional activities.

CATIA v5
Practical Studies Using Finite Element Analysis

Ionuţ Gabriel Ghionea

CRC Press
Taylor & Francis Group
Boca Raton London New York

CRC Press is an imprint of the
Taylor & Francis Group, an **informa** business

Designed cover image: Ionut Gabriel Ghionea

First edition published 2024
by CRC Press
2385 NW Executive Center Drive, Suite 320, Boca Raton FL 33431

and by CRC Press
4 Park Square, Milton Park, Abingdon, Oxon, OX14 4RN

CRC Press is an imprint of Taylor & Francis Group, LLC

© 2024 Ionuț Gabriel Ghionea

ISBN: 978-1-032-71164-5 (hbk)
ISBN: 978-1-032-73121-6 (pbk)
ISBN: 978-1-003-42681-3 (ebk)

DOI: 10.1201/9781003426813

Typeset in Times
by KnowledgeWorks Global Ltd.

Access the Instructor and Student Resources/Support Material: https://www.routledge.com/9781032711645

To the past that shaped me, to the present who has patience with me and to the future that awaits me.

To all those who unconditionally believed in me: family, friends, colleagues, editors and readers.

Contents

Preface

CATIA v5 is the world's leading 3D CAD engineering and design software for product excellence. It is used to design, innovate, simulate, analyse and manufacture products in a variety of industries. CATIA is taught at thousands of academic institutions around the globe to train today the great engineers of tomorrow.

This book is more than an introduction to *CATIA v5* Finite Element Analysis, providing a practical approach to the subject. The basic concepts of FEA in *CATIA v5* are explained and augmented with examples and figures for a thorough understanding of the subject.

The author wrote this book to be part of a series of tutorial books that covers both basic and advanced concepts and functionalities of CATIA v5, the modern CAD software solution.

Thus, this tutorial book is intended to be used by students from faculties with a mechanical or industrial engineering profile, but also by design and control engineers from various industries (automotive, aerospace, military, heavy machinery, medical technology, etc.). These users need to work and verify their 3D parts and assemblies by applying various methods. Among them, the finite element method (FEM) is a very important tool because it provides information on how the stresses are distributed in the component parts, how the loads are applied and the values and orientations of the resulting displacements.

Thus, FEM should be used for different calculations in all the engineering fields. This method and the programs that include it have become fundamental components of the modern CAD systems. It is true that FEM and other simulation methods are indispensable in all engineering activities where high performance is necessary.

The purpose of this book is not to be a finite element simulation course. Readers should have a minimum knowledge of FEM and strength of materials to understand certain concepts that are presented in the introductory chapters and throughout the applications. Also, the author assumes that readers have a good experience in modelling parts and establishing constraints for mechanical assemblies.

The aim of this book is, however, to present the FEM tools and ways of working in CATIA v5 to help the reader understand the steps to follow in order to create a correct and appropriate computation model for each studied problem.

The book is organized in a logical manner, with chapters that cover both theoretical concepts and practical issues addressed through the use of modelling, assembly and finite element analysis. The applications presented are clearly written and easy to understand, with step-by-step instructions and ample explanations, illustrations and figures. Many of the tutorials start from the beginning, including the parametric modelling of the part and the interpretation of FEM analysis results.

Moreover, the author helps readers download important files that contain the 3D models of the parts and assemblies considered in the applications of this book. CATIA version V5R21 or higher is required to open these files.

Basic parametric modelling is presented (in short) in this book. It is very important for readers to know how to create variants of the studied problems, but also these become ideas to continue the applications with other values for the applied loads, changes in the dimensions of the parts and certain constraints that can be applied during the FEM analyses.

The results are presented in different forms, through values, tables, various representations of models, so that the reader can understand how each problem was approached and the correct interpretation of the results.

Whether the readers are beginners or have experience in using CATIA v5, following all written tutorials will help them to understand, upgrade and improve their knowledge and then to apply proved 3D modelling and simulation methods, to become familiar with many new operations and options.

Based on the author's 20+ years of teaching experience, he structured and wrote this book focussing on FEM simulation of many interesting and common type parts. Each tutorial is a challenge for the reader and gradually presents different and important FEM options and tools, carefully explained and accompanied by clear figures, with annotations for a better understanding of the context. Thus, the author used numerous graphical representations, drawings, dialog boxes, icons of the tools, etc. Text and figures support the reader in understanding the approach and highlight all important selections (geometric elements or options).

Although many theoretical aspects are briefly mentioned to solve the considered FEM problems in an easy-to-understand manner, the book does not present all available commands and their options. Therefore, the reader is encouraged to further explore important options encountered in the dialog boxes and then to search for some new solutions.

From students to engineers, all are advised to open and follow the pages of this book with concern and perseverance, to patiently go through all the explanations of the presented tutorials, to explore the proposed FEM problems and then to successfully apply the knowledge acquired in their professional activity.

The author made a consistent effort and passionately created all the written and video tutorials presented in this book, with great attention to detail. Several people, experts in other CAD programs and good engineers, improved the book with ideas and valuable observations. Many friends and colleagues from university and industry supported the author with patience and interest in writing this book. Taylor & Francis/CRC Press publisher also guided the author with interest and professionalism to create the manuscript that will provide a great experience for the book's readers.

All the people mentioned above, regardless of their contribution level, acted as a team that made possible the writing, proofreading, editing, printing and publishing of this book. The author thanks everyone for their time and feedback.

We all hope that this book, with its content, will rise to the level of exigency set from the very beginning and be useful to all those who will have the curiosity and need to open it and expand their knowledge.

The author also challenges readers to share other solutions for the presented problems, links to their own video tutorials and to contribute with their ideas and suggestions towards improving the content of the future editions of this book.

Ionuţ Gabriel Ghionea
Bucharest, Romania
November 28, 2023

About author

Ionuţ Gabriel Ghionea is an Associate Professor, member of the Manufacturing Engineering Department, Faculty of Industrial Engineering and Robotics, National University of Science and Technology Politehnica Bucharest since 2000. In 2003 he completed an internship to prepare his doctoral thesis at the École Nationale Supérieure d'Arts et Métiers in Aix-en-Provence, France, and received a PhD in industrial engineering in 2010.

Ionuţ Gabriel Ghionea published, as first author or co-author, 12 books in the field of computer-aided design for mechanical engineering and over 125 papers in scientific technical journals and conference proceedings. He is a member of the editorial board and review panel of several international and national journals and conferences.

In addition, he participated in research projects with applicability in industry and education. He is one of the most known and appreciated CAD trainers in Romania, carrying out this activity since 2002.

Ionuţ Gabriel Ghionea is considered to be one of the main and first didactic promoters of the *CATIA v5* program in the Romanian academic environment. He created and recorded hundreds of video tutorials, wrote books and practical works for students and taught CAD courses for companies in Romania and abroad.

In December 2020 he became a *CATIA Champion,* a recognition issued by the Dassault Systèmes for his long-term commitment in promoting and using *CATIA v5.*

CATIA Champion
CATIA Certified Professional Part Design Specialist
Contact: ionut76@hotmail.com
https://upb.ro/en, http://www.fiir.pub.ro
http://www.tcm.pub.ro
www.linkedin.com/in/ghionea-ionut-gabriel
http://www.catia.ro
https://orcid.org/0000-0002-9062-0064

1 Introduction

INTRODUCTION

The main aim of any engineering activity is the creation of products, in all their diversity, followed by complex production processes. The main stages of a product development consist of defining a general concept, writing the technical project, establishing the manufacturing technology, making the experimental model, its mass production, ensuring a service and maintenance network. During the main steps of creating a product, the *finite element analysis* is very important and it's used in the validation phases, before having the physical prototype.

This introductive chapter briefly presents the importance of the finite element method and its role in the conception and development phases of a product.

1.1 INTRODUCTION TO FEM

The *finite element method* (FEM) is one of the best methods for performing various calculations and simulations in many engineering fields, especially when a mechanical part or assembly needs validation. This method and, of course, the software solutions that incorporate it, have become basic components of modern *computer-aided design* (CAD) systems. Analyses performed by FEM are, today, indispensable in all high-performance engineering activities. Using the *FEM* makes it possible to numerically solve differential equations that appear in engineering and mathematical modelling. Typical problem areas of interest include the traditional fields of structural analysis, heat transfer, fluid flow, mass transport and electromagnetic potential.

The FEM is a general numerical method for solving partial differential equations in two or three space variables. To address and solve a problem, the FEM subdivides a large system into much smaller and simpler parts that are called *finite elements*. This is achieved by the construction of a mesh for the simulated model/object; it has a finite number of points.

The FEM formulation of a boundary value problem finally results in a system of algebraic equations. The method approximates the unknown function over the domain. The simple (but many!) equations that model/rule these finite elements are then assembled into a much larger system of equations to deal with the entire problem. The FEM then approximates a solution by minimizing an associated error function via the calculus of variations.

Studying or analysing a phenomenon with FEM is often referred to as *finite element analysis* (FEA).

1.2 BASIC CONCEPTS

FEM allows detailed visualization of areas where structures are weak, bend or twist, indicating the distribution of stresses and displacements and providing a wide range of simulation options for controlling of both modelling and analysis of a system, no matter its complexity.

FEM helps entire projects to be virtually built, refined and optimized before the manufacturing phase. The mesh is created automatically or by the user. It must be carefully applied and used to give the best results. This involves a good discretization strategy, meaning a clearly defined set of procedures that cover the creation of finite element meshes, the definition of basis function on reference elements (also called shape functions) and the mapping of reference elements onto the elements of the mesh. Examples of discretization strategies are the *h*-version, *p*-version, *hp*-version, isogeometric analysis, etc. with certain advantages and disadvantages. A reasonable criterion in selecting a discretization strategy is to realize nearly optimal performance for the broadest set of mathematical models in a particular model class.

Usually, a higher number of elements in a mesh may conduct to a more accurate solution of the discretized problem. However, there is a value at which the results converge and further mesh refinement does not increase accuracy; it only needs more time to be computed.

The subdivision of a complex model into smaller and simpler parts/elements has some advantages: accurate representation of complex geometries, inclusion of material properties, fast and easy representation of the whole solution, determination and capture of local effects.

Typical appliance of the FEM consists of dividing the domain of the problem into a complex group of subdomains, each one being represented by a set of element equations to the original problem, systematically recombining all sets of element equations into a global system of equations for the final calculation. This global system of equations has known solution techniques, and can be calculated from the initial values of the original problem to obtain a numerical answer. The element equations are simple equations that approximate at the local level the complex equations to be studied. These are often partial differential equations. The FEM uses approximation and it is commonly introduced as a special case of the Galerkin method. As it is known, in mathematics, in the area of numerical analysis, Galerkin methods convert a continuous operator problem, such as a differential equation that has a weak formulation, into a discrete problem by applying linear constraints determined by finite sets of basis functions.

The process consists of setting an integral of the inner product of the residual and the weight functions and then equalling the integral to zero. Thus, it is a procedure that minimizes the error of approximation by fitting trial functions into the partial differential equations. The residual is the error caused by the trial functions and the weight functions are polynomial approximation functions that project the residual. The process eliminates all the spatial derivatives from the partial differential equations. These equations may be approximated locally with a set of algebraic equations for steady-state problems or with a set of ordinary differential equations for transient problems. The element equations are linear if the underlying partial differential equation is linear, and vice versa. Algebraic equation sets used in the steady-state problems are solved using numerical linear algebra methods, whereas ordinary differential equation sets used in the transient problems are solved by numerical integration using standard techniques such as Euler's method or the Runge–Kutta method.

Further, a global system of equations is generated from the element equations through a transformation of coordinates from the subdomains' local nodes to the domain's global nodes. This spatial transformation includes appropriate orientation adjustments as applied in relation to the reference coordinate system. The process is often carried out by FEM software using coordinate data generated from the subdomains.

The application of FEM in practice, known as FEA, is used in engineering as a computational tool for performing various analysis. FEA uses mesh generation techniques for dividing a complex problem into smaller elements by using a software system fitted with FEM abilities. The complex problem is usually a physical system with the underlying physics such as the Euler–Bernoulli beam equation, the heat equation, or the Navier-Stokes equations expressed in either partial differential equations or integral equations, whereas the divided small elements of the complex problem represent different local areas in the physical system.

1.3   TYPES OF FINITE ELEMENT METHODS

Many engineering disciplines from automotive, naval, aerospace, biomechanical, consumer goods, etc. industries use integrated FEM in the design, test and development of their products. Several modern CAD packages include specific FEM abilities to set up mechanical, thermal, electromagnetic, fluid and structural working environments. In a structural simulation, FEM is a great help in producing stiffness and strength visualizations and in reducing weight, materials and costs.

Being used in such diverse engineering fields, FEM has developed in several branches and directions, the best-known being:

AEM – the *applied element method* combines features of both FEM and *discrete element method* (DEM);

A-FEM – the *augmented-finite element method* is introduced by Yang and Lui whose goal was to model the weak and strong discontinuities without the need for extra degrees of freedom (DoF) as in *partition of unity method* (PUM) stated;

GFEM – the *generalized finite element method* uses local spaces consisting of functions, not necessarily polynomials that reflect the available information on the unknown solution and thus ensures good local approximation. Then a partition of unity is used to link these spaces together to form the approximating subspace. The effectiveness of GFEM has been shown when applied to problems with domains having complex boundaries, problems with microscales or with boundary layers;

MFEM – the *mixed finite element method* is a type of FEM in which extra independent variables are introduced as nodal variables during the discretization of a partial differential equation problem;

hp-FEM – the *variable-polynomial method* combines elements with variable size *h* and polynomial degree *p* in order to achieve fast, exponential convergence rates;

hpk-FEM – this method combines elements with variable size *h*, polynomial degree of the local approximations *p* and global differentiability of the local approximations ($k - 1$) to achieve best convergence rates;

XFEM – the *eXtended finite element method* is a numerical technique based on the *GFEM* and the *PUM*. *XFEMs* enrich the approximation space so that it can reproduce the challenging feature associated with discontinuities, singularities, boundary layers, etc. For some problems, such as embedding of the problem's feature into the approximation space, can significantly improve convergence rates and accuracy. Moreover, addressing problems with discontinuities with XFEMs suppresses the need to mesh and re-mesh the discontinuity surfaces, thus attenuating the computational times, costs and projection errors associated with conventional FEMs, at the cost of restricting the discontinuities to mesh edges;

SBFEM – the introduction of the *scaled boundary finite element method* came from Song and Wolf in 1997 and has been one of the most profitable contributions in the area of numerical analysis of fracture mechanics problems. It is a method which combines the advantages of both the finite element formulations and procedures and the boundary element discretization;

S-FEM – the *smoothed finite element method* is a particular class of numerical simulation algorithms for the simulation of physical phenomena. It was developed by combining mesh-free methods with the finite element method;

SEM – the *spectral element method* combines the geometric flexibility of finite elements and the accuracy of spectral methods. Spectral methods are the approximate solution of weak-form partial equations that are based on high-order Lagrangian interpolants and only used with certain quadrature rules; and

CPFEM – the *crystal plasticity finite element method* is a versatile advanced numerical tool developed by Franz Roters to describe the mechanical response of crystalline materials on all length scales from single crystals to engineering parts. CPFEM based on slip (shear strain rate) can calculate dislocation, crystal orientation and other texture information to consider crystal anisotropy during the routine and it is applied in the numerical study of material deformation, surface roughness, fractures, etc.

1.4 FEM INTEGRATION IN CAD SOFTWARE SYSTEMS

CAD is a creative activity, with many implications in other disciplines. To solve complex problems of analysis for parts and assemblies, the design engineer must have all the information necessary to formulate the problem numerically in a very accurate mode.

If only the qualitative conditions are given, without the quantitative information, it is expected that an unsatisfactory solution will be obtained, even if only from some points of view.

The main purpose of such an approach is to obtain the best solution for a set of conditions imposed. Thus, the engineer will design a virtual system and study its behaviour.

Generally, in engineering, and especially in the fields of machine manufacturing, mechanical devices, equipment and installations, the basic component of a system analysed by FEM is a rigid structure. It is defined as a mechanical assembly with a very clearly established functionality, such as dealing with various loads, moments, and accelerations, ensuring a certain functionality or movement between some subassemblies, getting a static and/or dynamic stability, guaranteeing a rigidity imposed by the design engineer, etc.

Strength, stability and durability are all part of the required characteristics of a part or assembly that consists of components or subassemblies.

The calculations performed by FEM represent a very important stage in the design of a product, but can be performed only after clarifying other aspects, such as requirements of the client, imposed costs, delivery times, available materials and technologies, product durability, production volume, manufacturing capabilities, available resources, etc.

Thus, for a certain product some restrictions can be taken into account, such as maximum number and value of static and/or dynamic loads, maximum values of deformations, different safety coefficients (for buckling, breaking or fatigue), imperfections in design, assembly or operation, vibration frequencies, product life, weight, material used for parts, moments of inertia, stiffness at various stresses, static and/or dynamic stability, behaviour at different simultaneous loads, etc.

Also, some issues will be taken into account, such as different breaking modes and overloads that appear during transport, assembly or operation whose values, frequencies and cause are not known in advance.

In the field of CAD and computer-aided manufacturing (CAM), the FEA method is becoming an important component of an integrated process.

It should be noted that in the sequence CAD-FEM-CAM, there is an iterative process of design-calculation-manufacturing. In this process, the synthesis and analysis operations of the prototype and of the model for the calculation with finite elements are performed successively (Figure 1.1).

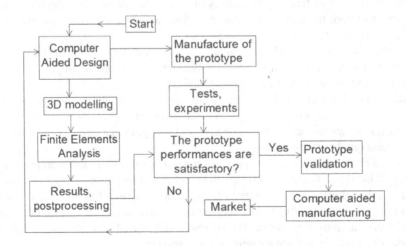

FIGURE 1.1 Simplified diagram of CAD-FEM-CAM integration in the creation of products.

At each iteration of the process, improvements are made to the prototype or computational model, until the desired or closest performance is achieved.

The FEA of a three-dimensional (3D) model is, in fact, a numerical verification calculation. Thus, for a certain dimensionally defined geometry, for a given load and well-specified restrictions, the values of displacements, stresses, reactions in supports, frequencies, etc. are obtained.

It is not, however, obvious in most cases how the model needs to be modified in order to meet the set of requirements in the best way possible. Although there are many attempts, a general automatic optimization technique is not yet available to solve any problem of any kind, but an optimal design methodology can be developed.

The CAD solutions that incorporate FEM contain special optimization procedures that allow the determination of a good solution by automatic computation of the optimal values for some design parameters, to satisfy a set of conditions imposed on an objective function, defined by the user. Also, the designing engineer can permanently change the geometric shape of the 3D model, the material from which it will be manufactured and, of course, the loads to which it will be subjected in this stage of development.

In conclusion, this FEM powerful design tool has significantly improved both the standard of engineering designs and the methodology of the design process in many engineering applications and fields. The introduction of FEM has substantially decreased the time to conceive products from idea and concept to the production line and market launch.

Benefits of FEM include increased accuracy, enhanced design of multiple solutions, better insight into critical design parameters, virtual prototyping, fewer hardware prototypes, a faster and less expensive design cycle, increased productivity and a shorter time to obtain the product.

2 Basic Concepts in the Finite Element Analysis Method

INTRODUCTION

This chapter presents the main basic concepts in the *finite element analysis* method that are met in the *CATIA v5* program and the process is described briefly. The structure, the computational model, the network of nodes and elements, the finite element, advantages and disadvantages of using such simulation tools are all presented. These are all important in establishing the mesh for each three-dimensional (3D) model created by the user, in the computational phase, in viewing the result and drawing conclusions.

2.1 ABOUT FINITE ELEMENT ANALYSIS AND ITS BASIC CONCEPTS

One of the major advantages of the finite element analysis (FEA) method is the simplicity of the basic concepts. It is very important for the user of a finite element method (FEM) program to understand and apply these concepts correctly because they include certain hypotheses, simplifications and generalizations. If some of these are ignored, significant errors may occur in product modelling and then in their FEM analysis. This chapter briefly presents the most important basic concepts:

- *Structure.* In order to obtain a high efficiency, in the field of FEA, the concept of 'structure' is used in a more general and simpler way than usual. In FEM, 'structure' is a set of bars, plates, shells and (solid) volumes. Thus, for example, the landing gear of an airplane, the mechanical hand of a robot, the casing of an engine, a network of pipes, etc. can be considered structures.

  Defined in this way, the notion of structure implies the acceptance of Bernoulli's planar section hypothesis for bars and Kirchhoff's rectilinear normal hypothesis for plates and shells. The simplification consists in replacing the real external forces with loads (axial forces, shear, bending or torsional moments) with some that are statically equivalent to them. Of course, this equivalence is not valid in the theory of elasticity. In the analysis of structures, the concept of 'concentrated force' can be introduced, without producing fields of stresses, deformations or displacements with singularities, as they are found in the theory of elasticity, when the application of a concentrated force in a point of the elastic half-space (theorem of Boussinesq) causes infinite stresses and displacements at that point.

  Also, such a structure concept allows the use of unit displacement, unit force and minimal mechanical work of displacement (Maxwell and Castigliano's theorems – 'The partial derivative of the strain energy, considered as a function of the applied forces acting on a linearly elastic structure, with respect to one of these forces, is equal to the displacement in the direction of the force of its point of application'). These theorems have a very good applicability in the field of strength of materials and the theory of structures, but not in the theory of elasticity.

- *Computational model.* The first step in FEA is to create a computational model of the analysed structure. The FEM models are approximate mathematical models, but at present there are no general algorithms and methods to develop a unique model that approximates 100% or with a known error and as small as possible, the real structure.

  Generally, in practice, several models can be imagined and developed for the same structure, all being correct, but with different performances. This development of the model is based on the intuition, experience and imagination of the engineer, because the model

DOI: 10.1201/9781003426813-2

must be well enough designed to effectively synthesize all available information about the structure analysed.

- *Network of nodes and elements (mesh). Discretization.* The model of the analysed structure consists of lines (axes of the structure bars), of flat or curved surfaces (the median surfaces of the component plates of the structure) and volumes (massive bodies of the structure). At this stage of development, the model is continuous, with an infinite number of points, like a real structure. The purpose of FEA is, first of all, the discretization of the model. This means obtaining a clear and finite mesh, transforming the real, continuous structure into a discrete model, with a certain number of points named nodes. This operation is accomplished by 'covering' the model with such a mesh, being justified by the idea that, from an engineering point of view, the loads and deformations in a certain number of points inside the structure are sufficient to accurately characterize its mechanic behaviour.

  The FEM defines the unknown (displacements and stresses) only in the nodes of the created model and calculates their values at the respective points. This is the main reason the discretization process must be performed in such a way as to obtain a sufficiently large number of nodes in the areas of interest, and, thus, to obtain a satisfactory approximation of the structure geometry, the edge areas and the loading conditions.

- *Node.* The points resulting from the discretization operation are part of a mesh network and they are named nodes. These points are defined as the unknown primary nodes, whose values are the results of FEA. Depending on the type of the unknowns, the model obtained by FEM can take several names, as follows: if the unknowns associated with the nodes are displacements, a displacement model will result, and if the unknowns are stresses, there will be a stress model. For the displacement model, it is admitted that the deformed shape of the structure, as a result of some stress, is defined by the displacements of all nodes in relation to their network before deformation.

  Each node can have a maximum of six components of the displacement, called nodal displacements, in relation to a global mark: three components u, v, w of the linear displacement and three rotations φx, φy and φz.

  Thus, the geometric degrees of freedom of the structure as a whole can be defined. Some of the degrees of freedom of the model must be eliminated because some nodes are connected to the supports, having zero displacements or have known imposed values, and it is no longer necessary to calculate them.

- *Finite element.* The discretization process results in the division of the model structure into a certain number of fragments or elements. Those are called finite elements and connect all characteristic points (nodes) that lie on their circumference. This 'connection' is a set of equations named shape functions. Each finite element has its own set of shape functions that connect all of the nodes of that element). Adjacent elements share common nodes (the ones on the shared edge). This means that the shape functions of all the elements (Figure 2.1) in the model are 'tied/connected' thanks to those common nodes.

Generally, there are several types of finite elements and they can be divided into one-dimensional (1D) elements (called beams or bars), two-dimensional (2D) elements (called shells or plates) and three-dimensional (3D) elements (called solids), represented in Figure 2.2. The 1D element is just a line connecting two nodes. Two-dimensional elements are usually triangular (*tri*) and quadrilateral (*quad*) in shape. Three-dimensional solid elements can be based on triangles and quads as well.

Thus, any 3D model can be discretized into a number of prismatic and tetrahedral elements that are connected by common nodes, such as the tips of tetrahedra or prisms. From this point of view, a finite element can be seen as an independent component, interacting with the other elements only in nodes.

The study of the real structure is replaced by the study of the complex set of finite elements obtained by discretization, which becomes an approximation of the real structure, being a calculation model of it. For the results of the analysis to be as accurate as possible, the process of

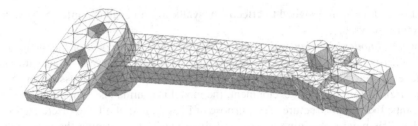

FIGURE 2.1   Example of a discretized 3D model.

discretization of the structure must be as efficient as possible, which involves compliance with rules and requirements for discretization, the development of the calculation model and the use of appropriate finite elements. In principle, the dimensions of finite elements can be as small as they are, but they must always be finite, and their dimensions cannot be close to zero.

The inherent limitations of FEM programs do not allow the design of a general finite element, with universal utility, and in order to be implemented and used as a calculation model, the finite element must be previously designed in all its details, completely defined geometrically, physically, by mathematics, etc.

From an informational point of view, a finite element is a model that must be able to accurately process as much information as possible for an imposed set of conditions. This assumes that the element of a certain geometric shape has as many nodes as possible, each node has many geometric degrees of freedom, and the interpolation functions are as complex as possible, with a large number of parameters. Of course, as the complexity of the finite element increases, so do the computational difficulties, for each concrete situation the best solution is sought when creating a finite element of a certain type.

For any finite element, it must be taken into account the hypothesis that the displacements within it vary according to an a priori known law, determined by means of an interpolation function. The consequence of this approach is that, locally, where the finite element will be placed, following the discretization process, the state of displacement of the structure will be approximated by the interpolation law implemented in the respective element.

The interpolation functions most often take the form of polynomials, the choice of the degree of the polynomial and the determination of the values of its coefficients must ensure the best possible approximation of the exact solution, unknown at the moment, of the given problem.

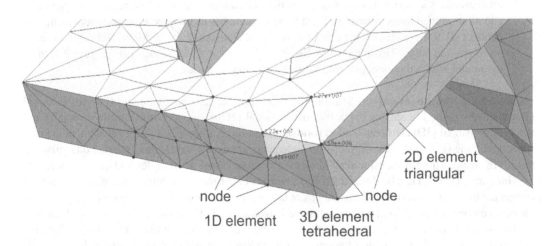

FIGURE 2.2   Detail of a discretized 3D model with nodes and elements.

Finite elements are classified according to the following criteria:

- *Type of analysis.* Finite elements can be defined on a discretization network that includes some mathematical procedures for various analyses: linear elastic, nonlinear, heat transfer, fluid mechanics, high-frequency electromagnetism, etc.;
- *Functional role.* The finite elements used to model a structure must be able to ensure its functional role as well as possible. Thus, the elements are of the point type (mass or arc element), of the line type (straight or curved bar type elements, in plane or in space), of the surface type (flat or curved plate elements, thick or thin, in plane or in space) or of volume type (spatial, 3D elements for solid structures, with a variable number of nodes). There are also finite elements with a special functional role: rigid, contact, friction, connection, defined by the rigidity matrix, etc.;
- *Geometric shape.* The finite elements generally have simple shapes, such as a straight line or arc of a circle, triangle, tetrahedron, hexahedron, etc. Also, some geometric features can be constant or variable (bars sections or plates thicknesses);
- *Number of nodes.* For some of the elements, a given shape can have several variants in terms of the number of nodes, so each discretization variant will use a certain number of nodes, the higher it is, the more finite elements will better cover the surface of the 3D model;
- *The number of degrees of freedom of each node.* The nodes of the elements have attached, implicitly, some degrees of freedom out of the six possible ones, so the user can also operate with the total number of degrees of freedom for a finite element, defined as the product between the number of nodes and the number of degrees of freedom per node;
- *The degree of the interpolation polynomial.* Each finite element incorporates interpolation polynomials of a certain degree, starting with the first degree. Of course, the higher the degree of the polynomials, the higher the amount of information with which the element operates and therefore the more efficient/precise it is; and
- *Material characteristics.* In the practice of FEA, the material of the model can be homogeneous and isotropic or with an anisotropy of a certain type. Also, the elastic and physical constants of the material may be temperature and/or load-dependent.

Any finite element is a set of conditions and assumptions; it must be regarded as a whole and used as such. Among the parameters that define the element results include its behaviour on request, the type of stresses state, the interaction with other elements, etc.

In a FEA, a 3D model has far more than a single finite element, and these elements share nodes. This means that one single node can have several 'coupled equations sets' it's involved with. Simply, each node can belong to more than one finite element, and each element will put its own shape functions onto the node.

According to Figure 2.3, element A shape functions are 'independent' from element B shape functions, because it's a different set of coupled equations. This means that both elements A and B will 'interpret' what is going on in the common nodes (2 and 6) in a different way.

When a computer-aided design (CAD)-FEM program is applied on a 3D model, it creates all the 'single element' shape functions and inserts them into the stiffness matrix for each element separately. When the FEM solver finishes the simulation task, it results in many small stiffness matrixes, one for each element. But these matrixes are not 'connected' together yet.

This connection happens, due to the common nodes between elements. Solver assumes that deformations/displacements in any given node should be the same, regardless from which element shape functions are used.

The assembly of the global stiffness matrix, created from the small element stiffness matrixes, is possible because the deformation of nodes is 'constant' across elements.

It means that, first, the solver gets the deformations at the nodes. Then knowing what is the deformation between nodes, using element shape functions, it can calculate the strain in the element. If

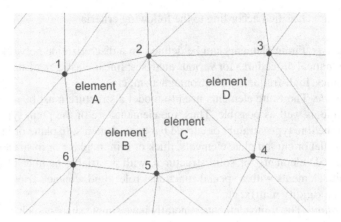

FIGURE 2.3 Adjacent finite elements that use the same nodes.

nodes 2 and 4 get closer to node 5 it means that the element C is compressed in a specific direction. Because it is known how much closer those nodes got together and what their relative position was, it is possible to know the 'reduction' of the element C. Using these displacement values, strains and stresses are calculated for the complete structure of the 3D model.

2.2 THE FINITE ELEMENT ANALYSIS PROCESS

The most common finite element technique is displacement-based. In this approach, displacement is assumed to be an unknown quantity. The problem is solved using FEM to find out displacements. The overall process can be subdivided into several main steps:

- *Pre-processing* (modelling of a finite element). In this step, the real, physical structure is converted into a 3D model (Figure 2.4), then into an equivalent finite element model (Figure 2.5) and problem at the same time. The user defines the material properties for this model. Physical forces, moments, accelerations, etc. are converted into equivalent loads (Figure 2.6) and the physical boundary conditions (Figure 2.7) are converted into equivalent conditions for the finite elements;
- *Computation* (solving the finite elements model). The finite element problem stated in the pre-processing step is solved to find out the unknown displacement values;
- *Post-processing* (to view the simulation results). Using these displacement values, strains and stresses are computed for the whole structure. The engineer can view and study the deformation of structure, variation of strains and stresses (Figure 2.8) all over the structure; and
- *Mesh refinement* (multiple iterations). The first solution provides an initial estimation of stress/strain values. If this is considered to be sufficiently accurate, no other iteration (computation) is needed. To get a more accurate solution (Figure 2.9), the mesh requires refinement by a new iteration of computation. A number of mesh refinement and multiple computation iterations are performed until the required solution accuracy is obtained.
     Figure 2.9 presents a refined mesh of a simple two-part assembly, with a larger number of finite elements that better cover the solids. After computation, new values are obtained, different from those in Figure 2.8. Also, the representation of the stressed areas is better and more precise.
- *Creating reports*. Once the required accuracy level is achieved, various data such as displacement, principle stress and *von Mises* stress can be obtained in a concise form. So, a report file is generated using text, tables with values, figures, etc. a fragment is represented in Figure 2.10.

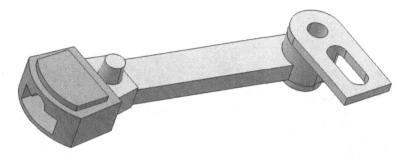

FIGURE 2.4 Three-dimensional model of the two-part assembly to be analysed.

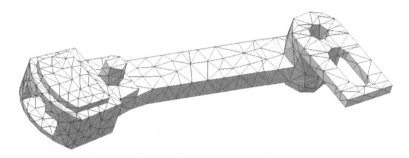

FIGURE 2.5 Finite element model representing the assembly.

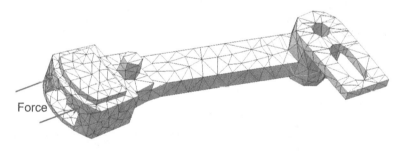

FIGURE 2.6 Applying a force load on the finite element model.

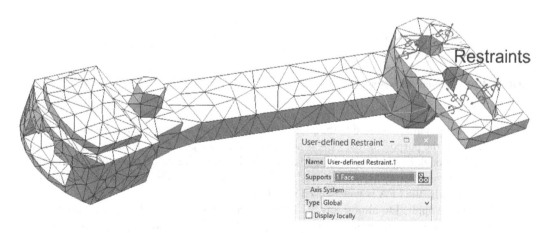

FIGURE 2.7 Applying boundary conditions to the finite element model.

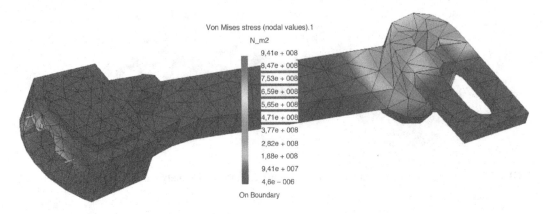

FIGURE 2.8   Representation of the von Mises stresses with values.

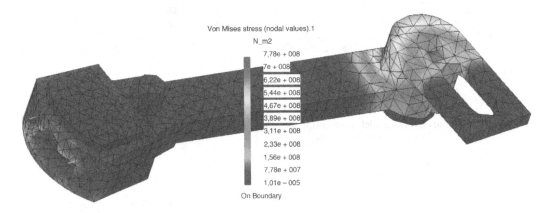

FIGURE 2.9   Representation of the von Mises stresses with values after the mesh refinement.

Equilibrium				
Components	Applied Forces	Reactions	Residual	Relative Magnitude Error
Fx (N)	-1.2806e-009	3.9940e-009	2.7134e-009	3.8474e-012
Fy (N)	-5.0000e+003	5.0000e+003	-5.5934e-010	7.9309e-013
Fz (N)	3.7253e-008	-3.8295e-008	-1.0420e-009	1.4775e-012
Mx (Nxm)	8.5714e-001	-8.5714e-001	-3.7570e-011	8.1395e-013
My (Nxm)	4.3876e-009	-4.4435e-009	-5.5931e-011	1.2117e-012
Mz (Nxm)	2.3182e+002	-2.3182e+002	-7.8160e-011	1.6933e-012

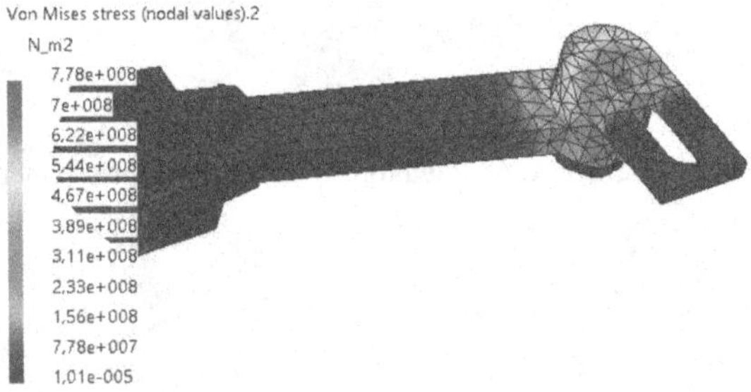

FIGURE 2.10   Fragment of a generated report.

The reports are presented to the project manager who will decide if the results are satisfactory, if the purpose of the FEA is achieved and if changes in the geometry of the analysed 3D model, its material and the loads to which it is subjected are necessary. Through several complex iterations, the expected results are obtained and the 3D model is validated.

This is why the FEA is widely used in industry:

- FEA can be applied to almost any 3D model placed in a system and having an arbitrary shape including various boundary and loading conditions. This flexibility is not possible or very difficult with classical analytical methods;

- The user becomes able to verify a proposed product or structure in a virtual manner long before manufacturing or construction. The product or the structure is asked by the customer who is not always an engineer, so his specifications may not be complete and/or engineering correct. By FEA, these requirements may be adjusted and verified in several iterations, starting with the CAD phase to manufacturing with minimum costs and in a very short time. This minimizes the product life cycle time significantly; and
- The user can evaluate the advantages and effectiveness of various product design alternatives without having any kind of experimental test setup. As an example, with FEA software tools, it is possible to optimize the product for minimum weight and volume with small costs, thus increasing product life and improving its reliability.

2.3 ADVANTAGES, DISADVANTAGES AND LIMITATIONS OF THE FINITE ELEMENT ANALYSIS METHOD

Currently, the FEM is almost widespread in assisted engineering design and has multiple applications in mechanical research, heat transfer, hydraulics, biomechanics, etc.

Among the many advantages of FEA, these are some of the most important:

- FEA is general, it is an approximate numerical calculation method that can be used to solve problems of deformable structure mechanics, fluid mechanics, heat transfer, etc. The loads applied in these systems can be static, dynamic, periodic, stationary, non-stationary, transient, etc.;
- *Flexibility.* In order to approach a certain concrete problem with FEM, there are no restrictions derived from the method. The elaboration of the calculation model of the given problem can be with minimal constraints, the user experience being very important. The flexibility of the FEM ensures the very easy elaboration of the calculation model and its complex use, regarding the boundary and loading conditions and the analysis options. The target is to obtain new, improved variants of the initial model/product, fully satisfying the various requirements of the user;
- *Simplicity of basic concepts.* To use FEA procedures in the CAD environment it is not necessary for the user to have special knowledge of mathematics or computer science (although it would still be very useful for understanding the phenomena!), but only to be a good engineer to understand the basic concepts of FEM;
- *Use of computers.* The FEM requires a very large number of calculations and interpolations, almost impossible to perform without the help of a high-performance computer. The way the method was conceived and implemented made it possible to run FEM programs on personal computers, making it very accessible to engineering companies; and
- *Existence of programs that have implemented FEM.* At present, there are many software programs available on the market dedicated to the FEM, but also CAD-dedicated programs in which FEM has been implemented. These modern programs allow the modelling and analysis of any mechanical structure and of high complexity from all points of view (geometric shape, dimensions, stresses, variants of analysis, etc.).

Among the disadvantages and limitations of the method are the following:

- *The method is approximate.* The analysis with the help of FEM is done for a calculation and virtual model and not on the real structure, the obtained results representing an approximation of the states of displacements, stresses, temperatures, etc. The truth is that the exact solution of an analysed problem cannot be estimated with precision, there are some deviations of the values computed with FEM from the real values, and are still unknown;

- *The computation model is subjective and maybe arbitrary.* The user of this method has to elaborate the model in the best way, according to his knowledge. The FEM is not imposing any restriction, being just a tool in his hands. The accuracy of the model thus depends on the user's experience and intuition. Many companies, that analyse with finite elements in the field of activity, use norms and rules to create good models with many inputs and details, based on their experience after the tests validated by the real practice;
- *The elaboration of the computation model is quite laborious.* To achieve such a complete and viable model, the user of an FEM program must have a solid knowledge of mechanical engineering and material's strength. Often, the activity involves advanced CAD abilities to obtain the model, according to certain strict rules;
- *Limitations of FEM programs.* Regardless of the degree of generality and performance of such a program, there are some limitations which allow the use of only certain types of finite elements and procedures for analysis. However, the software companies, based on the comments received from customers and following developments and evolutions in the field, offer a new version or various updates to the current one every few months; and
- *Resources of computer systems.* Running FEM programs in optimal conditions requires high-performance computing systems with good processors and serious storage space, but they can also be overcome in the case of very complex models.

In conclusion, the application of FEA in all engineering fields that require FEM simulations is a modern trend and often an irreplaceable phase to shorten the time to obtain products, from idea, conception, design, testing and validation, prototyping, to series production, and launch on the market and good sales.

FEA requires experienced specialists, with knowledge in related fields of engineering, with initiative and desire for innovation, who are able to understand and correctly interpret the results obtained by FEM simulating, and then apply them to optimize the product.

3 CATIA Generative Structural Analysis

Interface, Toolbars, Theoretical Aspects

INTRODUCTION

This chapter presents the interface of the *CATIA v5 Generative Structural Analysis* workbench, its toolbars and the most used tools, different ways to set up restraints, to apply loads, to create good meshes of finite elements, to create reports, to save data, etc. The tools are explained using figures and brief examples, values, details, comments and theoretical aspects. Thus, the user will learn how to deal with parts and assemblies. This chapter is also a good starting point to understand how to solve the applications presented in the Chapter 4.

3.1 INTERFACE OF THE GENERATIVE STRUCTURAL ANALYSIS WORKBENCH

The *CATIA v5 Generative Structural Analysis* workbench is accessed from the *Start → Analysis & Simulation → Generative Structural Analysis* menu, as shown in Figure 3.1.

The interface is very intuitive, as in any other workbench of the *CATIA v5* program. The tools are organized simply and logically on the specific toolbars, but, unlike other workbenches of the program, they are not found in the menus.

This interface (Figure 3.2) contains a work area that contains the specification tree, the toolbars and a three-dimensional (3D) representation of the part/assembly to be analysed by the finite element method (FEM).

In the interface, some icons may or may not be active depending on the current working phase, such as the pre-processing phase (finite element geometry is defined, also the restraints and loads for the analysed model), processing phase (the numerical solution of the problem is offered, on the basis of the computations) and the postprocessing phase (the results of the performed analysis are displayed).

CATIA v5 provides and helps the user to work with and understand several types of finite element analysis: *Static Case, Frequency Case, Buckling Case, Combined Case, Static Constrained Modes, Preprocessing Case, Solution Case* and *Envelop Case*.

The first step in approaching a new analysis is to choose its type from the *Insert* menu (Figure 3.3), but, when launching the workbench, the static analysis type *(Static Case)* is predefined/selected by default. As an immediate result, the specification tree (Figure 3.4) is completed with the *Static Case* feature.

By expanding the *Static Case* feature (click on the icon with the + symbol), its features are displayed: *Restraints, Loads, Static Case Solution* and *Sensors* (Figure 3.2). These features have particularly important roles in finite element analysis. Of course, at the beginning of the analysis, the features do not contain values, the user must set the restraints and the loads for the *CATIA v5* program to compute/calculate the solutions and activate the sensors.

For example, Figure 3.2 shows three possible solutions *(Deformed Mesh, von Mises Stress* and *Estimated Local Error)*. They can be active and/or inactive simultaneously, the result of the analysis being presented on screen.

DOI: 10.1201/9781003426813-3

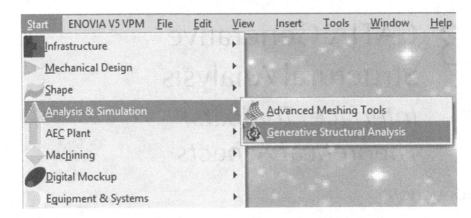

FIGURE 3.1   Accessing the *CATIA v5 Generative Structural Analysis* workbench from the *Start* menu.

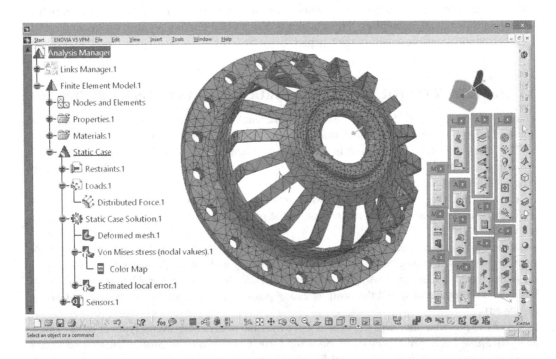

FIGURE 3.2   *CATIA v5 Generative Structural Analysis* workbench interface.

3.2 ESTABLISHING THE PARAMETERS FOR FINITE ELEMENT ANALYSIS

The main condition for a 3D model to be analysed using the FEM is that the user establishes a material for the model, according to its functional role.

For example, if the 3D model was created in the *CATIA v5 Part Design* workbench, before saving and inserting it in an analysis, the user must click on the *Apply Material* icon. As a result, the *Library* dialog box opens (Figure 3.5). It contains the standard library of available materials, all with a set of physical and mechanical parameters, which will be used in the analysis.

Having the 3D model thus parameterized with a material, the *CATIA v5 Generative Structural Analysis* workbench is launched and the type of static analysis is established.

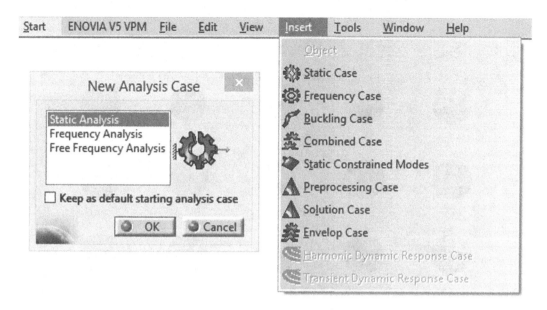

FIGURE 3.3 The *Insert* menu and its options to start an analysis.

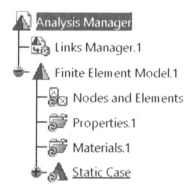

FIGURE 3.4 The specification tree for a finite element model.

By choosing the *Static Case* option from the *Insert* menu, the program opens the dialog box with the same name (Figure 3.6), used to create a new set of restraints, loads and masses or to use an existing one (as a reference) from a previous analysis.

Of course, that analysis and the current one must be in the same file, with the user able to select items of the current set from the previous set.

Pressing the *OK* button the user creates a new set of parameters in the specification tree, as part of the *Static Case* feature.

Next, the user will double-click the *Static Case Solution* feature in the specification tree to open the *Static Solution Parameters* dialog box, shown in Figure 3.7. There are four finite element analysis methods, as follows:

- *Auto*: The program will automatically choose one of the following three methods;
- *Gauss*: Direct method, recommended for processing small and medium models;
- *Gradient*: Method of iterative processing of very large models, does not require much memory, but uses a lot of the processor computing power; and
- *Gauss R6*: fast *Gauss* method, recommended for large models.

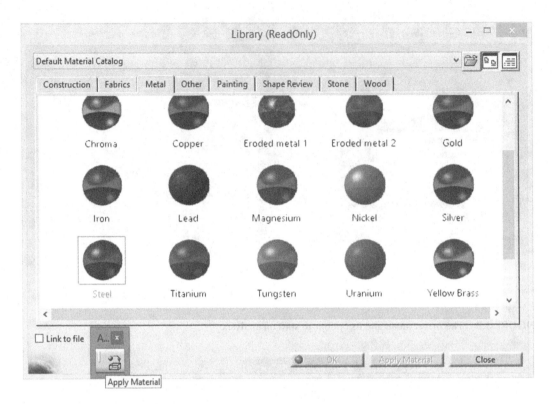

FIGURE 3.5   Dialog box with the list of available materials organized in a library by categories.

To perform the discretization of the 3D model and, thus, to define the network of nodes and elements (mesh), the user clicks on the *Octree Tetrahedron Mesher* icon on the *Model Manager* toolbar, displaying the dialog box (Figure 3.8). The size of the finite element (in the field *Size*), the maximum tolerance between the real model and the discretized model used in the analysis (in the field *Absolute sag*), the type of the element (*Element type*), but also other settings are established.

The minimum size of the finite element is 0.1 mm and size of the tolerance is 0.001 mm. The element type can be *Linear* (four nodes, one *Gauss* point and three degrees of freedom – translations) or *Parabolic* (ten nodes, four *Gauss* points and three degrees of freedom – translations). It follows that the *Parabolic* discretization is more accurate, but also more resource consuming.

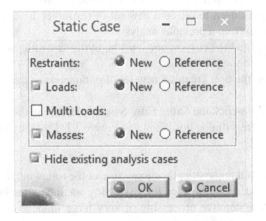

FIGURE 3.6   Establishing the set of restraints, loads and masses.

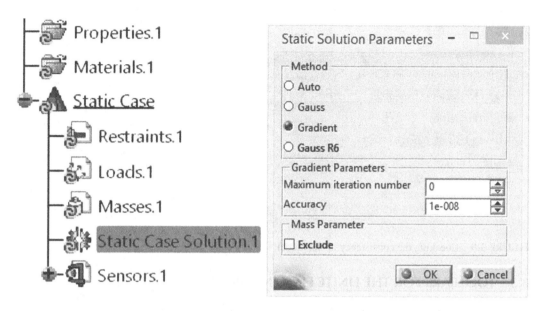

FIGURE 3.7 Choosing the method of analysis.

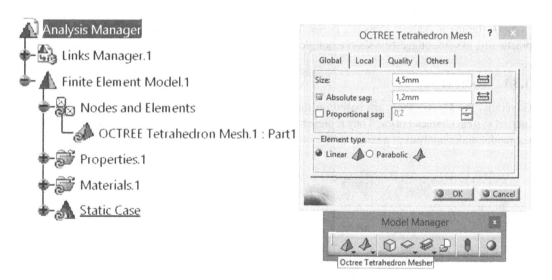

FIGURE 3.8 Defining the network of nodes and elements.

The definition of the finite element can be done globally, for the whole 3D model, but also locally, only for a certain area of it, about which the user wants to know more information and whose values are more accurate.

After applying these parameters, the user can check the consistency (correctness) of the model by clicking on the *Model Checker* icon on the same *Model Manager* toolbar. As a result, the dialog box with the same name opens (Figure 3.9). The figure shows the name and type of the model, the fact that it was discretized, its material and status.

After checking, if the definition of the model receives the status *OK* the user can advance to the next stages of finite element analysis, to create restraints and loads to obtain its solutions. The tools needed for the analysis are available on the toolbars presented in the next subchapter.

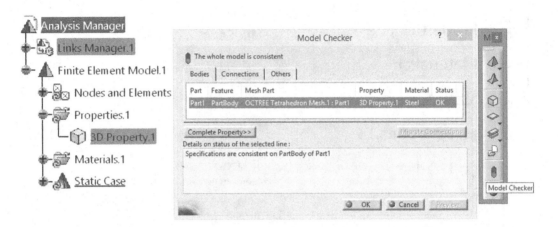

FIGURE 3.9 Checking the consistency of the model.

3.3 TOOLBARS FOR THE FINITE ELEMENT ANALYSIS

Numerous toolbars are available to the user, through the interface or through the menu *View →
Toolbars* (Figure 3.10), as follows: *Model Manager, Restraints, Loads, Masses, Adaptivity, Analysis
Supports, Virtual Parts, Compute, Analysis Tools, Image, Groups, Solver Tools, Analysis Results*
and *Connection Properties*.

Next, the main analysis tools are presented: theoretically, through the possibilities of use and
practically, in some suggestive short examples.

3.3.1 TOOLBAR *MASSES*

Generally, the *Masses* tools are based on the theory of distributed masses, used to model purely
inertial system characteristics. They represent scalar fields of mass, of a certain intensity, applied to
the model under analysis at a single point, on an edge or on a surface.

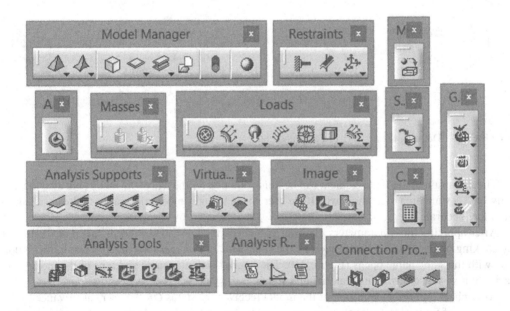

FIGURE 3.10 Toolbars used in finite element analysis.

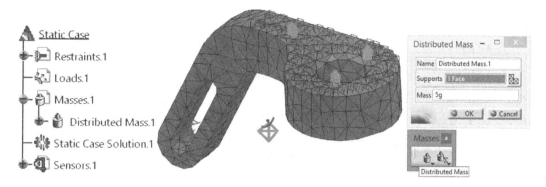

FIGURE 3.11   Applying a mass distributed on the face of a part.

This toolbar has six icons: *Distributed Mass, Line Mass Density, Surface Mass Density, Distributed Mass and Inertia, Combined Masses* and *Assembled Masses*. The unit of mass is in accordance with the international system of measurement, in kilograms (by default).

To exemplify the use of *Masses* tools, the user clicks on the *Distributed Mass* icon, the dialog box with the same name is displayed, the *Name* field contains the name of the load, in the *Supports* field are selected the elements to which the mass will be applied and in the field *Mass* is entered its value in grams.

Of course, the user can create a multiple selection of the elements to which the mass will be applied by holding down the *Ctrl* key during the selection. In Figure 3.11, together with the *Distributed Mass* dialog box, it is shown a support type part, having a distributed mass on its upper flat face. Graphic elements of cylindrical shape and green colour are observed on the selected face.

3.3.2   Toolbar *Virtual Parts*

A virtual part is considered to be a structure that does not have a geometric support, thus representing bodies for which no geometric shape is available, but with an important role in the analysis of a model (part or assembly).

Virtual parts are used to transmit action (forces, restraints, etc.) at a distance, they are considered rigid bodies, unless an elasticity is explicitly introduced by means of a spring. The point from which the virtual part transmits action to the model is called the reference point, handle point, or handle node. Such a point is established using the existing geometry of the model (edge or face). In fact, the whole virtual part is substituted by this point.

The *CATIA v5* software package allows the creation of several types of virtual parts: *Smooth, Contact, Rigid, Rigid Spring* and *Smooth Spring*, a short description of each is given below:

- *Smooth*. A rigid body without mass that connects a specified point to a certain geometry of the model, without stiffening the body or bodies to which it is attached. This virtual part takes into account the elastic deformation of the models to which it was attached;
- *Contact*. A rigid body without mass that connects a specified point to a certain geometry of the model, without stiffening the body or bodies to which it is attached, but prevents interference with it. Also, this virtual part takes into account approximately the elastic deformation of the models to which it was attached;
- *Rigid*. A rigid body without mass that connects a specified point to a certain geometry of the model, simultaneously stiffening the body or bodies to which it is attached. Rigid virtual parts do not take into account the elastic deformation of the models to which they were attached;
- *Rigid Spring*. An elastic body that connects a specified point to a certain geometry of the model, it behaves like a spring with six degrees of freedom, serially with a rigid body without mass, simultaneously stiffening the body or bodies to which it is attached. Like

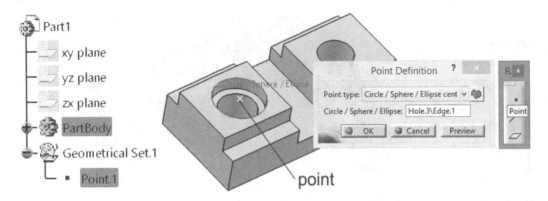

FIGURE 3.12   Creation of a point in the centre of the hole at the top surface.

the *Rigid* virtual part, it does not take into account the elastic deformation of the models to which it was attached; and

- *Smooth Spring.* An elastic body that connects a specified point to a certain geometry of the model, it behaves like a spring with six degrees of freedom, serially with a rigid body without mass, without stiffening the body or bodies to which it is attached. This virtual part takes into account the elastic deformation of the models to which it was attached.

For example, it is considered a bridle type part, on which a *Smooth* virtual part will act. The *Virtual Parts* tools definition requires the selection of a previously created point in the workbench in which the 3D model was drawn.

Thus, in the *CATIA v5 Part Design* workbench, using the *Point* tool on the *Reference Elements* toolbar, a point will be created in the centre of the hole, on the upper flat face of the part. Figure 3.12 shows the *Point Definition* dialog box, how to create the point (*Point Type* field – *Circle/Sphere/ Ellipse center* option), the selection of the circle (the projection of the hole on that face), whose centre will be taken and a preview the position of the newly created point.

Once the point is created, in the *CATIA v5 Generative Structural Analysis* workbench, the user clicks on the *Smooth Virtual Part* icon to choose the face of the virtual part and the reference point. Thus, the support face is the contact face between the virtual part and the 3D model. For example, a force applied at the reference point will act on the part.

In the dialog box displayed in Figure 3.13, in the *Supports* field the user chooses the upper flat face of the part, and in the *Handler* field the point previously created, as a reference point of the

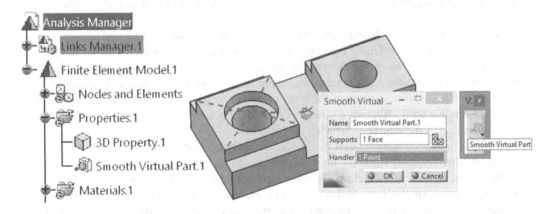

FIGURE 3.13   Creating the virtual part.

virtual part. If the point is missing or the user does not select it, *CATIA v5* will automatically consider the centre of that face (in this case the intersection of the diagonals of the rectangular face).

At this point, forces and moments, restraints can be placed to simulate a rigid body acting on the analysed model of the bridle.

Also, the specification tree in Figure 3.13 is completed with the *Smooth Virtual Mesh* and *Smooth Virtual Part* features under *Nodes and Elements* and *Properties*.

3.3.3   TOOLBAR *RESTRAINTS*

Restraints (or constraints) must be defined to represent a structural interface between the 3D model under analysis and the assembly of which it belongs. If the model is not correctly constrained, many numerical problems (singularities) will result, and finite element analysis will not be possible.

The tools belonging to this toolbar add restraints to the analysed model, thus having the role of taking over some degrees of freedom (translations and rotations). The program provides the following tools:

- *Clamp.* Takes all degrees of freedom for a certain selected geometry (curve or surface type). Any point belonging to such a geometry will remain fixed during the analysis;
- *Surface Slider.* Creates a restraint that allows points placed on a deformable surface to slide along a second rigid surface, coinciding with the first surface. At each point on the deformable surface, the program automatically generates a restraint that takes over the translation degree of freedom in the direction of the normal to the surface at that point;
- *Slider.* Creates a restraint of type prismatic coupling applied to the reference point of the virtual part, so that it moves along an axis;
- *Sliding Pivot.* Creates a restraint of type cylindrical coupling applied to the reference point of each virtual part, so that it performs two movements simultaneously: translating along a considered axis and rotating around it. The restraint takes over all degrees of freedom of the point, except those of translation and rotation for the respective axis;
- *Ball Joint.* Creates a restraint of type spherical coupling applied to the reference point of the virtual parts so that it rotates around a coincidentally fixed point. All degrees of freedom of type translation of the point are taken over;
- *Pivot.* Restraints of type conical coupling (hinge) applied to the reference point of the virtual parts, so that it rotates around an axis. All degrees of freedom of the point are taken over, except for one rotation;
- *User Defined.* Creates generic restraints allowing the fixing of any combination of nodal degrees of freedom on any type of geometry. Thus, there are three translation degrees of freedom for each node of continuous finite element and a combination of three translation degrees of freedom with three of rotation for each node of structural finite element;
- *Isostatic.* Creates statically defined restraints to support a body. The program automatically chooses three points and takes some degrees of freedom, so that the body can no longer perform translation and rotational movements, however, without overconstraining it.

For example, it is presented the creation of two restraints, one of the *Clamp* type, the other of the *Sliding Pivot* type. The *Clamp* restraint requires a model of one part created in a solid modelling workbench, and the *Sliding Pivot* restraint requires a virtual part. The model resulting from the example in Figure 3.13 will be used.

The user clicks on the *Clamp* icon, and in the dialog box with the same name displayed (Figure 3.14), in the *Supports* field, he chooses one flat surface *(1 Face)* placed on the side of the rectangular part.

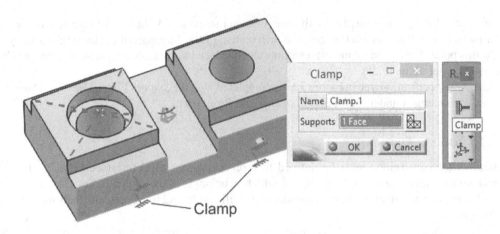

FIGURE 3.14   Applying a *Clamp* restraint.

In the *Name* field, the user has the option to enter a name for the created restraint, instead of the default one provided by the program. The symbols of this restraint are also displayed on that surface.

For the other restraint, the user clicks on the *Sliding Pivot* icon and a dialog box with the same name (Figure 3.15) opens. In the *Supports* field, he chooses the virtual part *(Smooth Virtual Part)*, previously created on the model.

The *Name* field contains the editable name of the restraint, in the *Axis System* area the user can define his own axis system or he can keep the global one, provided by default by the *CATIA v5* program.

The *X*, *Y* and *Z* fields of the *Released Direction* area contain the values corresponding to the components of the sliding pivot direction, relative to the selected axis system. These fields can take values in the range [–1, 1].

In the current considered example, the user entered the value 1 for the *Z* direction, the pivot axis becoming vertical, oriented in the positive direction of the respective axis. The direction of the pivot can also be set with the compass, the values of the fields correspond to the directions of its axes.

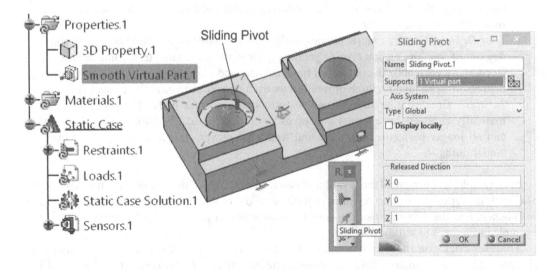

FIGURE 3.15   Applying a *Sliding Pivot* restraint.

The figure is completed with this restraint symbol, and the *Clamp* and *Sliding Pivot* features are added to the specification tree in *Static Case → Restraints.1*.

3.3.4 TOOLBAR *LOADS*

These tools add various loads to the models that will be analysed by the FEM, representing an essential step in the analysis process. These virtual loads simulate the real loads to which the model will be subjected during its operation.

CATIA v5 allows loads such as pressure, distributed force, concentrated force, moment, acceleration, etc. providing the following tools:

Pressure is defined as a load, expressed in N/m^2, that creates a uniform pressure, applied on a certain surface, so that the directions of the forces exerting the pressure are permanently perpendicular to surface. Applying a pressure involves using the *Pressure* icon and the dialog box with the same name opens. In the *Name* editable field the user can enter the name of the pressure load, in the *Supports* field he chooses the surface of the model on which the pressure will be applied, and in the *Pressure* field enters its actual value (positive or negative).

Thus, in the example in Figure 3.16 it is presented a part with a cavity that can be *Clamp* restrained at its base planar surface. Inside the cavity there are several surfaces, the user loads them with a pressure of 40 N/m^2. This load can be applied simultaneously, with the same value, on multiple surfaces of the model by holding down the *Ctrl* key during selection. The specification tree is completed with the *Pressure.1* feature under *Static Case → Loads.1*.

Distributed Force is, in fact, a static equivalent force system with a resultant of a real force, applied at a certain point to a model or to a virtual part. The user specifies the magnitude (size and direction) of the force and the values of its components in the directions of the *X*, *Y* and *Z* axes. Any change in these values automatically leads to the updating of the magnitude and of the components. The values, expressed in Newtons (N), can be positive or negative, the minus sign having the meaning of changing the direction of the vector of the respective component.

Figure 3.17 considers an example, using a hinge model. The user clicks on the *Distributed Force* icon to open a dialog box with the same name. In the *Supports* field, the user chooses two curved surfaces of the model, uses the global axis system (*Axis System* area) and then sets the value of the force vector and its components in the three directions *X*, *Y* and *Z*.

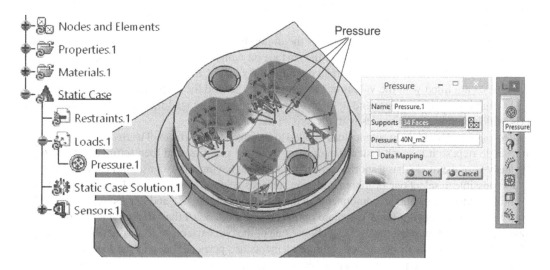

FIGURE 3.16   Applying a *Pressure* load.

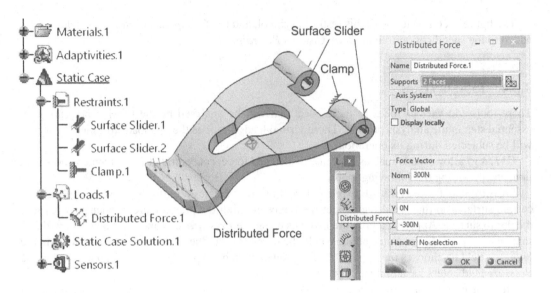

FIGURE 3.17 Applying a *Distributed Force* load.

The application point of this load (*Handler* field) is computed by default by the program as the centroid of the selected face or faces, in the case of models with their own geometry, in the reference point for virtual parts or it can be defined by the user in the *CATIA v5 Part Design* workbench by adding a point anywhere on the considered surface (Figure 3.18).

Visually, as in the case of pressure loads, the distributed forces are represented by arrows on the considered surface, the meaning of an arrow is that of the force vector. After loading, the specification tree is completed with the *Distributed Force* feature.

Moment is a static force system, equivalent to a resultant torque (N × m) distributed on a model with defined geometry or on a virtual part, being applicable on elements of type point, vertex, edge and face. The user specifies three components for the direction of the resulting moment, but also its value, any change having the effect of automatically updating them. The values can be positive or negative, thus determining the vector direction of the respective component.

In Figure 3.19 is an example of applying a moment load to a shaft. A *User-defined Restraint* is applied to one end of the shaft to take over all the degrees of freedom of the finite elements nodes at that end.

The user clicks on the *Moment* icon and the dialog box with the same name will be displayed, in which he selects in the *Supports* field the surface on which the *Moment* load is applied. Also, a global axis system is used, the *Norm* field contains the value of the resulting moment, and the X, Y and Z fields are filled with the values of the components around the respective axes.

As can be seen in the figure and in the *Moment* dialog box, the loading takes place around the Y axis, which in this case coincides with the axis of the shaft.

Bearing Load simulates a contact load applied to the revolution surfaces. The *Bearing Load* is done in a single step, so it is much faster than creating a virtual part first and then setting the load. Also, the analysis of the model has a shorter duration, while also having a low consumption of resources.

It is possible to select one or more revolution surfaces belonging to the model, then to specify the resulting contact force and its components in the directions of the X, Y and Z axes, using a global or user system. The *Bearing Load* has more options (see Figure 3.20) than the previous loads, as follows: in the *Orientation* list the user can choose one of the two variants of force distribution (*Radial*, when all force vectors in the nodes of finite elements are perpendicular to the loaded surface or *Parallel*, when they are parallel to the resulting vectors).

The *Angle* field contains the value corresponding to the angle at which the force is distributed (180° by default), but it is possible to enter values in the range (0°… 360°). Low values for this angle

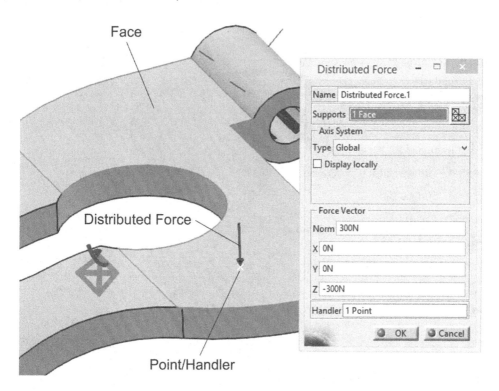

FIGURE 3.18 Applying force in a chosen point on the planar face of the part.

FIGURE 3.19 Applying a *Moment* load.

mean a concentrated load, only in a certain area, and high values a diffuse load, distributed over a larger area of the surface.

The *Profile* field contains three variants (*Sinusoidal, Parabolic* and *Law*), specifying how the force intensity will vary depending on the previously established angle.

The *Distribution* field, through its two variants *(Outward* and *Inward)* allows the user to specify the force distribution, according to Figures 3.21 and 3.22.

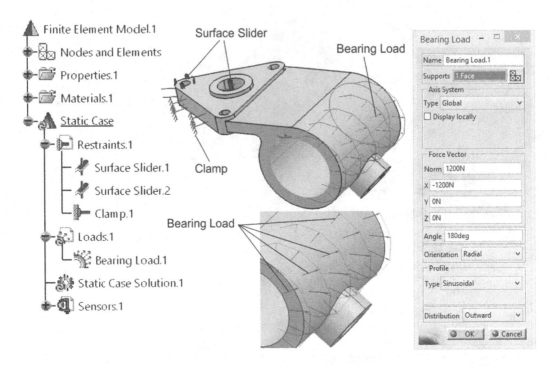

FIGURE 3.20   Applying a *Bearing Load*.

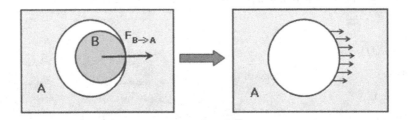

FIGURE 3.21   *Outward* distribution.

Considering element A as the analysed model and element B the virtual part, from Figure 3.21 it is observed that B acts on A, and from Figure 3.22 that A acts on B. The orientations of the force vectors for the *Parallel* variant are also presented schematically. However, in both cases, the *Bearing Load* acts and will only deform element A, the model.

For example, Figure 3.20 shows a model, a support part with a cylindrical surface and a hole, to which three restraints were attached: two *Surface Slider* and one *Clamp*. Then, through the *Bearing*

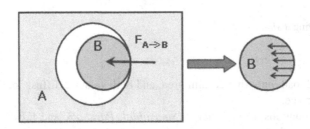

FIGURE 3.22   *Inward* distribution.

Load dialog box, a load was created on the cylindrical inner surface, with the parameters displayed in the respective fields, in the *Force Vector* area.

Also, the corresponding graphic elements (arrows) are arranged on the loaded surface, and the specification tree is completed with the *Bearing Load* feature.

The *Imported Force* and *Imported Moment* tools allow the user to add new forces and moments from text files with tabular data and from *Microsoft Excel* files. The data files must be created before the import and should contain 3D coordinates of the points where the loads will be applied together with their values on the coordinate axes.

Acceleration represents concentrated loads of acceleration field type (N/kg or m/s^2) with uniform intensity, applied to the analysed models or to virtual parts. The user will specify the values of the three components of the acceleration field, but also the resulting value. Any subsequent changes will automatically update them.

The example of Figure 3.23 is considered a fork-type part, with a *Clamp* restraint, applied to the left end, on some flat surfaces. The user clicks on the *Acceleration* icon and, in the available dialog box, he can change the load name *(Acceleration.1)*. The user selects the body of the model and the *Supports* field is filled in with the value *1 Body (on publication)*.

In the fields of the *Acceleration Vector* area, the values of the acceleration vector and its components are set. The plus or minus signs of the value influence the direction of the vector of that component. In the specification tree is added the *Acceleration* feature and its corresponding graphic elements that are placed on the model body.

Rotation Force represents concentrated loads of types of acceleration fields, induced by uniform or accelerated rotational movements, applied to the analysed models or virtual parts. Thus, the user must specify an axis of rotation, the values of angular velocity (rad/s) and angular acceleration (rad/s^2), the program automatically calculating the linear variation of the distribution of the acceleration field.

A new part of type bladed rotor is opened and in the *CATIA v5 Generative Structural Analysis* workbench, the user clicks the *Rotation Force* icon to display the dialog box with the same name (Figure 3.24), in which he selects the model body, thus completing the *Supports* field with the value *1 Body (on publication)*.

For the *Rotation Axis* field, the circular edge of the rotor hub is chosen, the force symbol is displayed together with an axis, then in the *Angular Velocity* and *Angular Acceleration* fields the user enters the values for velocity and acceleration, respectively.

Following the current load, a centrifugal force is created which will act on the analysed model, keeping its geometric shape in the area immediately adjacent to the axis of rotation.

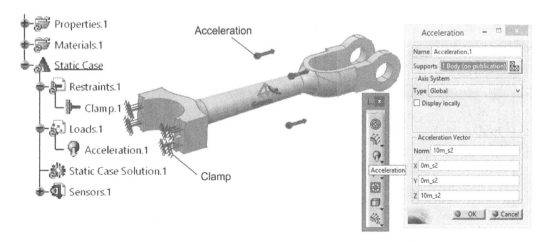

FIGURE 3.23   Applying an *Acceleration* load.

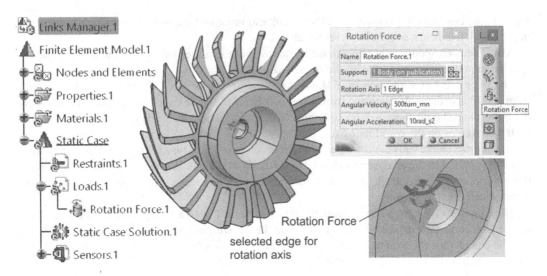

FIGURE 3.24 Applying a *Rotation Force* load.

Line Force Density represents concentrated loads (N/m) of types of linear traction forces field, having a uniform intensity, applied on the edges of the analysed models. Thus, only the edges of an existing geometry are allowed to be selected, not the spline lines or curves created later on the model surface.

The user will specify three components of the linear force field in the directions of the X, Y and Z axes, along with its resulting value. Changing any of the four values automatically leads to updating the other three.

The example in Figure 3.25 shows the hinge type model, with *Pivot* and *Slider* restraints applied against a virtual part (*Rigid Virtual Mesh* under *Nodes and Elements* feature). A linear force of 1000 N/m is applied to the right edge, the minus sign representing the direction of the force vector. The specification tree is completed with the feature *Line Force Density*.

Surface Force Density represents concentrated loads (N/m²) of the traction force field type, with uniform intensity, applied on the surfaces of the analysed models. The user sets the three components of the force field in the directions of the X, Y and Z axes, along with the resulting value. Changing one value automatically leads to changing the other three.

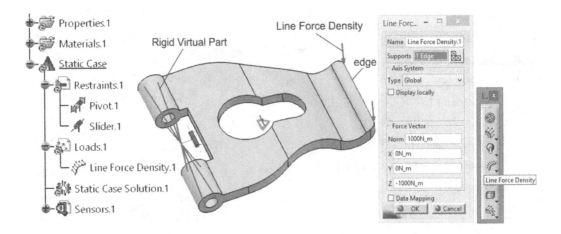

FIGURE 3.25 Applying a *Line Force Density* load.

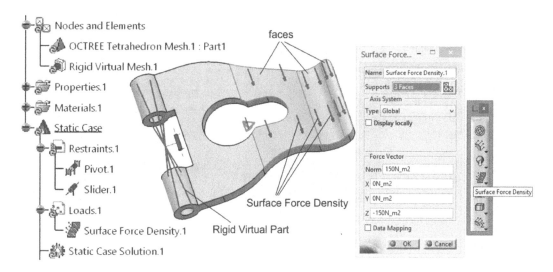

FIGURE 3.26   Applying a *Surface Force Density* load.

It is easy to observe that this type of loading is similar to the one obtained by using the *Pressure* icon/tool, presented earlier, except that the pressure is exerted only in a direction perpendicular to any point on the considered surface.

The example in Figure 3.26 shows the *Surface Force Density* dialog box, which contains three surfaces of the model in the *Supports* field. The surfaces are also highlighted on the figure. In the *Force Vector* area, the loads are established by the user, but also their direction (positive or negative values).

Volume Force Density are loads (N/m³) of concentrated traction force field type, with uniform intensity, applied on the whole body of an analysed model. The user sets the three components of the force field in the directions of the *X, Y, Z* axes and the resulting value. Changing one value of the four leads, of course, to changing the other three.

The example in Figure 3.27 contains a sphere type part with a *Surface Slider* restraint applied in the central hole. The load is placed on the whole solid body, so the field *Supports* contains the value *1 Body (on publication)*. The specification tree is completed with the *Volume Force Density* feature.

Force Density is part of the force loads group, with the difference that its unit of measurement is N, compared to N/m, N/m², respectively, N/m³. This particularity allows it to be applied to any edge, surface or volume geometries of the analysed models. However, in use, only geometries of the same type can be selected in a multiple selection for a load in the *Supports* field.

Enforced Displacement is a load applied to the geometries of the analysed models, which assign non-zero values to the displacements of the points in some constrained directions.

Thus, the user will select an applied restraint in the field with the same name (Figure 3.28), depending on its type, some *Translation* and *Rotation* fields become available (representing the degrees of freedom taken over). Multiple restraints selection is not possible for an *Enforced Displacement* load, even if they are of the same type. During the analysis, the constrained points will have a forced translational or rotational motion according to the specified values.

Temperature Field creates temperature fields, in degrees Kelvin (⁰K) on the bodies of the analysed models.

Temperature Field from Thermal Solution imports temperature fields in degrees Kelvin on the bodies and surfaces of the analysed models using an already existing solution, which was obtained with the help of a *CATIA v5* partner product.

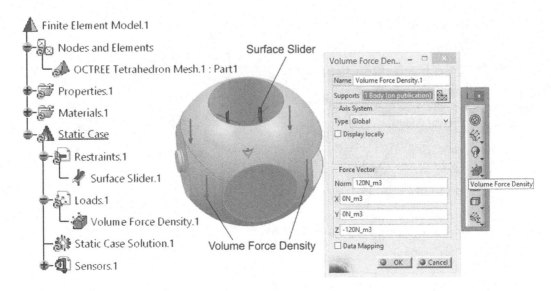

FIGURE 3.27   Applying a *Volume Force Density* load.

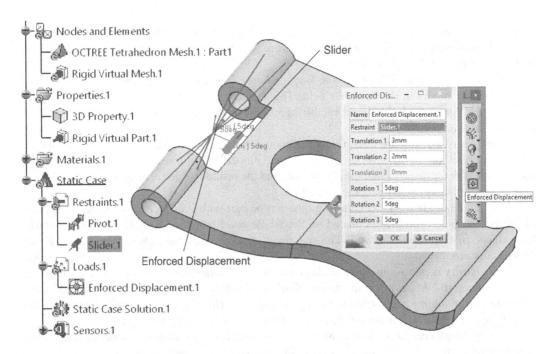

FIGURE 3.28   Applying an *Enforced Displacement* load.

3.3.5   Toolbars *Adaptivity and Compute*

With the *New Adaptivity Entity* tool, the user has the possibility to refine the network of nodes and elements (mesh) in order to obtain the required accuracy for the network (the whole analysed model) or only for a specified area of it (edge or face). The criteria for refining the network are based on predictive error estimation: they determine the distribution of a local error field for a given case of static analysis. The role of this refinement is very important to obtain a correct analysis model, usable in industry applications.

FIGURE 3.29 *Compute* dialog box.

The *Compute* tool has one of the most important roles in the finite element analysis process of a model: it triggers/starts this process, but only for the completely constrained/restrained and loaded model. Thus, throughout the process, the user has to define a set of parameters (material characteristics, restraints, various loads, etc.), following the steps previously described in this chapter. Then, during the computation, the program transforms and interprets these parameters in conditions applied to the model, allowing the access of the user to visual and numerical data.

For example, on the hinge part the user applies three restraints: two *Surface Slider* and one *Clamp*, then a *Distributed Force* load. An estimated computation of the model is performed by clicking on the *Compute* icon, opening the dialog box with the same name (Figure 3.29). As a result, the *Static Case Solution* feature is updated in the specification tree (Figure 3.30).

This first computation contains a number of errors that can/should be reduced by refining the network of nodes and elements. Thus, by clicking on the *New Adaptivity Entity* icon, the *Global Adaptivity* dialog box is displayed (Figure 3.31) in which the user selects the whole model (*1 Mesh part*) in the *Supports* field. The model selection is made in the specification tree by clicking on the *OCTREE Tetrahedron Mesh* entity.

In the *Current Error (%)* field, the program displays a value of the actual error, resulted from the first computation applied to the considered model. The value obtained in this case, of 56.5038% is unacceptable, and the user has, however, the possibility to enter in the field *Objective Error*

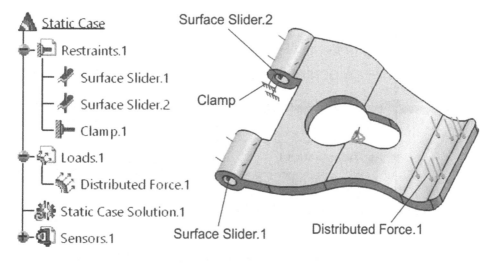

FIGURE 3.30 Example of model ready for finite element analysis and its specification tree.

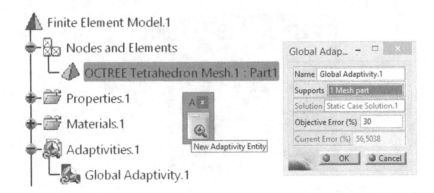

FIGURE 3.31 *Global Adaptivity* dialog box and model selection in the specification tree.

(%) a much lower value (30%), suitable for a model with a real correspondent. The *Adaptivities* feature is added to the specification tree and the user should try to reduce the computation errors of the analysis.

In industrial practice it is considered that the model is correctly refined if it meets an error rate close to 10%.

Reducing errors is not always possible for the whole model, and the user is more interested in certain areas after applying loads and restraints.

It is therefore considered useful to create a local refining, only for certain edges and/or faces. In the example of the considered hinge part, the stress area is close to the applied restraints. In the specification tree, the user right-clicks on the *Global Adaptivity* feature, and from the context menu that appears, selects the *Local Adaptivity* option (Figure 3.32). This one is added to the specification tree and the next step is to select the faces in the restraints area.

In the *Local Adaptivity* dialog box (Figure 3.33), in the *Supports* field, the user selects five faces of the model that will be refined more compared with the rest of the model in order to obtain a 20%

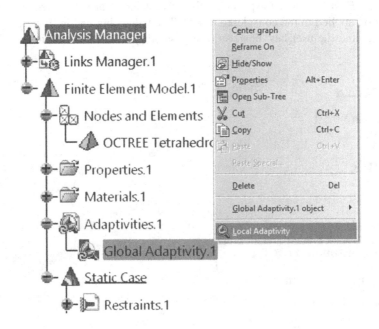

FIGURE 3.32 Creation of a local refining.

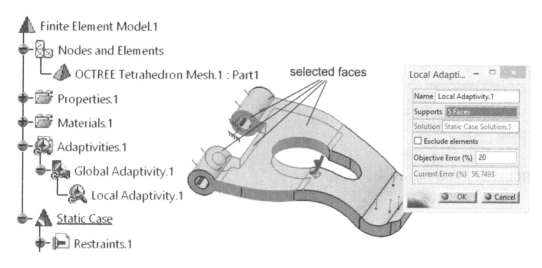

FIGURE 3.33 Selection of faces for local refining.

error rate. If other cases, the selection could include some edges, the field *Supports* being completed with the respective entities.

In order for these refinements (global and local) to be taken into account in the finite element computation process of the model, the *Compute with Adaptivity* icon will be used. The *Adaptivity Process Parameters* dialog box (Figure 3.34) sets the number of computation iterations to which the model will be subjected during the analysis *(Iterations Number* field*)*.

The higher the number of iterations, the longer the computation time. Pressing the *OK* button displays the *Computation Status* information box (Figure 3.35). A suggestive progress bar is shown and the total computation time for each iteration.

For the considered part, following the three iterations, the imposed error was not reached, the user being informed through a warning box.

Of course, following the refining and computation, some reduction of the error rate was obtained both locally and globally (Figure 3.36). The information is displayed by double-clicking on the *Global* and *Local Adaptivity* features in the specification tree.

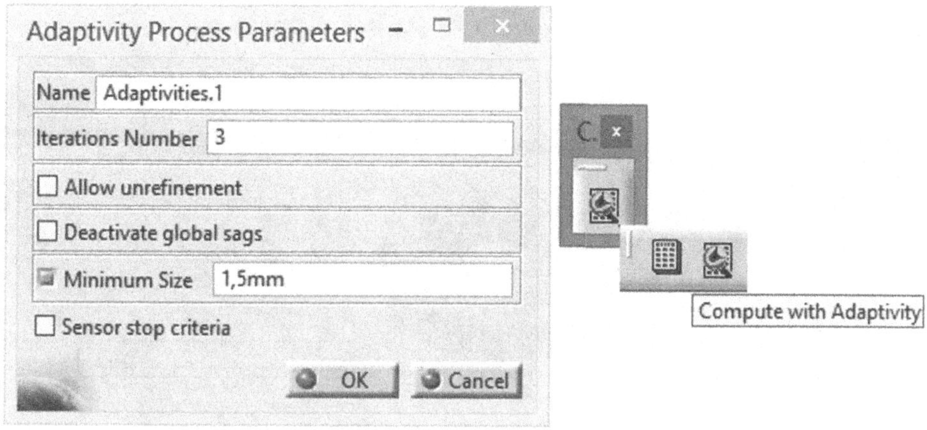

FIGURE 3.34 Establishing the iterations number.

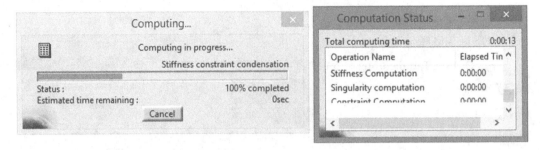

FIGURE 3.35   The computation progress.

FIGURE 3.36   *Global* and *Local* results.

The reduction is significant, but the imposed objectives (error rates) were not reached. As a result, several solutions can be adopted, such as accepting these error rates, continuing to refine the model through other iterations, choosing another method (Figure 3.7) or resuming the finite element analysis process by refining the model (Figure 3.8).

3.3.6   TOOLBAR *IMAGE*

The results of the finite element analysis are presented in the form of images in most programs that include FEM. *CATIA v5,* through the tools on the *Image* toolbar, offers the user the possibility to identify areas with problems, the safe ones, with maximum stresses, amplitude of deformations and displacements.

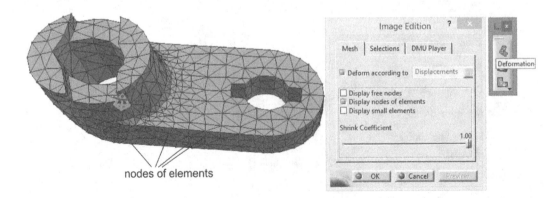

FIGURE 3.37   Deformed representation of the model.

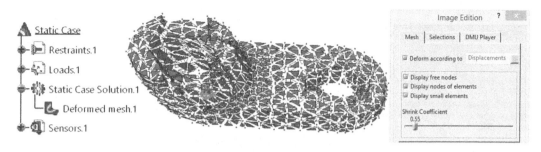

FIGURE 3.38 Changing how the elements are displayed.

Several variants of interpretation and presentation of the finite element analysis results are available, using these tools: *Deformation, Von Mises Stress, Displacement, Principal Stress* and *Precision*. By default, the program displays only one of these, but two or more can be placed on screen for comparative purposes.

The images provided by *Deformation* are used to visualize the analysed model in a distorted representation, as a result of all the imposed conditions (material, restraints, loads). By clicking on the tool icon, the model is displayed as shown, for example, in Figure 3.37. The network of finite elements that covers it should be observed, but especially its deformation.

The *Deformed Mesh* feature, currently active, is added to the specification tree (Figure 3.38). Double-clicking on it opens the *Image Edition* dialog box which contains three tabs: *Mesh, Selections* and *DMU Player* (Figures 3.37 and 3.38).

In the *Mesh* tab, checking the options has the following meanings: *Deform according to Displacements* allows viewing the results of the analysis in distorted representation, *Display free nodes* shows the unconstrained nodes, *Display nodes of elements* highlights the network of nodes (in red) belonging to the finite element network, *Display small elements* helps the user to see the smallest finite elements and the *Shrink Coefficient* slider is applied to obtain a different/smaller representation of finite elements for values <1 of the coefficient (Figure 3.38).

The images created after using *Von Mises Stress* show the field of the *Von Mises* model, representing, in fact, the values of a scalar field obtained from the volume of the deformation energy density (strain energy density) and used to measure the stress state created in the model.

This is often used together with the value of the yield strength of the material (in $N/m^2 = Pa$) to verify the integrity of the model under analysis, according to *Von Mises* criteria.

Before displaying the analysis result, the user must choose another view mode of the model. Thus, on the *View* toolbar, he expands the *Render Style* icons group (Figure 3.39) and chooses *Customize View Parameters*. Alternatively, he can access it in the *View → Render Style* menu.

In the opened *View Mode Customization* dialog box, the *Shading* and *Material* options should be checked in the *Mesh* area, the result is a different view mode on the deformed model (Figure 3.40).

In this *Von Mises* view mode, the dark blue areas are the least stressed, and the blue, yellow, green and red areas highlight the most stressed, the value of which the user is interested because the load on the part is very high.

The graphic presentation, through colours, of the stresses values greatly facilitates their examination. Graphic results are accompanied by captions or additional information. To interpret the model in Figure 3.40, for example, the user must know that each colour type represents a range of values of a specific solution, shown in an adjacent palette (Figure 3.41).

Also, the extreme values are either at the first, or at the last colour range in the palette. The order of the colours can be changed using the *Inverse* option in the *Color Map Edition* dialog box. The display of the dialog box is obtained by double-clicking on the colours and values palette that accompanies the *Von Mises* representation of the model.

Double-clicking on the *Von Mises Stress* feature in the specification tree displays the *Image Edition* dialog box, which contains the *Visu, Selections* and *DMU Player* tabs (Figure 3.42).

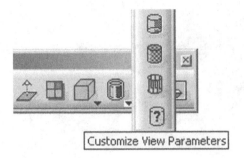

FIGURE 3.39 Changing how the model is displayed.

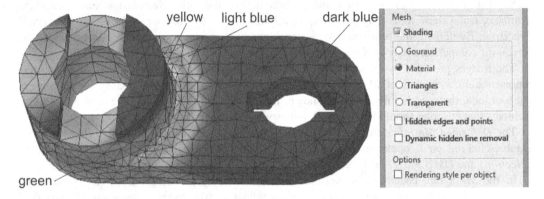

FIGURE 3.40 *Von Mises* representation of the model.

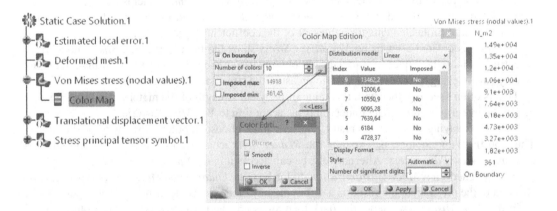

FIGURE 3.41 *Von Mises* colours and values palette of the model.

In the *Visu* tab, the options have the following meanings: *On deformed mesh* allows the user to view the results of *Von Mises* analysis in such a representation, *Average iso* graphically displays stresses values in nodes, *Discontinuous iso* also graphically displays stress values in elements, each of them having a single colour, *Symbol* visualizes the results with the help of cube symbols (by default), but the shape, type and their dimensions are established by clicking the *Options* button (*Tetra* representation, for example, in Figure 3.43). The last type, *Text,* serves to effectively display the stress values on the model surface.

The user is interested in knowing the actual values of stresses, displayed in the colour palette, because they may exceed the value of the yield strength of the material, in which case the

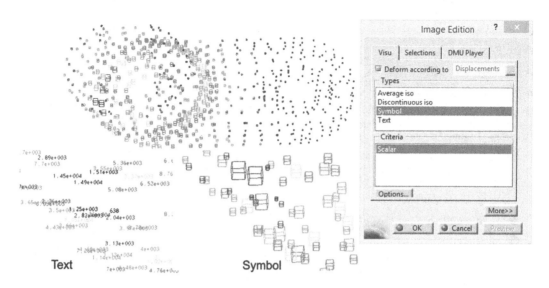

FIGURE 3.42 *Image Edition* dialog box with two details on *Symbol* and *Text* representations.

deformation is no longer elastic but plastic, so there is the possibility of cracks in the structure of the analysed model, and, of course, in that of the real model.

Displacement is used to visualize the displacements field (in millimetres) of the network nodes, as a result of imposing restraints and various loading conditions on the model. The field is symbolized as vectors (arrows), each with a magnitude and a direction.

Figure 3.44 graphically shows the displacements of the nodes using vectors of different colours, which signify their values, the directions are established by the arrowhead, as it results from the adjacent detail.

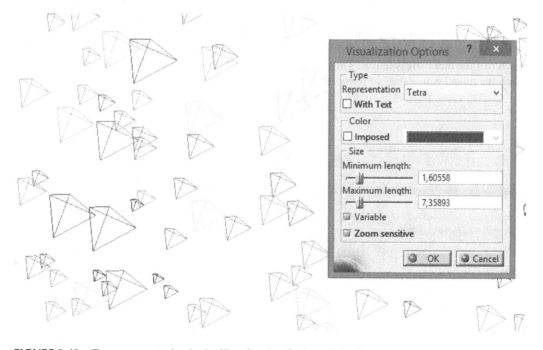

FIGURE 3.43 *Tetra* representation in the *Visualization Options* dialog box.

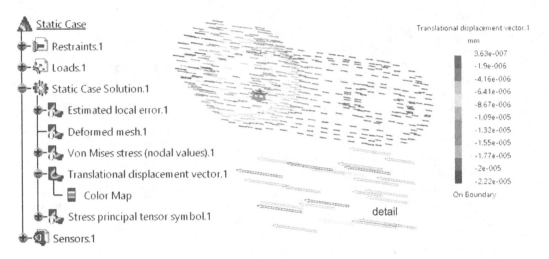

FIGURE 3.44　Vector symbolization of displacements.

The *Translational displacement vector* feature is also added to the specification tree. Double-clicking on it opens the *Image Edition* dialog box which contains three tabs: *Visu*, *Selections* and *DMU Player* (Figure 3.45).

The *Visu* tab is very similar to the one previously presented for *Von Mises*, the options having the following meanings: *Deform according to Displacements* allows viewing the analysis results in distorted representation, *Average iso* displays the displacements in nodes, *Symbol* presents the displacement vectors, their shape, type and magnitude, which are set by clicking the *Options* button. *Text* is used to display the values of the model's displacements. In the *Visualization Options* dialog box (Figure 3.45), the user has chosen to display the values (*With Text*) near the double arrow representation:

Also, the user has the possibility to find out the displacements of a vector by clicking on any of the arrows of the field, the result is the display of the three components of the displacements in the *X*, *Y* and *Z* directions, according to the considered reference system.

The images provided by using *Principal Stress* allow the visualization of the fields of the components of the main stress tensor. Thus, at each point, the tensor indicates the relative directions in which the model is in a state of tension/compression, but also the respective values.

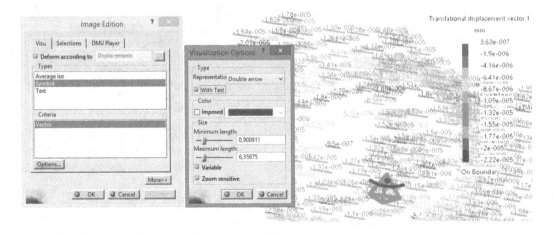

FIGURE 3.45　Values of vectors displayed next to their arrows.

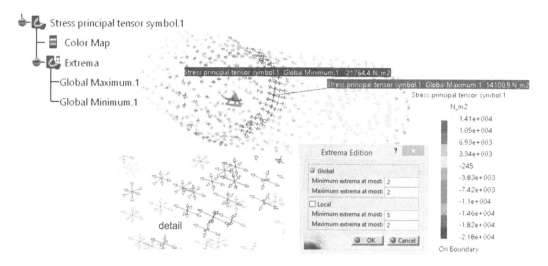

FIGURE 3.46   Symbolization of the principal stress tensor components.

Each point of the tensor contains a set of three directions, symbolized by lines with pointing arrows, these being the main directions of stress. The arrowheads (outwards or inwards from the model) indicate the direction of the principal stresses, their colour offering quantitative information on each value (in N/m^2).

Figure 3.46 shows the visualization of the fields of the tensor components for the analysed model, the colour palette and corresponding values, but also a detail. Also, the *Global Minimum extrema* and *Global Maximum extrema* are displayed by flags in the respective nodes of the network. The *Stress principal tensor symbol* feature is added to the specification tree. Double-clicking on it opens the *Image Edition* dialog box, its content and the included options being similar to those presented earlier. Thus, it is possible to visualize the stress in the nodes and in the elements of the model, but also different ways of representing the fields of the tensor components.

Precision is the tool used to display the deformation energy error distribution map (in J) for a static analysis. This map provides qualitative information on how the estimated errors are distributed across the model. The program evaluates the validity of the analysis and provides a conclusion on it, an estimated error rate indicating a valid solution.

Thus, if the error is relatively large in a particular area of interest, there is a possibility that the results of the analysis for that area will not be taken into account by the user, the model must be refined, followed by a new computation. The purpose of using this tool, together with the analysis and refinement of the model, is to obtain the best possible accuracy for the considered model.

Figure 3.47 graphically shows the deformation energy error map, the corresponding colour palette and values, as well as the *Color Map Edition* dialog box obtained by double-clicking on the palette itself or on any of the palette values. By pressing the *More>>* button, the user expands the basic options to choose the distribution mode (*Linear, Histogram* or *Logarithmic,* two of them being presented in figure), the value display mode (*Scientific, Decimal* or *Automatic*) and the number of significant decimals used to estimate the errors.

The *Information* tool on the *Analysis Tools* toolbar is often used together with the *Precision* tool. By clicking on its icon and selecting the *Estimated local error* feature in the specification tree (Figure 3.47), the *Information* box opens (Figure 3.48), in which more data are displayed regarding the extreme values of the deformation energy errors, the estimated accuracy, the characteristics physical properties of the model material, as well as the estimated error rate.

During an analysis process, the user can choose how the graphical display of the results involves the use of one or more tools on the *Image* toolbar. By default, only one of the existing images is active, but with the context menu of each one, they can be activated or deactivated, respectively.

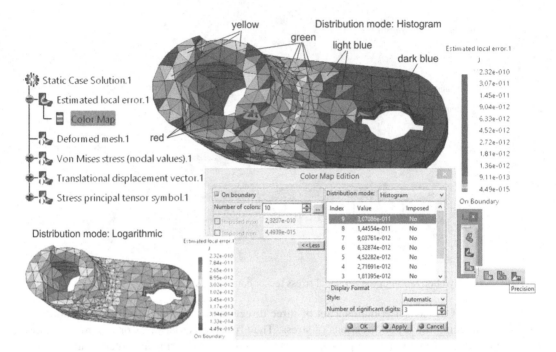

FIGURE 3.47 Distribution of deformation energy errors.

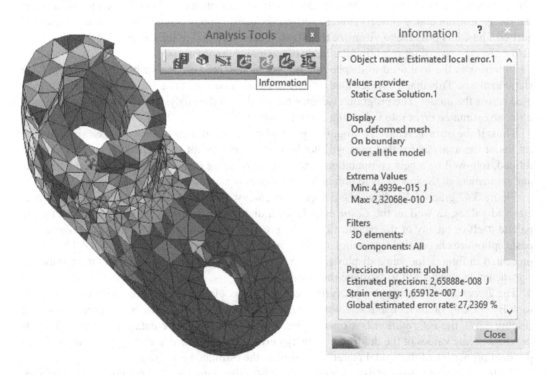

FIGURE 3.48 Conclusions on estimating errors.

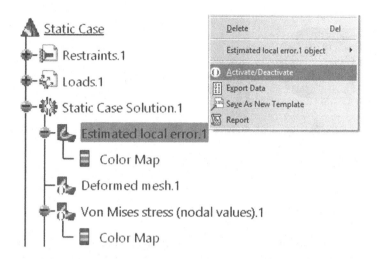

FIGURE 3.49 *Activate/Deactivate* option applied to an image.

Thus, the example in Figure 3.49 shows that the specification tree contains three such images, two inactive (*Deformed mesh*, *Von Mises stress*) and one active (*Estimated local error*). To activate/deactivate them, the user right-clicks on each feature/image in the specification tree. As a result, a context menu is displayed, from which the user chooses the *() Activate/Deactivate* option, the result being the overlapping display of all active images.

Figure 3.50 presents such an example with four active images and one being inactive, but the solutions are difficult to understand even if the palettes with values and colours are also displayed.

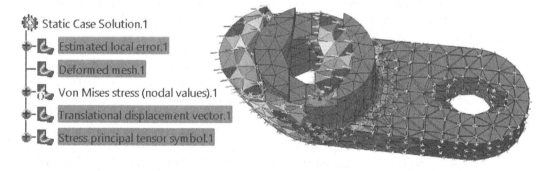

FIGURE 3.50 Overlapping four images/solutions.

3.3.7 Toolbar *Analysis Tools*

This toolbar contains the tools needed to analyse the results represented by images. Thus, the user can create animations, sections through the model, find out information on existing stresses, two images can be analysed comparatively, etc. Seven icons/tools are available, namely: *Animate, Cut Plane Analysis, Amplification Magnitude, Image Extrema, Information, Images Layout* and *Simplified Representation*. All are also available in the *Tools* menu.

The *Animate* tool provides a continuous sequence of successive frames on the result of the model analysis, based on a previously created/computed image/solution. Each frame presents the result

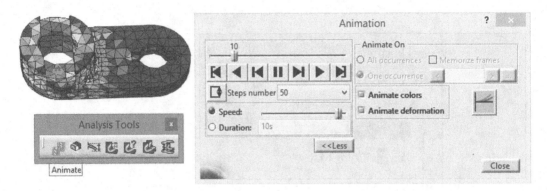

FIGURE 3.51   Controls for animation.

obtained, with a certain amplitude. The frames follow one another quickly, giving the user the impression of movement. Of course, the role of this tool is only to facilitate the understanding of the behaviour of the whole system consisting of the model, restraints, loads, etc., the displayed deformations being deliberately exaggerated, just for clarity. These are not representing the real deformations of the model.

Clicking on the *Animate* icon and selecting a graphic result (image), which is active in the specification tree, the user opens the *Animation* dialog box (Figure 3.51), it contains the standard animation controls, a slider in the *Speed* area to increases/decreases the running/playing speed. The animation can be displayed once or several times using the *Change Loop Mode* button, its refinement (number of key-frames) being set by the value in the *Steps number* field. The minimum number of frames is 5 and the maximum number is 50. For a higher number of frames, the animation is a lot smoother, but slower.

By pressing the *More* ≫ button, the options of this dialog box are extended, the user thus is able to choose the animation type (*Animation Mode* button) between asymmetric (default) and symmetric, but also the interpolation of stresses and displacements values, so that a smooth animation can be made between the chosen key-frames.

Figure 3.52 shows three frames taken during the playing/running of an animation to highlight the way in which the analysed model is exaggerated deformed. Of course, the deformation has the effect of changing the positions of certain nodes and changing the dimensions of the finite elements in the stress zones. A *Deformed Mesh* image was used for the example shown.

With *Cut Plane Analysis,* by default, the user has access to view only the images outside the model, observing the results of the analysis on its surfaces, but there is also the possibility to have access inside the model, in certain section planes.

The tool creates a plane that the user manipulates with the help of the compass (rotations, movements), in order to visualize the results in real time. By dynamically changing the position and orientation of the sectioning plane, the inside of the model is revealed in two ways: as a portion of it, remaining after removing the volume above or below the section plane, or as a single area, a slice of the model, actually placed in this plane (Figure 3.53).

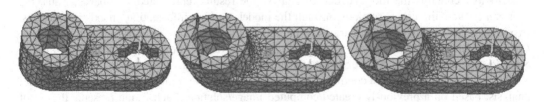

FIGURE 3.52   Examples of frames during an animation for the *Deformed mesh* solution.

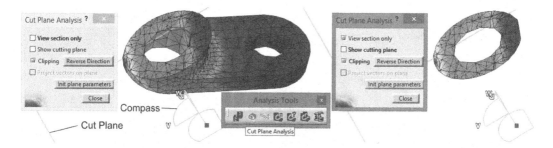

FIGURE 3.53 The model sectioned with a plane and just the section placed in that plane.

Figure 3.53 also shows the *Cut Plane Analysis* dialog box with its options. Checking/unchecking the *View section only* option leads to the presentation modes mentioned earlier. If the user clicks on *Show cutting plane*, *CATIA v5* displays/hides a transparent rectangle inside the frame that symbolize the section plane. The frame is always visible and only the rectangle can be hidden. Using the branches of the compass, the frame (and the rectangle) can be manipulated, in order to visually cut the model. The *Clipping* option and its *Reverse Direction* button present the cut model above or below the section plane.

The section plane remains active as long as the *Cut Plane Analysis* dialog box is opened, closing it leads to the disappearance of the plane and to the view of the whole model.

Amplification Magnitude is used to establish the maximum amplitude of finite element displacements when the analysis results are graphically presented. Of course, these displacements are exaggerated to show how the model is affected. *Amplification Magnitude*'s icon is active only if a fully defined system analysis computation has been performed, option *Deform according to* is checked (Figure 3.42) and there is at least one active image result in the *Static Case Solution* feature (Figure 3.54).

Clicking on the *Amplification Magnitude* icon, the user opens the dialog box with the same name, in which he can check the *Scaling factor* option to set a constant scaling factor or *Maximum Amplitude* to actually enter a value, in millimetres, through which the maximum amplitude of the displacements is displayed.

Depending on the selection, value fields and a slider are available, thus: for *Scaling factor*, the slider and the field *Factor* are operational, and, for *Maximum Amplitude*, only the *Length* field is displayed. Figure 3.54 present the slider and a value for factor. The *Set as default for future created images* option, available in both selection cases, allows the scaling factor or maximum amplitude parameter to be applied to all currently and future created image results.

Image Extrema is the tool that simplifies the location finding of points where the result's field is maximum or minimum. The user can ask the program to find one or both global extreme values and a certain number of local values. The availability of this tool requires the existence of *Estimated*

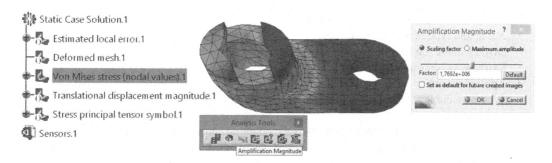

FIGURE 3.54 Determining the value of the displacements maximum amplitude.

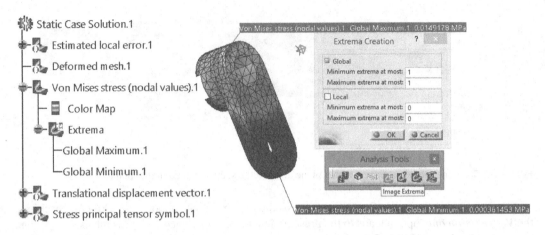

FIGURE 3.55 Locating the extreme values for the selected image.

Error, Von Mises Stress or *Translational displacement magnitude* features to be active in the specification tree. *Image Extrema* cannot be applied, for example, to the *Deformed mesh* feature.

The user clicks the tool icon, then selects the image for which he wants to find the extreme values. These values identified by the program have different units of measurement, depending on the choice of image in the specification tree.

For example, in Figure 3.55, the extreme values are global, measured in MPa, because the *Von Mises* feature was selected. Also, the values could be displayed in N/m^2 (1 N/m^2 = 1 Pa), depending on the user's choice in the menu *Tools → Options → Parameters and measure → Units* tab. Many other units of measurement can be selected in this tab for all the parameters of the *CATIA v5* projects.

The *Global Maximum* and *Global Minimum* features are added to the specification tree, the two values are, also, located on the model by indicators, next to the points/nodes where they were reached.

Checking the *Global* option in the *Extrema Creation* dialog box will cause the detection of extreme values at the global level, meaning all entities of the model whose values are equal to one of the two extremes. Checking the *Local* option specifies finding the minimum and maximum values at the local level, on a radius of two nodes around the one identified with extreme value.

Information is a tool that is useful to find and display certain information based on the image results obtained from the analysis. They provide graphical information, that are easy to understand, but for a correct analysis it is necessary to apply the *Information* tool, and after clicking the icon, an information box is displayed.

Thus, Figure 3.56 shows an example of an information box. Generally, these boxes contain different data, depending on the selection of the image type result from the specification tree, namely:

- Name and type of image result used;
- Display mode (on the edges of the model and/or on its entire surface);
- Extreme values (min. and max.);
- Applied filters;
- Type of accuracy (local or global);
- Estimated accuracy;
- The deformation/strain energy, defined as the potential energy in a part body at the moment of an elastic deformation, that is equal to the mechanical work necessary to produce this deformation;
- Estimated global error rate; and

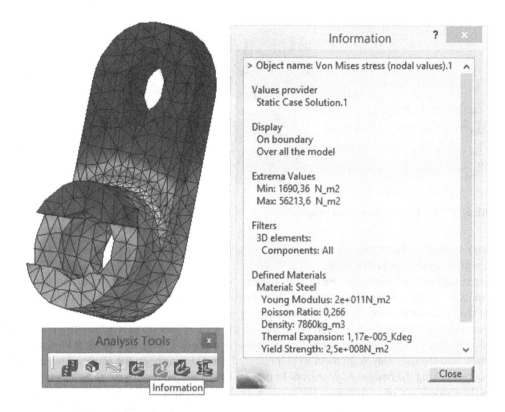

FIGURE 3.56 Information box.

- Material and its physical properties: Young's modulus, Poisson's ratio, density, coefficient of thermal expansion, yield strength. The latter is important and analysed because the stresses that cause cracks and ruptures are considered critical.

An important problem for the analysis using the finite elements of the considered models/parts and, then, for their real correspondent, is the comparison between the actual and the critical stresses.

As a result, in conditions of complete safety for the operation or use of real parts, mechanisms and assemblies, it is necessary that the actual stresses provided in the computation are much lower than the critical ones. For most critical conditions (breaking strength, yield strength, tensile strength, elastic limit, etc.) certain safety factors have to be defined.

The actual values of the safety coefficients must correspond to the total avoidance of the possibilities of reaching the state of critical stress, or, in certain cases, to limit the risk of this possibility to certain admissible levels.

Thus, the yield strength is defined as the maximum limit of the actual stresses under the safety conditions imposed by the permissible value of the safety coefficient. Yield strength also refers to an indication of maximum stress that can be developed in a material without causing plastic deformation. It is the stress at which a material exhibits a specified permanent deformation and is a practical approximation of the elastic limit. Yield strengths can only be considered as mechanical characteristics of the materials if they refer to parts with the same values imposed for the safety coefficients.

Depending on the value of the yield strength and, in general, on the data displayed in the information box, the user decides his next recommended actions: apply some changes to the model (its

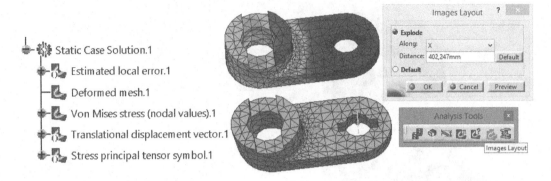

Static Case Solution.1

    Estimated local error.1

    Deformed mesh.1

    Von Mises stress (nodal values).1

    Translational displacement vector.1

    Stress principal tensor symbol.1

FIGURE 3.57   Relative positioning of images.

geometry, network refinement), restraints and/or loads, so that the deformations remain in the elastic range, with a convenient safety factor.

Images Layout, by default, the generated images, corresponding to the results of the finite element analysis, are positioned superimposed, therefore they cannot be viewed correctly. The user can change their relative position to analyse them comparatively.

Figure 3.57 shows that there are several images in the specification tree, resulting from the FEM analysis. Of these images, only two are active (*Deformed mesh* and *Von Mises stress*). Both results are important, and the user may want them to be presented simultaneously on the screen. This action is possible by clicking the *Images Layout* icon. In the available dialog box with the same name the user checks the *Explode* option and chooses the positioning direction (*X, Y, Z* axes or *XY, XZ* or *YZ* planes).

In the *Distance* field, the user enters the movement value in the selected direction. Pressing the *Default* button sets an automatic and optimal value in this field.

Relatively positioned images are displayed simultaneously on screen and the user can apply an animation to one of them, add a section with a plane and so on. Although, apparently, there are two or more models on the screen, in fact, there is only one, obtained from the applied analysis and represented by multiple images.

The images thus presented remain distinct until the user selects the *Default* option in the *Images Layout* dialog box. Images return to the overlay position, but remain active in the specification tree.

Simplified Representation is the tool that helps the program to simplify the representation of the model images, resulting from the analysis, during its manipulation by actions of movement, rotation, zoom, etc. The usefulness of such a simplified representation is obvious in the case of models with high complexity, with many loads, which contain several images displayed simultaneously, and whose manipulation is quite difficult, especially on computing systems equipped with a non-performing video card and low RAM.

Clicking on the *Simplified Representation* icon opens the dialog box with the same name, as shown in Figure 3.58. The *CATIA v5* program offers three simplified modes of representation: *None, Bounding box* and *Compressed.*

The *None* mode is considered as default, so no change is applied to the representation of the model image. *Bounding box* is the simplest mode, the analysed model is displayed in the image as a wire parallelepiped, which includes the volume of the part. Its geometry is not visually available (Figure 3.58). The *Compressed* mode allows the visual access to the model geometry and the images resulting from the analysis are available through three variants: *Low, Medium* and *High*. In the *Low* variant, the image intentionally loses most of its details and clarity, while *High* offers the highest visualization of details.

Figure 3.58 exemplifies, for the same model, two modes of representation: *Bounding box* and *Compressed*, the variant *Medium*. *Compressed* mode is not available for *Text* and *Symbol* image results (Figure 3.42).

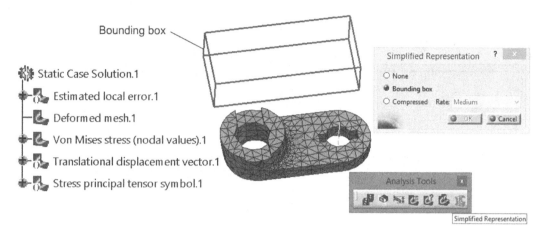

FIGURE 3.58 Choosing the simplified representation mode.

All modes of simplified representation become active only during movement, rotation, zoom of the model, its image temporarily changing its appearance, but immediately returning to the default display, with all the details, at the moment of ceasing the actions of manipulating the representation by the user.

Also, the specification tree is not adding new features; the simplified display modes have a local effect on one or more images displayed simultaneously, for the same analysed model.

3.3.8 Toolbar *Analysis Results*

Following the analysis with finite elements, various graphical representations are obtained, but also numerical data, based on which the user can generate the report at the end of a design project (drawing of sketches, 3D modelling, FEM analysis, etc.) for a product. The *Analysis Results* toolbar provides several tools for obtaining the report files.

Generate Report is the tool that creates such a report, consisting of a summary of the computation process results for an analysed model. The report is stored in an *.html* format file. This file displays numeric information and images, all saved in a specific folder on a computer. That folder will therefore contain an *index.html* file, along with a few other image files (*.jpg* format).

Figure 3.59 shows the *Report Generation* dialog box, obtained by clicking on the *Generate Report* icon. Several fields and options are available, as follows: in the *Output directory* field the user chooses the path (location) where the folder will be created and the report files will be saved, the *Title* field contains its name, the *Add created images* option allows adding the results report, as available and active images. In the *Choose the analysis case(s)* field the user selects the analysis for which the report is required.

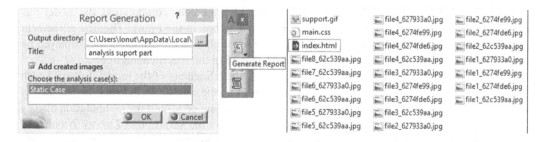

FIGURE 3.59 Choosing the files location and generating the report.

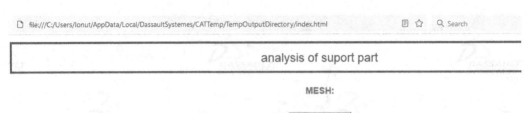

FIGURE 3.60   Example of a report web page.

Figure 3.60 shows some data of such a report. In fact, the report can be considered to be a complete web page, ready to be viewed on the computer on which it was created, to be displayed online or printed. The report data are similar to those obtained from using the *Information* tool, but are more complete and well synthesized to inform the user.

Generated Advanced Report is a tool that is practically an extension of the one previously presented, the generated data being much more detailed and varied.

Thus, in the example in Figure 3.61 the dialog box *Advanced reporting options* is displayed in which the user can select from the left panel the features on which he wants an advanced report to move them to the right list.

Compared with the report previously generated that referred only to the results of type image, through the advanced report, the user can found many other data on material, loads, restrictions, etc.

Historic of Computations – to be able to access the icon of this tool, at least two computations are required (applying the *Compute* tool twice). For the model in Figure 3.62, constraints and loads are considered so that the system is ready for analysis. Before the second computation, the model was refined according to the parameters displayed in the dialog box in the figure. The first refinement used the parameters: *Size* = 10 mm and *Absolute sag* = 2 mm. These parameters are displayed by double-clicking on the *OCTREE Tetrahedron Mesh* feature in the specification tree.

By modifying several times, by the same procedure, the refinement of the model, followed each time by a computation, several results are obtained, meeting the conditions to apply the *Historic of Computations* tool. Clicking its icon has the effect of displaying on screen the computation history graph for the analysed model. This graph (Figure 3.63) shows the variation of *Energy, Number of Elements* and *Number of Nodes*. It can be observed that the number of nodes increases from one iteration to another, due to the finite elements network refinement.

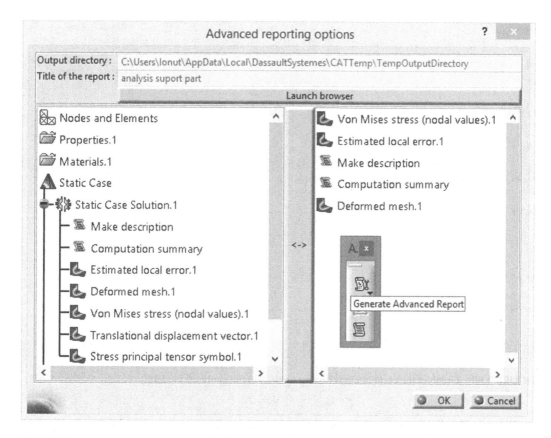

FIGURE 3.61 Choosing the components of an advanced report.

The figure shows that three computations/iterations were applied for the three previously mentioned elements of the graph. During the computation stages, the user can, however, add more elements to observe their variation. For this, in the specification tree there must be several results – images, active or inactive. The user will perform a right click on the *Sensors* feature, displaying the contextual menu in Figure 3.64.

Choosing the option *Create Global Sensor* opens the dialog box in the same figure, in which the user makes the selection, single or multiple, to create new sensors, depending on the comparisons he wants to highlight between the computation steps.

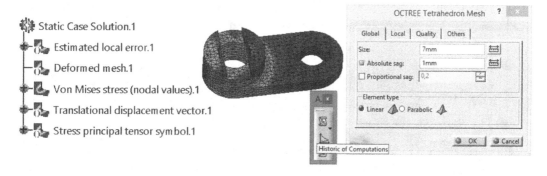

FIGURE 3.62 The model after the second refinement and its defining parameters.

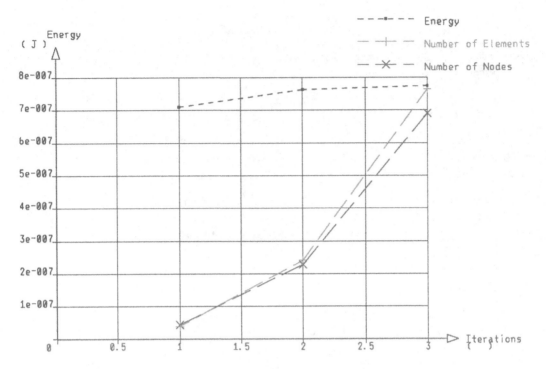

FIGURE 3.63 Graph of the computations history applied to the analysed model.

Changing the display mode of the graphs of the compared elements is done by double-clicking on their lines to open the *EditPopup* dialog box (Figure 3.65). The *Function Name* field contains the editable name of the element whose graph is edited. The user may check the *Point* and *Line* options so the graph will consist of lines and inflection points. The *Line* and *Point* fields allow changing the types of lines and points, as well as the colour used to display them on the graph.

The graph thus obtained is visualized, printed or saved in *.pdf* or *.xps* formats if the user right-clicks on the graph and chooses the *Print → Print Screen* option.

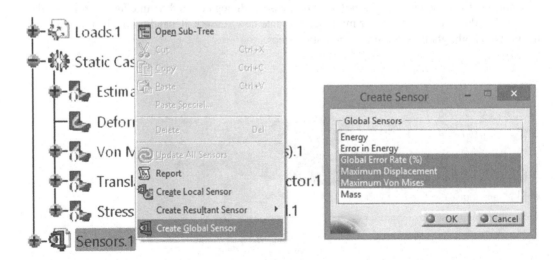

FIGURE 3.64 The context menu of the *Sensor* feature and the creation of sensors.

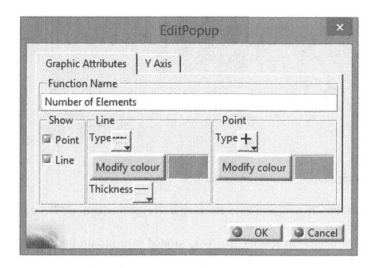

FIGURE 3.65 Changing the way graphs are displayed.

The dialog box that appears (Figure 3.66) contains all the options needed to print the graph if a printer is installed on the computer. Saving the document is frequently applied, the preferred format being *.pdf,* but prior installation of a special program for creating this type of files is necessary.

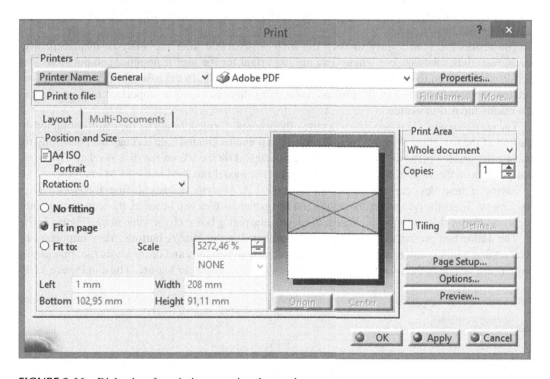

FIGURE 3.66 Dialog box for printing or saving the graph.

3.3.9 TOOLBAR *SOLVER TOOLS*

All results of the finite element analysis are temporarily stored in the computer memory to be quickly accessed by the program during computation operations, refinement, modification of model

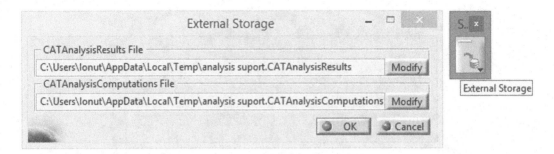

FIGURE 3.67 Setting the path on disk to save the files.

loads and restraints, etc. To save and keep them on a storage medium, *CATIA v5* uses the tools of the *Solver Tools* toolbar.

External Storage is used to specify the files, located in a certain folder, where the obtained data will be saved. Following an analysis process the obtained result is a file with the extension *.CATAnalysis.*

Between the model file (having the extension *.CATPart*) and the analysis file the program creates an important link, which is kept even after the work session has ended. Similarly, there is also a link between the analysis file and the output files located in the folder specified by applying the *External Storage* tool.

The data obtained after the analysis is completed are saved in two files, one for results (that has the extension *.CATAnalysisResults*) and one for computations (with the extension *.CATAnalysisComputations*).

The results file is necessary to keep the images generated after the analysis: displacements, loads, restraints, energies, etc. These data are important for the user if no other computations are performed based on them and they use a little space on the disk. In the case of other computations needed, the user has to restart the process, an activity that can consume important time resources, the results file is overwritten.

The computations file is necessary in case others are performed, based on the results obtained at the end of the first stage. The user benefits from the previous created data and has the possibility to make comparisons between them, but the space occupied by the file on the disk is relatively large, compared with the sizes of the other files related to the model involved in the FEM analysis.

Saving of these files occurs when the user saves the *.CATAnalysis* analysis file from the *File → Save As...* menu. To set the path where the results and computations files will be saved, the icon corresponding to the *External Storage* tool is clicked, displaying the dialog box with the same name (Figure 3.67).

The dialog box presents two fields, each editable using the *Modify* buttons. Very important is the fact that a link is automatically created between these two files and the analysis file. File destinations are visible in the specification tree, under the *Links Manager* feature. Thus, in Figure 3.68,

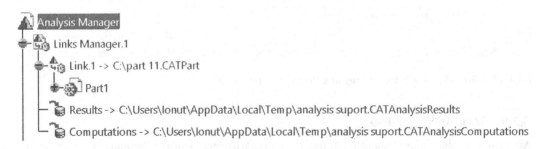

FIGURE 3.68 Displaying paths in the specification tree.

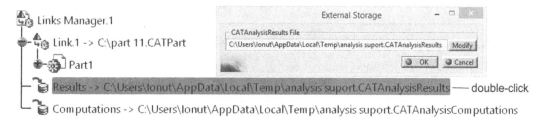

FIGURE 3.69 Editing destinations paths.

three destinations can be observed for three saved files: the model, the results and the computations.

Saving the analysis file to a destination path other than the initial one, previously established, is not allowed, this being the main proof of the link created between the files. However, if their destination paths are previously changed, the analysis file can be saved under a different name, in a different folder. Paths can be edited after they have been created by double-clicking on their names, and a dialog box will be displayed, like the one in Figure 3.69.

External Storage Clean-up is a tool that has the role of definitively removing the data from the results and computations files, to reset the current analysis and to start a new one or to free up important space on the storage disk.

Clicking the icon of the tool leads to the opening of the dialog box *External Storage Clean-up* in which the user chooses to remove only the computation files or them together with the results files (Figure 3.70).

Also, the figure shows the warning message automatically generated by the program, having a safety role, so that files that may be important for the user are not lost in a hurry.

Temporary External Storage, is used during an analysis process; the working data are stored in a temporary folder.

The save action, as previously presented, has the role of keeping the data stored temporarily, in two files, of results and computations, respectively. Thus, the user can work within the analysis process, all current information is temporary, but to become final it must be saved. Once the save is done and the analysis process is complete, the temporary data folder is emptied.

By default, temporary data resides in the temporary folder set by the operating system. However, it can be assigned a space quota, which cannot be exceeded, otherwise the operating system will perform actions to remove certain files.

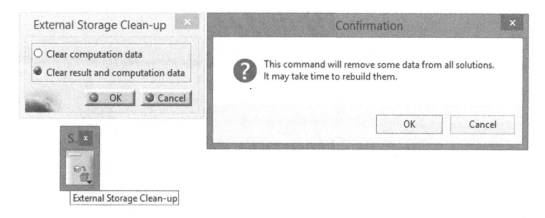

FIGURE 3.70 Removing files and warning message.

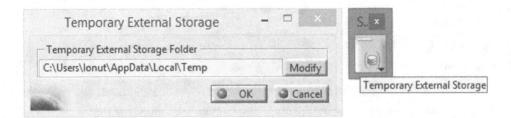

FIGURE 3.71   Setting the destination for the temporary files resulting from the analysis.

So that there are no unwanted data losses, resulting from the analysis process, it is recommended that the user define a new folder on one of the computer disks (Figure 3.71).

The way to choose the new folder for temporary data storage is similar to the one presented above, when saving results and computations files. Clicking on the *Temporary External Storage* icon brings up the dialog box with the same name.

3.3.10   Toolbar *Connection Properties*

The tools on this toolbar have a very important role in finite element analysis of assemblies. The user creates in the *CATIA v5 Assembly Design* workbench the required assembly constraints for the correct positioning of the components. With the help of the *Connection Properties* tools, the physical nature of the constraints can be defined. Thus, the assembly constraints are not sufficient for the analysis, but necessary, because they are the basis of the physical constraints created within the *CATIA v5 Generative Structural Analysis* workbench.

Slider Connection Property creates a connection/link between two parts, constrained to move together in the direction of the local normal in the common contact zone, behaving as if the sliding movement of one part relative to the other in the local tangential plane is allowed. The resulting elastic deformations are taken into account, and the tool assumes the previous creation of an assembly of two parts, between which there must be a surface contact constraint (*Contact Constraint*).

By clicking the *Slider Connection Property* icon, the dialog box with the same name opens. In the *Supports* field, the user chooses the contact constraint (Figure 3.72) between the two parts (the *Surface contact* feature in the structure *Links Manager → Link → Product → Constraints*, of the specification tree). The contact surface between the two parts is cylindrical.

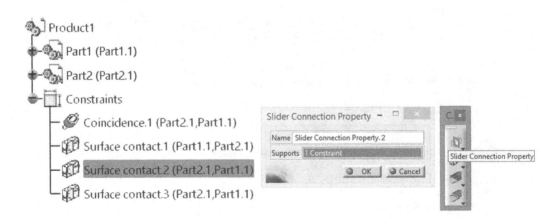

FIGURE 3.72   Selecting the contact constraint.

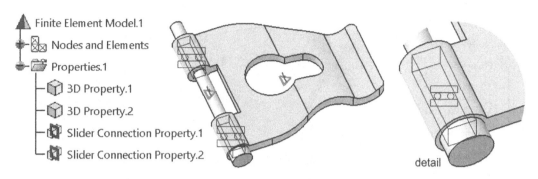

FIGURE 3.73 Representation of *Slider Connection* links.

In the *Name* field, the user enters the name of the *Slider Connection*. If the parts present several contact constraints, the user can also create respective links, these being represented alphanumerically in the specification tree, but also by symbols in the assembly (Figure 3.73).

The figure shows the place that two *Slider Connection* links take place in the specification tree and one of their corresponding symbols (in detail).

Contact Connection Property is used to create a connection between two parts, having the role to prevent their interference in the common contact area, behaving as if some relative movement of the parts was allowed, relative to each other, if they do not come into contact within a normal distance specified/imposed by the user. The *Contact Connection* tool considers the elastic deformation occurring in the area between the two parts.

The previously presented assembly is considered to have a contact constraint (*Surface contact.1*). In this case, the surfaces have a flat shape. The user clicks the *Contact Connection Property* icon to open the dialog box with the same name. In the *Supports* field, the contact constraint (Figure 3.74) is selected between the two parts, flat surfaces. The selection of the feature *Surface contact.1* can be observed, written in black on an orange background.

The *Name* field contains the name of the *Contact Connection* link, then in the *Clearance* field the user can enter a value, in mm, to establish the geometric distance between the surfaces of the parts (only those that are in contact). The value can be positive (meaning a clearance between the surfaces, they still have the possibility of approaching until contact is made) or negative (the parts are already too close, in which case a certain interference occurs). In the *Friction ratio* field, the user can enter a coefficient of friction between the considered surfaces. Thus, it is assumed that the user knows the materials from which the parts are made and how they are assembled in real conditions.

The created connection/link is represented in the specification tree, but also by a symbol in the assembly (Figure 3.75). In the detail on the right this symbol is more clearly identified.

Fastened Connection Property creates a connection between two parts, fastened together on the common contact area, so that they behave as a single body/part. From the point of view of the finite

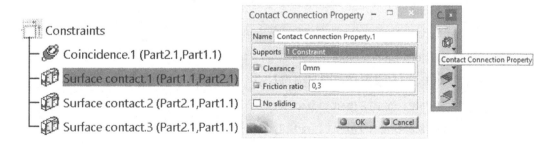

FIGURE 3.74 Selecting a contact constraint.

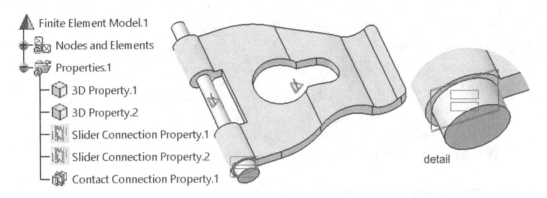

FIGURE 3.75 Representation of the *Contact Connection* link.

element analysis method, the *Fastened Connection* is equivalent to the situation in which two corresponding nodes, that belong to the two surfaces, merge to become a single node. The link created takes into account the elastic deformation occurring in the area between the two parts.

As an example, a simple assembly is considered, consisting of two components, between which there are three contact constraints, the respective surfaces having a planar shape.

By pressing the *Fastened Connection Property* icon, the dialog box with the same name opens, in which, in the *Supports* field, a contact constraint (Figure 3.76) is chosen between the parts. Optionally, two other *Fastened Connections* can be established based on the two remaining constraints, but the initially created connection is sufficient for fixing the respective parts together.

Figure 3.77 shows the specification tree completed with the *Fastened Connection Property* feature, the connection symbol in the detail on the right and two *OCTREE Tetrahedron Mesh* located within the *Finite Element Model* feature.

The existence of the two mesh elements in the specification tree is explained by the fact that in the analysis process an assembly consisting of two parts was inserted, transformed into a model by discretizing its components. The properties of the network of nodes and finite elements can also be modified at this stage by double-clicking on each mesh element, displaying a dialog box similar to the one in Figure 3.8.

Such completion of the specification tree with mesh elements is valid in all cases of creating connections for the analysed assemblies.

Fastened Spring Connection Property creates an elastic connection between two surfaces located on two different parts. This tool is similar to the *Fastened Connection Property*, but it's still slightly different: the selection can contain a contact constraint or a coincidence constraint, previously created in the *CATIA v5 Assembly Design* workbench.

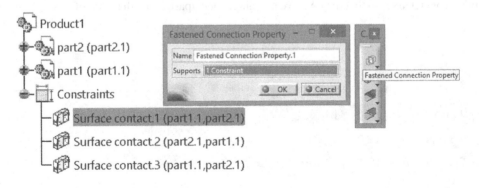

FIGURE 3.76 Selecting the contact constraint.

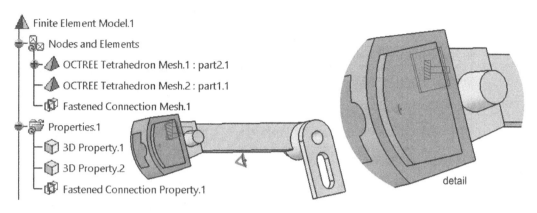

FIGURE 3.77   Representation of *Fastened Connection* link.

Also, in the *Fastened Spring Connection Property* dialog box, shown in Figure 3.78, the *Translation Stiffness* and *Rotation Stiffness* fields contain the imposed stiffness values for the assembly of the two components in the area where the connection was created. Thus, the stiffness is distributed to the involved elements and can be set interactively.

The values allow the assembly to elastically take on a certain intensity of the loads it is subjected to during the analysis process. The specification tree is completed with the *Fastened Spring Connection Property* feature.

Pressure Fitting Connection Property uses a surface contact constraint, created within an assembly, as a support for two parts assembled by pressing, between which there are interferences or overlaps, appearing in each of them. The *Pressure Fitting Connection* is similar to the one created

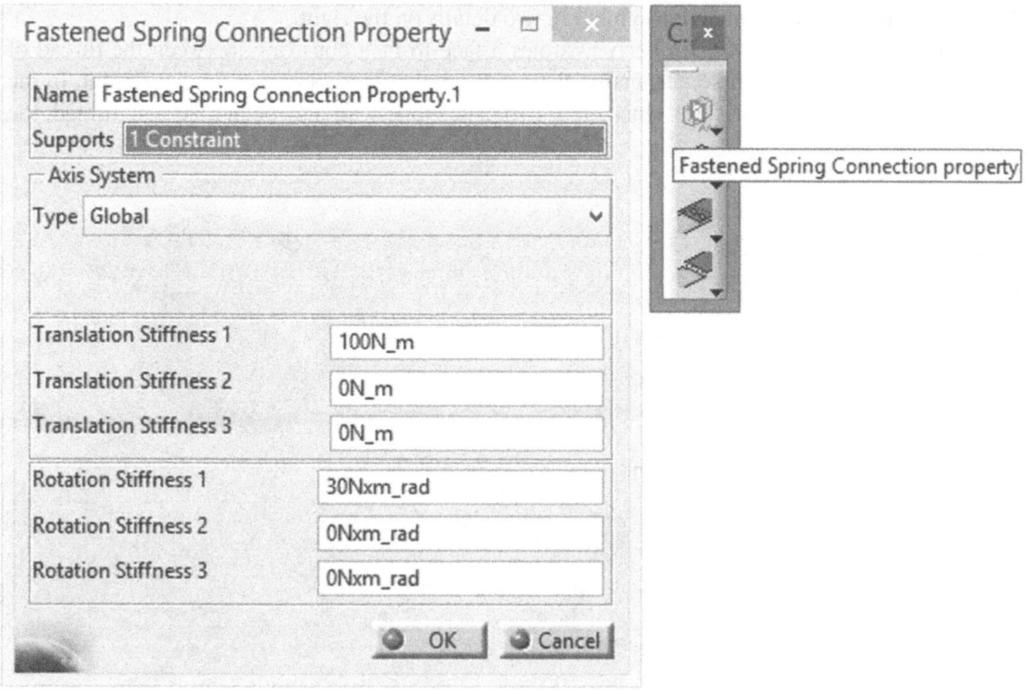

FIGURE 3.78   Imposing stiffness.

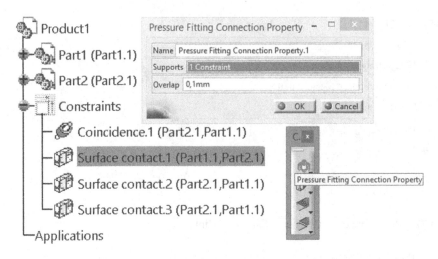

FIGURE 3.79   Selecting the contact constraint.

with the *Contact Connection* tool along the surface normal, the difference being in the tangential directions in which both parts are connected together.

The *Pressure Fitting Connection Property* dialog box (Figure 3.79) allows the selection of only one surface contact constraint in the *Supports* field. In the *Overlap* field the user enters a numerical value to establish the maximum allowed interference distance between the two parts. Generally, the value is considered to be positive or zero for contact constraints.

A negative value represents the existence of a space/tolerance between the surfaces of the parts supposed to be in contact, and they have the possibility to get closer until the contact is made. The default value of the field represents the current value of the distance between the respective surfaces.

Figure 3.80 shows the specification tree completed with the *Pressure Fitting Connection Property* feature and the connection symbol in two details on the right.

Bolt Tightening Connection Property assumes a face-to-face constraint between the thread of a screw and the thread of a hole, respecting the condition that these surfaces are coincident. In the following example, three parts are considered: two plates (one with a through hole – no thread, the

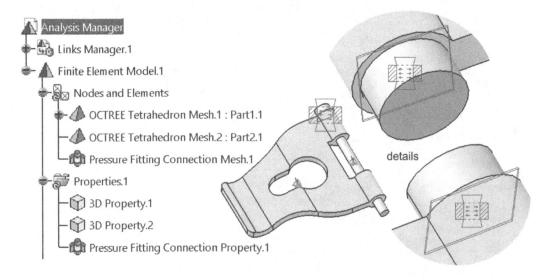

FIGURE 3.80   Representation of the *Pressure Fitting Connection*.

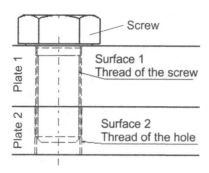

FIGURE 3.81   Assembling the plates with screw.

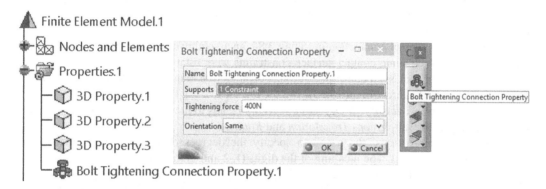

FIGURE 3.82   Establishing the parameters of screw tightening.

other with a threaded hole) assembled and fixed with a screw. Thus, Figure 3.81 shows the assembly, highlighting the surfaces of the threads.

In the example, the two surfaces are the supports for a surface contact assembly constraint.

Figure 3.82 shows the *Bolt Tightening Connection Property* dialog box. The *Supports* field contains a coincidence constraint, selected in Figure 3.83, between the screw and the bottom/second plate, having the threaded hole (fastening assembly). In the *Tightening Force* field, the user enters the value of the tightening force (central axial force) created by the screw, and the *Orientation* field allows the selection of its orientation.

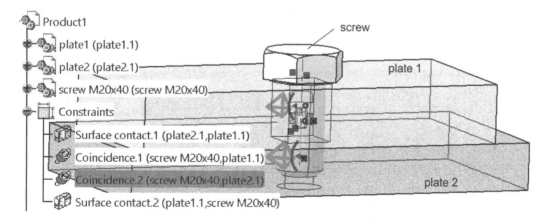

FIGURE 3.83   Selecting the coincidence constraint.

The *Bolt Tightening Connection* takes into account the pre-tensioning that occurs in threaded assemblies. Computations are then performed according to two conventional approaches.

In the first stage, the model of the assembly is subjected to tension forces relative to the tightening exerted by the screw, by applying opposite forces on its thread, respectively, on the threaded bore.

In the second stage, the relative displacements of these two surfaces (previously obtained) are imposed due to the various loads created by the user. During the respective stages, the displacements in the screw and the plate are in the direction normal to the screw axis. Of course, this type of connection takes into account the elastic deformations that occur.

Rigid Connection Property creates a connection between two parts to fix and stiffen them on the common contact area. From the moment of stiffening, these parts behave as if they are one, the connection between them is not taking into account the elastic deformations in the respective contact area. This is one of the main differences from the *Fastened Connection Property*.

Figure 3.84 shows the selection of a contact constraint between two parts and the *Rigid Connection Property* dialog box, displayed by using the icon with the same name. The *Supports* field contains the chosen constraint.

To exemplify this type of connection in Figure 3.84, an assembly consisting of two parts is considered, between which some contact surface constraints were established. Fixing and stiffening the parts are done for each such constraint, the specification tree being completed with three *Rigid Connection Property* features (Figure 3.85).

Checking the option *Transmitted Degrees of Freedom* leads to the expansion and completion of the dialog box with the sub-options *Translation* and *Rotation*.

Through these sub-options, the user can specify the degrees of freedom transmitted after a connection/link is created. The meaning of the digits (1, 2 and 3) that follow the translations and rotations is as follows: digit 1 corresponds to the *X* axis, digit 2 to the *Y* axis, and digit 3 to the *Z* axis.

Following the completion of the finite element analysis process and the establishment of the *Rigid Connection,* the two parts no longer influence each other from the point of view of deformations and stresses, the common area remaining unchanged.

Smooth Connection Property is a tool, from some points of view, similar to the *Rigid Connection Property*, previously presented. Through it, a connection is formed between two parts, fixed together on the common contact area. The difference lies in the fact that *Smooth Connection* roughly takes into account the elastic deformation in the respective contact area.

Pressing the *Smooth Connection Property* icon brings up the dialog box with the same name where the user can select a surface contact constraint established at the time of creating the assembly.

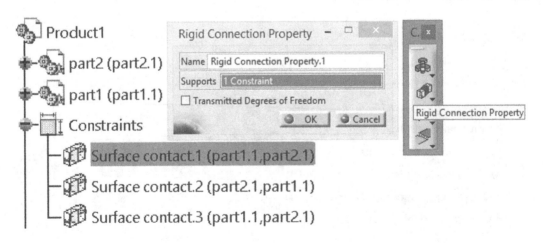

FIGURE 3.84 Selecting the contact constraint.

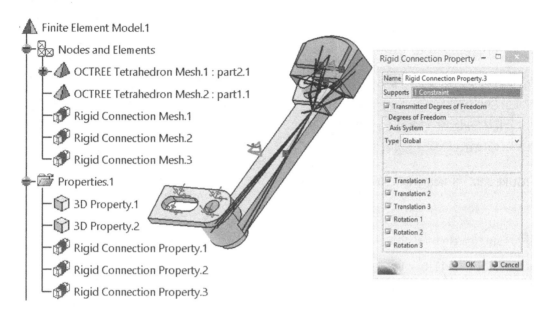

FIGURE 3.85 Creation of stiffening and fixing connections, establishing the degrees of freedom.

Figure 3.86 shows two such connections, located within the *Properties* feature in the specification tree.

Virtual Bolt Tightening Connection Property is applied to specify the interaction at the level of the considered contact zone between the parts in the assembly. Here is the similarity with the *Bolt Tightening Connection,* previously presented, with the difference that in this case the screw is no longer physically necessary, its role being taken by a virtual screw, its presence being simulated by a certain tightening force, set by the user.

It is considered an example of an assembly consisting of two plates, each provided with one hole: through (not threaded) and, respectively, threaded. In the case of the *Bolt Tightening Connection,* it is a screw passed through the two holes. In Figure 3.87 it can be seen that a surface contact constraint and a coincidence constraint between the axes of the holes were established between the two plates.

To create the connection, in the *Virtual Bolt Tightening Connection Property* dialog box (Figure 3.87) it is possible to select the coincidence constraint and set the tightening force (*Tightening Force* field). The specification tree is completed with the connection feature, represented in Figure 3.88 by the symbol of a screw.

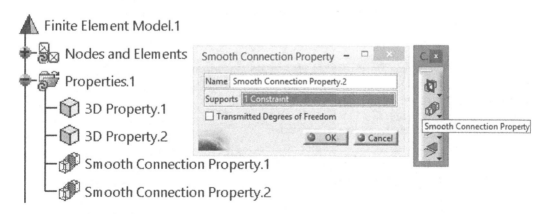

FIGURE 3.86 Creating a *Smooth Connection.*

FIGURE 3.87 Selecting the coincidence constraint.

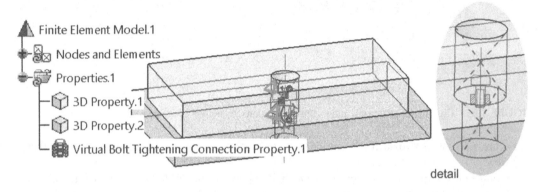

FIGURE 3.88 Representation of virtual screw tightening.

The use of the *Virtual Bolt Tightening Connection Property* tool does not require, as it follows from the explanations presented earlier, the previous creation of a virtual part (the screw), but it is implicitly present through the tightening force it exerts, set by the user.

In practice, there are many cases of assemblies that contain numerous screws, their 3D models must be considered during the analysis with finite elements, consuming important resources in the discretization and computation stages. It is, therefore, important to use such a tool that considers the tightening forces of the screws, without including the 3D models in the analysis.

Virtual Spring Bolt Tightening Connection Property is similar to the previous connection type, with the difference that it is elastic. The user has the possibility to establish in the dialog box in Figure 3.89, in the *Translation Stiffness* and *Rotation Stiffness* fields, the values required to ensure the stiffness of the assembly in the area where the connection was created. Of course, the values

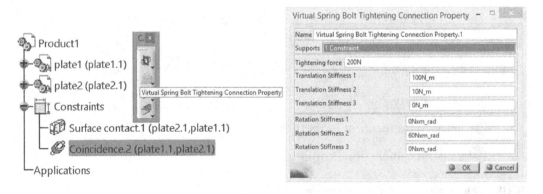

FIGURE 3.89 Establishing the parameters of elastic tightening with virtual screw.

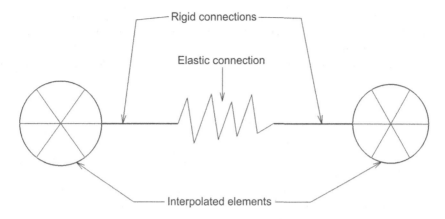

FIGURE 3.90 Components of the user-defined connection.

1, 2 and 3 that follow the translational and rotational stiffnesses correspond to the *X*, *Y* and *Z* axes, respectively. The value of the tightening force of the virtual screw is also indicated.

User-defined Distant Connection Property creates a special connection between two elements located at a certain distance from each other. The user must define the types of elements that can be part of this connection, that is divided into three areas: left, middle and right, as it results from the following schematic representation (Figure 3.90).

The scheme assumes two interpolated elements, an elastic connection (spring) and two rigid connections (rigid beam).

The element located in the left area is connected with the middle area by links of the type: *Smooth, Rigid, Spring-Smooth, Spring-Rigid* and *Contact-Rigid*.

Similarly, the element in the right area connects to the middle area also using the links: *Smooth, Rigid, Smooth-Spring, Rigid-Spring* and *Rigid-Contact*.

For the middle area are available the following connections: *Rigid, Spring-Rigid-Spring, Rigid-Spring-Rigid, Spring-Rigid, Rigid-Spring, Beam, Spring-Beam-Spring, Beam-Spring-Beam, Spring-Beam, Beam-Spring, Bolt-Rigid, Rigid-Bolt, Bolt-Beam, Beam-Bolt, Bolt-Rigid-Spring* and *Spring-Rigid-Bolt*.

Pressing the *User-defined Distant Connection Property* icon leads to the display of the dialog box with the same name, where it is possible to choose the connections for the three zones in the *Start, Middle* and *End* fields. Figure 3.91 shows in the *Supports* field a selected constraint and an example of establishing a variant of connections.

FIGURE 3.91 Variant of connections.

Depending on the variant chosen, if in the *Start, Middle* or *End* fields there are connections involving one of the *Contact* or *Spring* types, the *Component edition* button appears in the right part of the respective field (Figure 3.92).

Clicking this button allows the user to establish the parameter values for the contact connection or for the elastic connection.

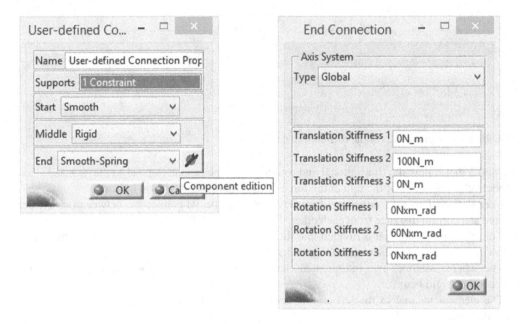

FIGURE 3.92   Displaying the *Component edition* button.

4 Simulation and Computation Applications Using Finite Element Analysis

INTRODUCTION

This chapter contains various finite element method (FEM) applications, different methods to establish meshes for the proposed three-dimensional (3D) models, how to impose restraints, apply loads, create connections between parts in assemblies, obtain and interpret results, create reports, etc. The parts used in these applications are presented by their two-dimensional (2D) drawings and brief explanations on how to 3D model them. Also, the proposed 3D models are available for download. By the end of this chapter, the user will understand the finite element analysis working methods and gain knowledge on creating finite element models in the *Generative Structural Analysis* workbench. It is assumed that the user has a good knowledge in computer-aided design (CAD) modelling of parts and assemblies, but also in strength of materials.

4.1 ANALYSIS OF A CONNECTING BEAM

This application presents a simple analysis on a connecting beam used in some mechanical assemblies to link different parts. The beam, shown in Figure 4.1 by its two-dimensional (2D) drawing, is clamped/fixed at one end and subjected to a load on two planar surfaces, at the other end.

There are two types of solid elements in *CATIA v5*: linear and parabolic, but both refer to tetrahedron elements, as explained in previous chapters. The linear elements are faster to compute, but less accurate. Usually, they are best used for simple parts with planar faces. The parabolic elements are more complex and require more computational resources from the computer's hardware. The obtained results are more accurate, this being the main reason of their use within curved surfaces.

The user will model the three-dimensional (3D) part using the *CATIA v5 Part Design* workbench and assign it the name *App1*. Thus, in the *ZX Plane,* a first sketch is drawn, symmetrical to the *V* axis, using the *Profile* tool. The lines are horizontal and vertical, according to Figure 4.2, and constrained to the dimensions from the 2D drawing.

The profile is extruded along the *Y* direction over a distance of 400 mm by applying the *Pad* tool (Figure 4.3).

The beam has two Ø36 holes placed at the left end of the 3D model. They can be obtained using the *Hole* tool or by cutting a circle with the *Pocket* tool. The holes together with the nearby flat face will be used to clamp the part at its end (Figure 4.4).

At the other end of the part, the user applies another *Pocket* tool using a sketch with a rectangle. The result is two flat surfaces on which the force loading the part will be applied. The rectangle cuts the part on a depth of 10 mm (Figure 4.5, right).

The user should save the 3D model of the part before continuing with the selection and application of the steel material. Clicking the *Apply Material* icon, the *Library* (of materials) is opened as a dialog box in Figure 4.6. This library contains several types of materials, from metals to wood and stone. Thus, in the *Metal* tab, *Steel* is selected at the bottom of the list, but the *Apply Material* button is not available until the user selects the *PartBody* feature from the specification tree.

The selected material is added to the part and the user is able to view the connecting beam in a more realistic manner. From the *View* menu, under the *Render Style* option he must select the

DOI: 10.1201/9781003426813-4

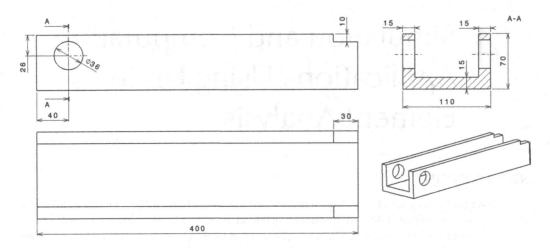

FIGURE 4.1   Two-dimensional drawing of the connecting beam.

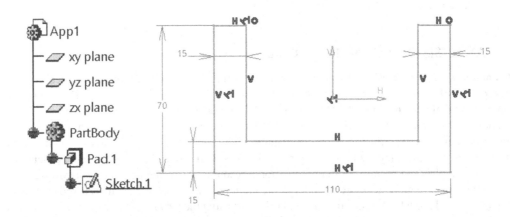

FIGURE 4.2   Drawing and constraining the sketch.

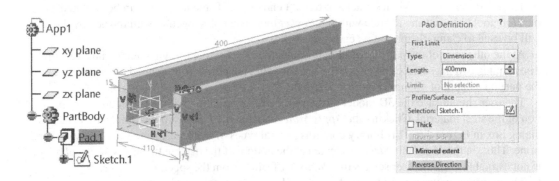

FIGURE 4.3   Extruding the sketch along the *Y* axis.

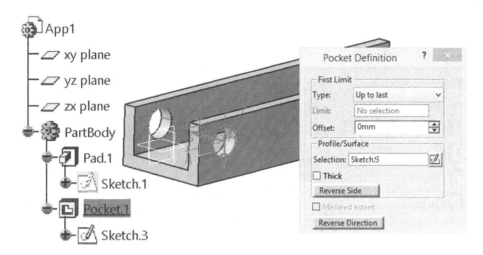

FIGURE 4.4 Obtaining the holes through the part.

Shading with Material sub-option, according to Figure 4.7. The colour of the part model becomes dark grey, with a slight glossy appearance. This setting only concerns the appearance of the part, its physical properties being already added with the *Steel* material.

The user has the possibility to view and edit, if necessary, the material properties assigned to the part by double-clicking on the material's name in the specification tree. The selection box from Figure 4.8 opens, which contains several tabs, which allow specifying a name for the material (since it is usually an alloy), its appearance in the *CATIA v5* program, its physical properties, the properties of the hatch that are visible in the *CATIA v5 Drafting* workbench, etc.

The physical properties of the material are very important for the way the part will behave under certain loads and restraints. They are displayed in the *Analysis* tab and the user is able to edit these properties.

The steps described earlier are for preparing the part model for finite element analysis. To start the analysis, the user accesses the *Generative Structural Analysis* workbench from the *Start →* *Analysis & Simulation* menu (Figure 4.9). The *New Analysis Case* selection box opens and the default choice, also selected by the user for this application, is *Static Analysis*.

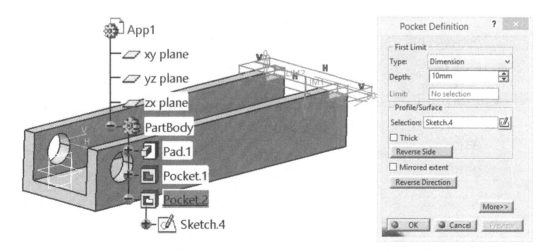

FIGURE 4.5 Cutting the part with a rectangle at its right end.

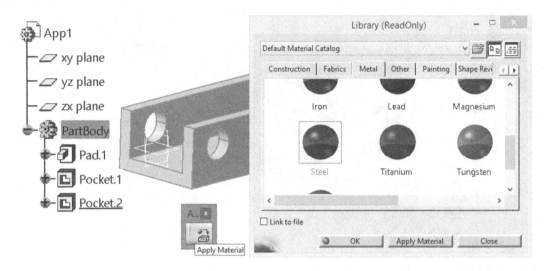

FIGURE 4.6   Selecting the material of the part from the library.

The specification tree structure becomes longer, new features, specific to the finite element method (FEM), being added. The bottom branches of this tree are empty by default in this moment, but as the user will work on *Generative Structural Analysis,* many specific features will be added. As an example, the user will add restraints, loads, will compute solutions, etc.

Figure 4.10 displays the specification tree at the beginning of the analysis, the 3D model, a mesh defined automatically on this geometry and the presence of a mesh symbol. This symbol means that the user may proceed with the analysis process even though the mesh is not displayed on screen. The symbol also reflects two important parameters of the mesh: size and sag.

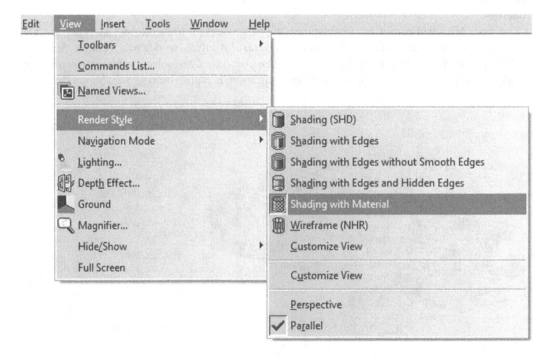

FIGURE 4.7   Changing the view mode of the part.

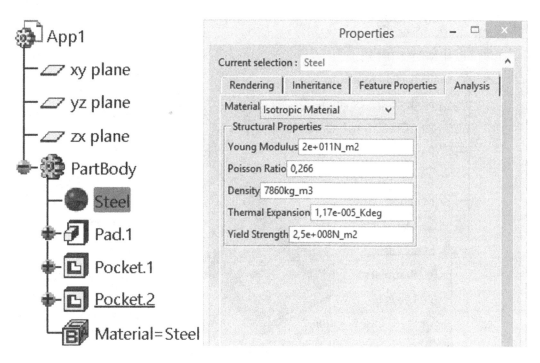

FIGURE 4.8 Accessing the physical properties of the material.

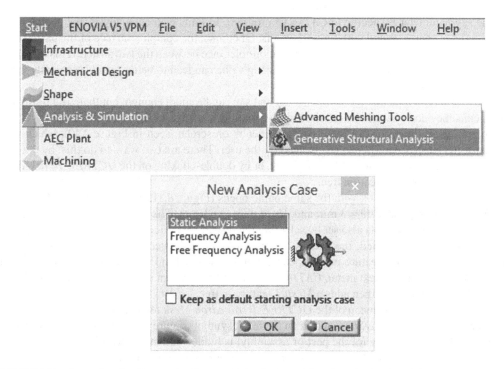

FIGURE 4.9 Accessing the *Generative Structural Analysis* workbench and selection of *Static Analysis*.

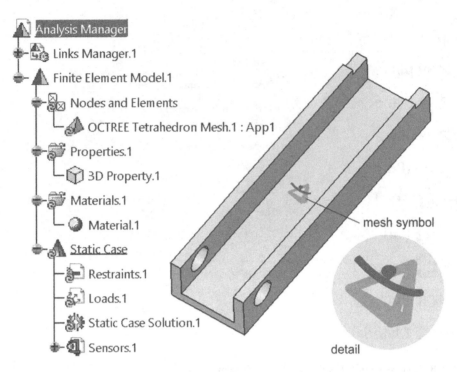

FIGURE 4.10   Displaying the specification tree and the mesh symbol.

A smaller element size means that the 3D model contains more elements, more nodes, more faces, etc. and, thus, more accurate results, but, at the same time, more resources are consumed, represented, mainly, by a longer computation time.

The sag parameter is specific to *CATIA v5* program. In finite element analysis, the geometry of a 3D model is approximated with finite elements. Of course, the complex surface of the part and its approximated mesh do not coincide. The deviation/tolerance between the two structures is materialized by the sag parameter. Therefore, a smaller sag value can lead to better results especially when the size value is also small.

The user should note that the values of size and sag can be small enough, but there is a limit, and this means they don't have to be infinitesimally small for better results.

The physical dimensions of these two parameters on screen (seen in Figure 4.10), which also influence the mesh accuracy, can be changed by the user. There are two ways to do this: by double-clicking on the mesh symbol visible on the part, or by double-clicking on the *OCTREE Tetrahedron Mesh* feature in the specification tree.

As a result, the dialog box with the same name opens (Figure 4.11). The user should change the values in the *Global* tab: *Size* = 3 mm and *Absolute sag* = 1 mm. The type of elements used in this analysis (linear/parabolic) is also set by selecting the respective option.

Once the mesh is defined, the user is able to view the generated structure, by right-click on the *Nodes and Elements* feature in the specification tree and select the *Mesh Visualization* option (Figure 4.12) from the context menu. *CATIA v5* needs a small amount of time to compute/update the mesh, displays a message, then adds a *Mesh.1* feature in tree.

From a similar context menu of the *OCTREE Tetrahedron Mesh* feature in the specification tree the user may apply the *Hide/Show* option on the mesh symbol (Figure 4.13). When the user displays the mesh, the 3D geometry (of the part or assembly) is hidden automatically. Figure 4.13 presents the hidden *Links Manager* feature, but it can be displayed again by the *Hide/Show* option in its context menu.

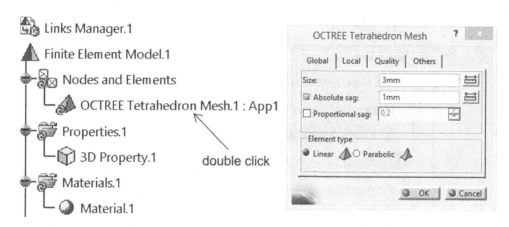

FIGURE 4.11 Setting the main parameters of the mesh.

The 3D geometry of the part and the mesh can be displayed together, but, usually, the finite element analysis does not require the 3D geometry to be displayed. The visualization of this geometry is important when the user needs to apply some changes to the part model. Of course, such a modification is followed by the resuming of the analysis with finite elements.

Figure 4.14 presents the representation of the analysed part when the user decided to show the 3D geometry and the defined mesh together.

The user should save the 3D model of the part before continuing with the phase of applying restraints and loads to the part mesh. Saving is done using the *Save Management* option from the *File* menu, which opens the dialog box in Figure 4.15. In this case, three files are required to be stored on disk, with the extensions: *CATPart, CATAnalysis* and *CATfct*. The first file contains the

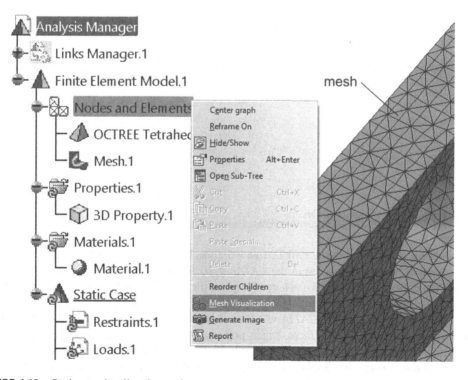

FIGURE 4.12 Option to visualize the mesh.

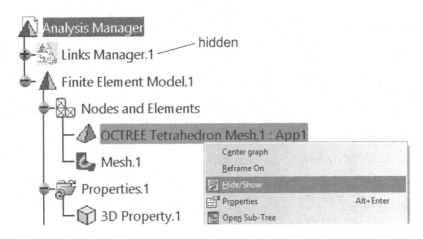

FIGURE 4.13 Hiding the mesh symbol.

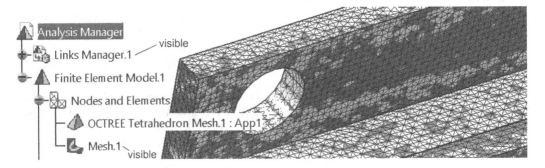

FIGURE 4.14 Displaying the 3D geometry and the mesh of the part at the same time.

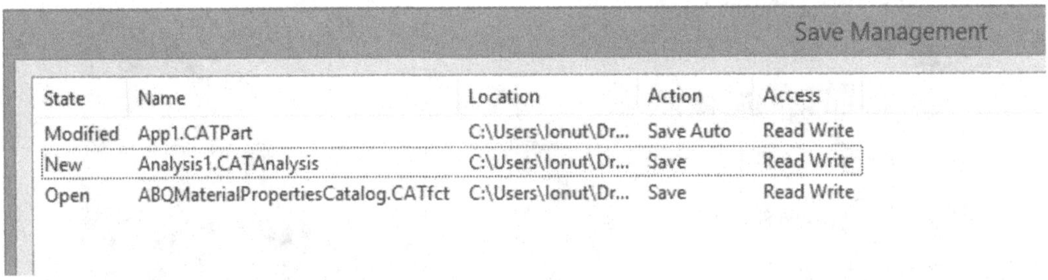

		Save Management		
State	Name	Location	Action	Access
Modified	App1.CATPart	C:\Users\Ionut\Dr...	Save Auto	Read Write
New	Analysis1.CATAnalysis	C:\Users\Ionut\Dr...	Save	Read Write
Open	ABQMaterialPropertiesCatalog.CATfct	C:\Users\Ionut\Dr...	Save	Read Write

FIGURE 4.15 Saving the analysis files.

part model as it was created in the *Part Design* workbench. The settings related to the mesh structure, loads and restraints are stored in the *CATAnalysis* file. *CATfct* is the extension of a *Feature Dictionary* and *Business Process Knowledge Template* file. A *CATfct* file contains the technological objects and behaviours that are defined for several *CATIA v5* workbenches.

The finite element analysis in *CATIA v5* is based on 3D geometry. Restraints and loads can be applied on this geometry such as, faces, edges, points and vertices, and the mesh has to be defined on the geometry. That means the geometry must be visible in order to apply restraints and loads.

Restraints refer to applying displacement boundary conditions in a specific area of the part's 3D model. In the present application, the user clicks on the *Clamp* icon within the *Restraints* toolbar.

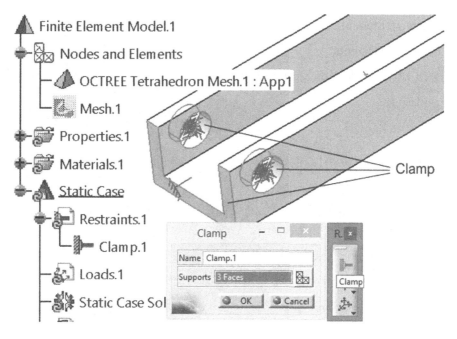

FIGURE 4.16 *Clamp.1* restraint added to three surfaces of the part's geometry.

The *Clamp* condition imposed to a surface means that the displacements of its nodes in all three directions are zero.

The dialog box with the same name is displayed and the user chooses three surfaces: the flat surface at the left side of the part and two cylindrical surfaces (of the holes), as shown in Figure 4.16. The *Clamp* symbol is represented on these three surfaces.

Also, the user should note that the *Mesh.1* feature in the specification tree is hidden, so only the 3D geometry of the part is displayed. By clicking the *OK* button, a new feature, *Clamp.1,* is added in *Restraints.1* under the *Static Case* feature in the specification tree.

To apply a force, the user has to press the *Distributed Force* icon on the *Loads* toolbar. The dialog box to set up the force opens (Figure 4.17) and the user selects two small flat surfaces in the *Supports* field, then specifies a value of −1000 N for the *Force Vector* in the Z direction field. The minus sign means that the force is applied opposite the Z axis. The *Norm* field contains the resultant value of the force if it has values in the three directions X, Y and Z. The symbol of the distributed force consists of a group of red arrows, which specify a certain direction and are positioned on the two surfaces.

In this case, the force is applied with respect to the global axis system, but it is also possible to use another axis system previously defined by the user when he modelled the part in 3D.

By clicking the *OK* button, a new feature, *Distributed Force.1*, is added in *Loads.1* under the *Static Case* feature in the specification tree.

The model proposed for analysis has to be checked before performing the computations and solving the problem created by the restraint and loading with the distributed force. For this, the user clicks the *Model Checker* icon on the *Model Manager* toolbar (Figure 4.18). The *OK* status in the last column is displayed along with the message *The whole model is consistent,* which means that there are no basic mistakes/errors in the mesh, restraint(s) and load(s).

So, the model is ready and the user should launch the solver. Thus, to run the finite element analysis, the *Compute* icon is pressed and the selection box is shown (Figure 4.19). The *All* option is set in the list, which means that everything is computed: mesh, restraints, loads, etc. as the *Static Case* requires. By clicking the *OK* button the analysis starts and the *Computing...* box is displayed showing the progress. Also, a second box, *Computation Resources Estimation,* becomes available

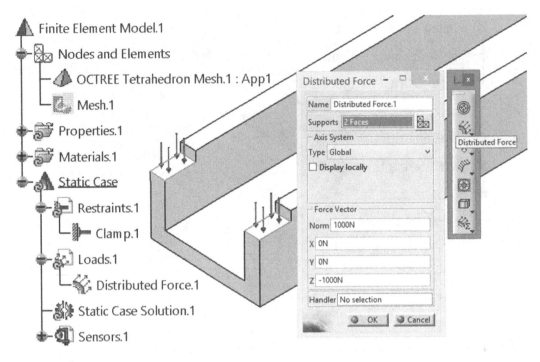

FIGURE 4.17 *Distributed Force* added to two surfaces of the part's geometry.

on screen providing information on the resources needed to complete the analysis. If the estimates are zero in the info box, then there is a problem in the previous step and the user should solve it.

The specification tree is changed/updated to reflect the location of saving the *Results* and *Computations* (Figure 4.20). As expected, the paths for the files can be modified by the user by double-clicking on each link.

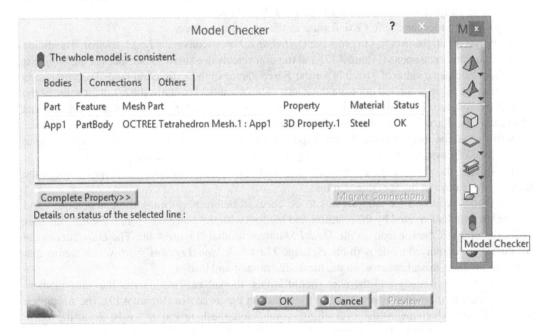

FIGURE 4.18 Checking the model.

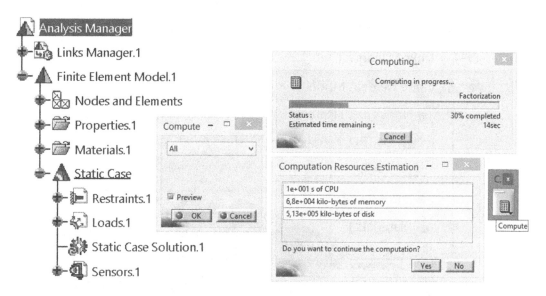

FIGURE 4.19 Computing the analysis, showing its progress and the resources needed.

After solving the finite element problem, the user advances to the post processing phase, to find and highlight the results. The main postprocessing toolbar is *Image,* shown in Figure 4.21. To view the deformed shape of the connecting beam the user presses the *Deformation* icon and the resulting shape is also represented in figure. Note that *Links Manager.1* is hidden in the specification tree, also *Mesh.1* is hidden, or it can be deactivated from its context menu (option *Activate/Deactivate*).

If the user decides to display the *Mesh.1* feature, both meshes (undeformed and deformed) are shown on screen (detail in Figure 4.22). The deformation image *(Deformed mesh.1)* can be very deceiving because the user could have the (wrong) impression that the analysed beam actually displaces/deforms to that extent. As presented in Chapter 3, the deformation is scaled considerably, in an exaggerated manner, for the user to view and understand the process. The displacements are scaled using the *Amplification Magnitude* icon in the *Analysis Tools* toolbar.

Once a new image is inserted, all the previous displayed images are deactivated (and hidden from screen). Thus, in order to see the displacement field of nodes (Figure 4.23), the user clicks on the *Displacement* icon in the *Image* toolbar. Also, the user should note that the *Deformed Mesh.1* feature is deactivated in the specification tree.

The implicit display for displacement vectors is as simple arrows; their colour and length represent the size of displacements. The palette indicates a maximum displacement of 0.054 mm. The specification tree adds a new feature, *Translational displacement vector.1,* double-clicking on it opens the *Image Edition* selection box (Figure 4.24).

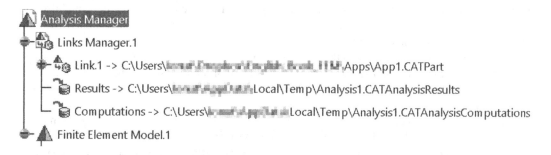

FIGURE 4.20 Locations for *Results* and *Computations* files.

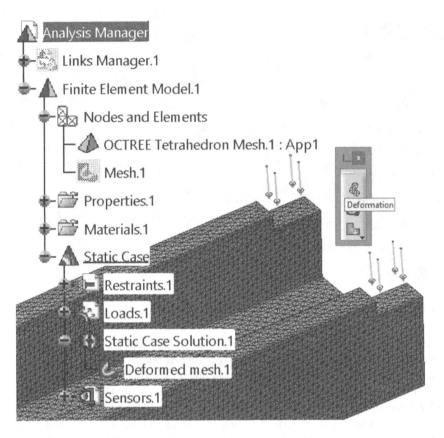

FIGURE 4.21   Showing the deformation of the connecting beam.

Many options are available in all tabs and under the *Options* and *More»* buttons. As an example, in the *Visualization Options* dialog box, the user is able to change the simple arrows with double arrows or points, with or without text (discrete values of displacements), dimensions of arrows, their property to be *Zoom sensitive,* etc. The contents of this dialog box depend on the selection in the *Image Edition* box.

For the next image, the user has to change the render style by using the *Shading with Material* icon in the *View mode* toolbar. The user must note that the finite elements are not showing in this render style, but he can also use the *Customize View Parameters* icon and check the *Edges and points* option in the *View Mode Customization* selection box (Figure 4.25).

The next step and image consists in applying the *Von Mises Stress* tool, its icon is, also, placed in the *Image* toolbar. *Von Mises Stress* provides information about the tensions/stresses that appear in the analysed part. This information is presented graphically by colours on the part, but also by values in the displayed palette. So, the user can compare the stress produced in the part with the yield strength of its assigned material.

After computing the restraints and the force applied on the connecting beam, *CATIA v5* displays the *Von Mises Stress* image in Figures 4.26 and 4.27 (detail around the hole). The user should note that the other two previous images, *Deformed mesh.1* and *Translational displacement vector.1,* are deactivated.

The part is visualized on screen, after the analysis, in a very colourful representation, from dark blue to green, yellow and red. The minimum stress is marked by the blue colour (9.74×10^3 N/m^2), the medium ones by green and yellow (4.6×10^6 N/m^2, 8.04×10^6 N/m^2), and the maximum stress, of course, by the red colour (1.15×10^7 N/m^2). These values are smaller than the yield strength of

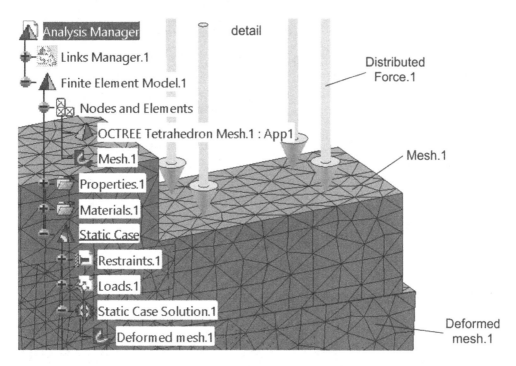

FIGURE 4.22 Showing together the undeformed and deformed meshes.

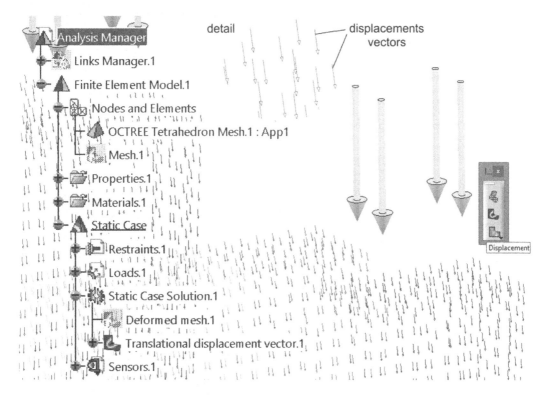

FIGURE 4.23 Showing the displacement vectors.

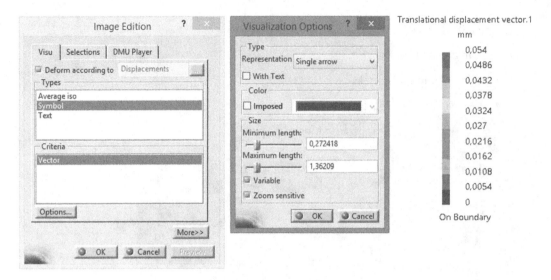

FIGURE 4.24   Options to view the displacement vectors and a palette with colours and values.

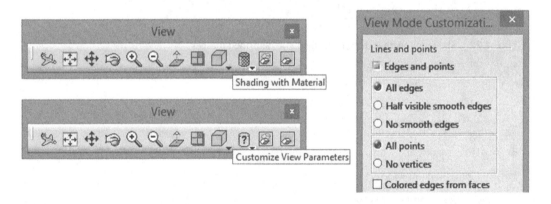

FIGURE 4.25   Options to display the part with its material and/or finite elements.

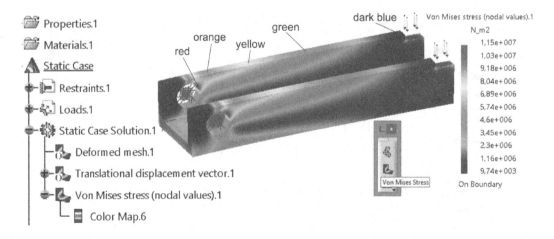

FIGURE 4.26   Stress distribution on the part model in *Shading with Material* visualization.

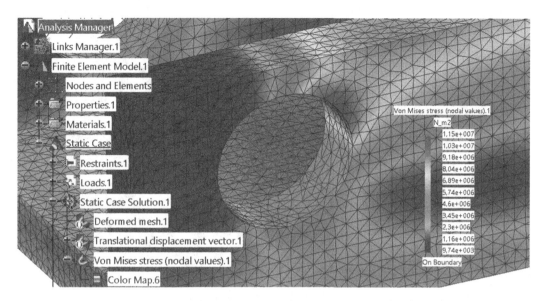

FIGURE 4.27 Stress distribution on the part model in *Customize View Parameters* visualization detail.

the chosen steel (2.5×10^8 N/m^2 in Figure 4.8), so the part has a normal behaviour, it works in the elastic domain, under the imposed conditions.

Double-clicking on any value of the palette or on the *Color Map* feature in the specification tree leads to opening the *Color Map Edition* selection box displayed in Figure 4.28.

The user can choose the number of colours used to display the *Von Mises Stress* image. To the right of the *Number of colors* field there is a button with three points, and by clicking it, several options become available. The *Smooth* option checked makes the colours pass smoothly from one to another. By unchecking it, the coloured areas displayed on the model are better separated, according to the Figure 4.29. The *Inverse* option changes the colours in the palette and on the model, so that the red colour is assigned to areas with minimum stress and dark blue to areas with maximum stress.

By clicking the *More≫* button, other options are available through which the user can choose the stress distribution mode (*Linear*, *Histogram* and *Logarithmic*), the precision of displaying the values in the palette, can impose a certain value, etc.

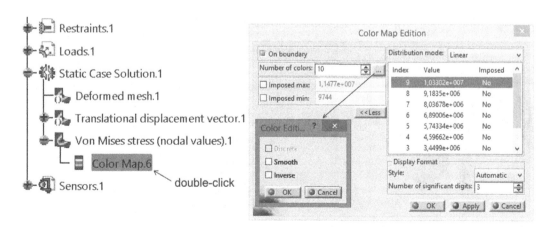

FIGURE 4.28 Options regarding the display of colours on the model and in the palette.

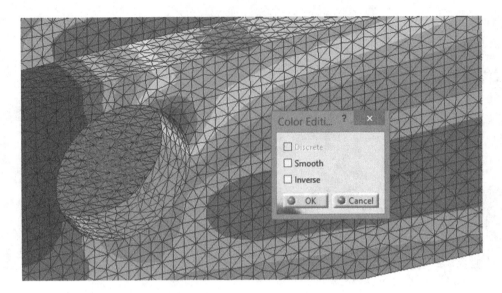

FIGURE 4.29   Displaying colours on the model with the *Smooth* option unchecked.

It is possible that the user wants to find out the stress in certain particular areas of the part, such as the area of the *Clamp* restraints or the area where the loading force is applied. By double-clicking on the *Von Mises stress (nodal values).1* image/result in the specification tree, a selection box is displayed, *Image Edition* (Figure 4.30). In the *Selections* tab, the user is able to move *Clamp.1* and *Distributed Force.1* from the upper field, *Available Groups*, to the lower field, *Activated Groups* using the simple arrow located between these fields. Figure 4.30 shows the *Von Mises stress* for the elements placed in the areas where the part is clamped and the distributed force is applied.

The image and values presented in Figures 4.27 and 4.29 show the stresses of the outer elements, but the user may see inside the part with the *Cut Plane Analysis* tool. Thus, the hidden stress results become available. Generally, the fracture in the part's volume starts from the outer surface, with the outer elements of the mesh. So, finding the results inside the volume is not that effective in analysis, but they are still important.

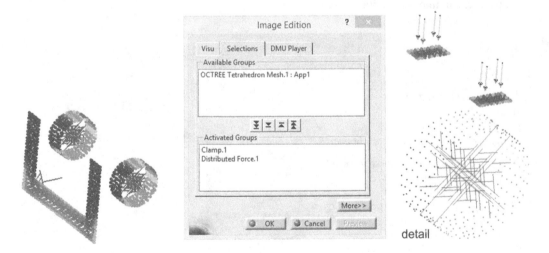

FIGURE 4.30   Displaying *Von Mises stress* only for clamping and loading force areas.

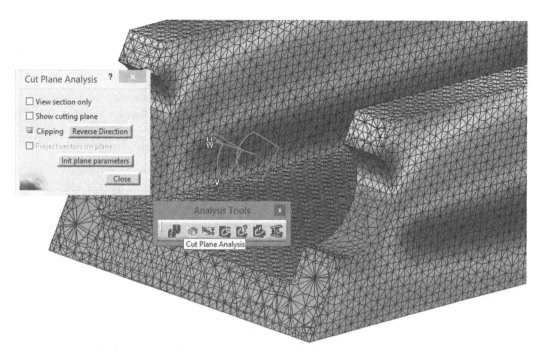

FIGURE 4.31 Displaying *Von Mises stress* inside the part's volume.

Figure 4.31 shows the part being cut near the clamping areas and a selection box, *Cut Plane Analysis*, opened by using the tool with the same name in the *Analysis Tools* toolbar. The compass is, also, shown; it is used to move and rotate the cutting plane. This plane was displayed during these manipulations, then hidden by unchecking the *Show cutting plane* option. The *Clipping* option is checked to remove a volume (that is cut from the model) above the cutting/sectioning plane.

The user should also note that during the manipulation of the cutting plane, the *Von Mises stress* palette with values is locked, and this information is placed under the palette's title.

The location and magnitude of the extremum values of an image/result (like *Von Mises stress*) can be identified using the *Image Extrema* tool in the *Analysis Tools* toolbar. The user opens the *Extrema Creation* dialog box (Figure 4.32) and, if the default values are maintained (1 in both fields under *Global*), the global maximum and minimum are displayed at their location pinpointed on the part. These values are also saved and placed in the specification tree under the *Extrema* feature.

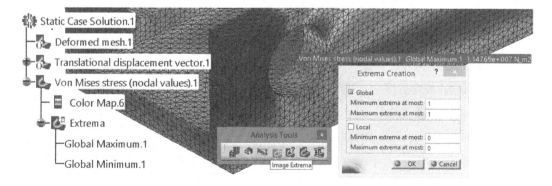

FIGURE 4.32 Displaying *Von Mises stress* extreme values on the analysed part.

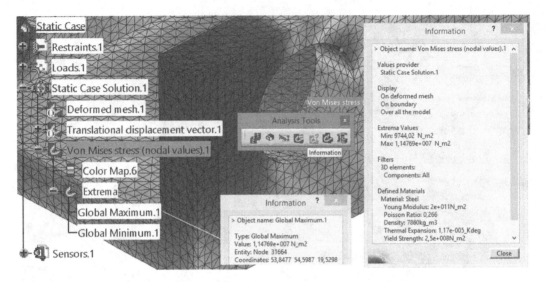

FIGURE 4.33 Displaying *Von Mises stress* information.

The user is able to display an information box about a selected image (Figure 4.33). In this figure, the *Von Mises stress* image was selected, which is active in the specification tree and in another selection, the *Global Maximum* extrema. The *Information* box is only shown for active images. Thus, the user should note that, in this case in Figure 4.33, the selection of another image (*Deformed mesh* or *Translational displacement vector*) is not allowed because they are deactivated (presence of the ( ) symbol).

If the selection is an extrema, the *Information* box displays the type, value, number and coordinates of the respective node.

For this part, three images/results were generated and the user may display one or more of them side by side, with the condition to be active in the specification tree.

Once the *Images Layout* icon is pressed, a dialog box with the same name is available on screen (Figure 4.34) and the user has to enter a distance (between the images) and to select an axis or a plane as a positioning direction. Thus, the images are displayed close to each another, but this status is not saved in the specification tree as a feature.

Before saving the files of the considered part and of the computed analysis, the user may take advantage of many other tools to see and understand the behaviour of the part under the restraints and the imposed load.

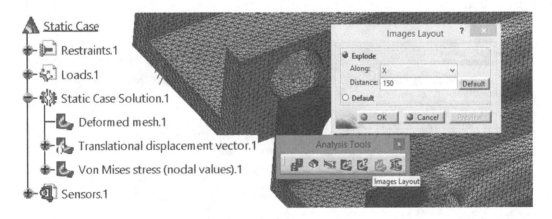

FIGURE 4.34 Displaying multiple images side by side.

Thus, the *Animate* tool shows an animation of how the stress propagates in the part's volume. Also, it is recommended to generate and save an HTML-based report which summarizes the features and results of the finite element analysis model of the part. This web page is full of information and images, yet easy to understand and to upload on a dedicated site to share the results.

4.2   ANALYSIS OF A PLATE SUPPORT

In the application, the finite element analysis of a plate support will be performed, with the 2D drawing represented in Figure 4.35. To fulfil its functional role, the 3D model of the part is provided with two flat support surfaces (left–right) and a connecting flat surface (middle), on which a load will be applied.

The user starts to create the part's 3D model using the *CATIA v5 Part Design* workbench and assign it the name *App2*. Thus, in the *XY Plane*, a first sketch is drawn. This is a *Centered Rectangle* of 120 × 45 mm, symmetrical to both *H* and *V* axes of the first sketch coordinate system, with the intersection of the rectangle's diagonals in the origin point of the sketch. The lines are horizontal and vertical and constrained to the dimensions from the 2D drawing.

Figure 4.36 presents how the profile is extruded along the *Z* direction over a distance of 12 mm by applying the *Pad* tool.

Another 18 × 45 mm rectangle is drawn at the base of the *Pad.1* plate and extruded in the negative direction of the *Z* axis for a distance of 13 mm. The *Pad.2* feature is created according to Figure 4.37, then it is copied symmetrically to the *YZ* plane to obtain the *Mirror.1* feature.

In the middle of the upper flat surface of the part, the user draws a circle having a diameter of Ø18 mm. Based on it, a *Pocket.1* feature is created through the part (Figure 4.38). In this case, the *Hole* tool was not used because it is possible that this cutout will no longer be cylindrical in another future version of the part. In such a case the user only has to replace the circle in *Sketch.3*.

At the base of the part, under the *Pad.1* plate and between the *Pad.2* and *Mirror.1* features, a line is drawn in *Sketch.4*. This is used to create a stiffening feature *Stiffener.1* (Figure 4.39), with a very important role in the strength of the part. It has a thickness of 6 mm and it is symmetrical with respect to the *ZX* plane.

The stiffening feature is copied symmetrically to the *YZ* plane and *Mirror.2* is obtained. Further, the user applies some fillets in the area of these stiffeners, of radii R6 and R2 mm, the features

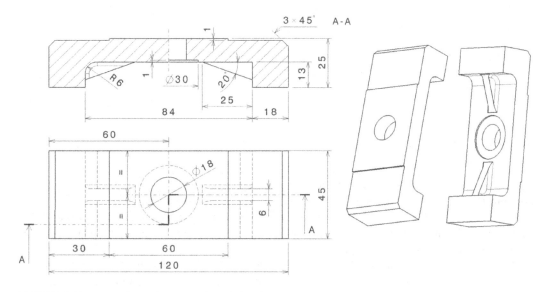

FIGURE 4.35   Two-dimensional drawing of the plate support.

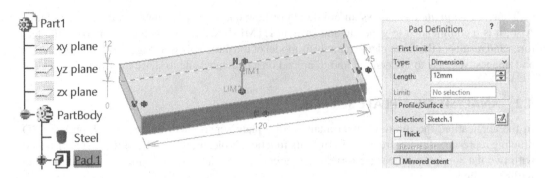

FIGURE 4.36 Extruding the first sketch along the Z+ axis.

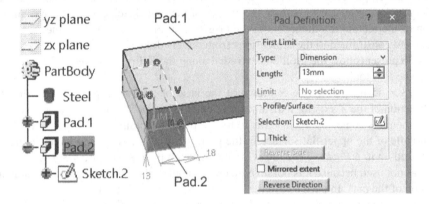

FIGURE 4.37 Extruding the second sketch along the Z– axis.

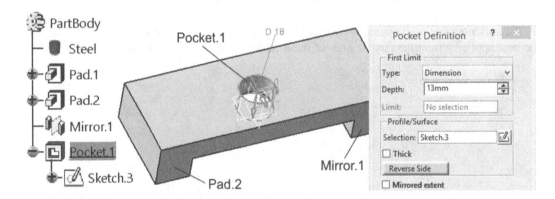

FIGURE 4.38 Creating the *Pocket.1* feature.

EdgeFillet.1 and *EdgeFillet.2* are added in the specification tree. Figure 4.40 also shows the Ø30 mm *Pocket.2* feature created at the base of the part, concentric with *Pocket.1*.

On the top flat part of the part, in *Sketch.6,* the user draws a rectangle to create a *Pocket.3* cut on the depth of 1 mm (Figure 4.41). One side of the rectangle is 30 mm from the *YZ* plane, and the other sides exceed the edges of the part.

Pocket.3 is copied symmetrically with respect to the *YZ* plane and the *Mirror.3* feature is obtained, then a 3 × 45° *Chamfer.1* is applied to the end edges of the upper flat face (Figure 4.42).

If the 3D model of the part is correct, it should have a volume of 81317.53 mm³.

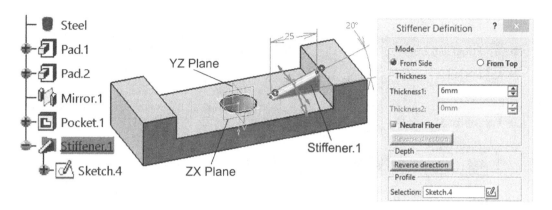

FIGURE 4.39 Creating the stiffening feature.

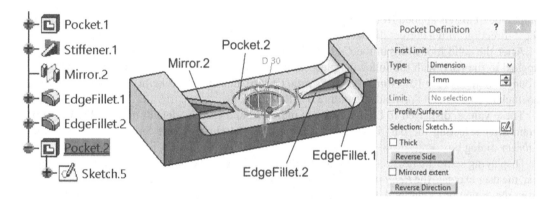

FIGURE 4.40 Mirroring the stiffening feature, adding fillets and cutting the *Pocket.2* feature.

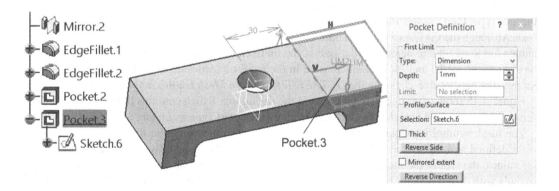

FIGURE 4.41 Creating the *Pocket.3* feature.

Let's consider that the user wants to test this part under a 300 kg load laid on the middle top flat surface. This mass is converted into a force of 2940 N by multiplying it with the gravitational acceleration. This application presents only the plate support and not the floor or any other part with which it comes into contact.

Knowing this, the analysed part requires a boundary condition on the contact faces with the floor. These faces are placed at the bottom of the part, being created by *Pad.2* and *Mirror.1*. A force applied on the top face of the plate support may reach an equilibrium status when the same

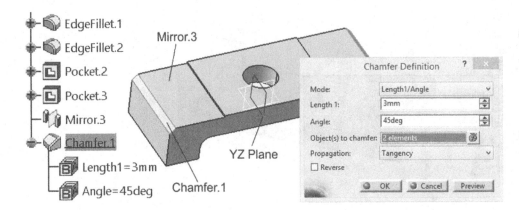

FIGURE 4.42 Creating the last features of the part, including the chamfering of two edges.

amplitude of load is applied from an opposite direction. The plate support is maintained static along the *Z* direction with a correct displacement restraint on the bottom faces.

After the solid modelling in the *CATIA v5 Part Design* workbench is done, a steel material is applied to the part, with the following physical properties, that are important during the analysis: Young's modulus (2×10^{11} N/m²), Poisson's ratio (0.266), density (7860 kg/m³), thermal expansion coefficient (1.17×10^{-5} K), yield strength (2.5×10^8 N/m²).

These values, displayed in Figure 4.43, are indicated and used by default in the *CATIA v5* program after selecting the part in the specification tree and choosing the *Steel* material from the *Library* dialog box.

Displaying some analysis results requires a different view of the model. Thus, on the *View* toolbar, the user expands the *Render Style* group of icons and chooses *Customize View Parameters*, then from the dialog box (Figure 4.44) that appears, he checks the *Shading* and *Material* options. As a result, the model acquires and displays a dark grey colour, with metallic reflections, specific to this visualization mode.

The user accesses the *CATIA v5 Generative Structural Analysis* workbench and sets the *Static Analysis* type, then the specification tree simultaneously displaying the feature with the name *Static Case*. Although the *CATIA v5* program implicitly defines the network of nodes and elements, it is recommended to edit it and determine the size of the finite elements, the maximum tolerance between the discretized model and the real model used in the analysis (*Absolute sag*), the element type, etc. For this, the user double-clicks on the *OCTREE Tetrahedron Mesh* feature in the specification tree.

Figure 4.45 shows the specification tree and the dialog box which contains the *Size* of the finite elements (5 mm), the *Absolute Sag* (1.25 mm) and the *Element Type* as *Linear*.

The mesh symbol is displayed on geometry (Figure 4.46) meaning that the part's mesh is correctly defined and the user is able to proceed with the analysis phases. By changing the *Size* and *Sag* values, the size of the mesh symbol is also modified. This symbol is, in fact, composed of two symbols with different colours: green for *Size* and blue for *Sag*, but it is selectable as a single entity. Note that the user may also double-click this symbol to open the dialog box in Figure 4.45.

PartBody\Material	Steel
`Steel\Steel.1.1\Young Modulus`	2e+011N_m2
`Steel\Steel.1.1\Poisson Ratio`	0.266
`Steel\Steel.1.1\Density`	7860kg_m3
`Steel\Steel.1.1\Thermal Expansion`	1.17e-005_Kdeg
`Steel\Steel.1.1\Yield Strength`	2.5e+008N_m2

FIGURE 4.43 List of physical properties of the steel material.

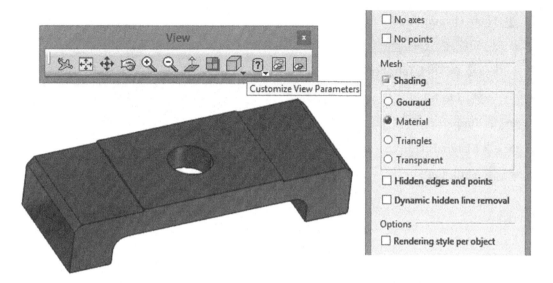

FIGURE 4.44 Changing the view mode of the part.

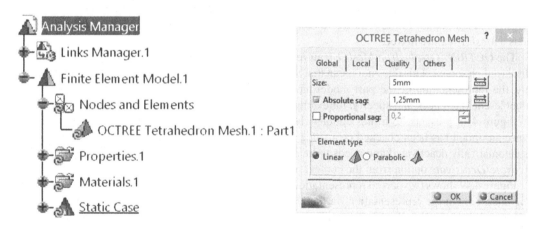

FIGURE 4.45 Discretization of the part model.

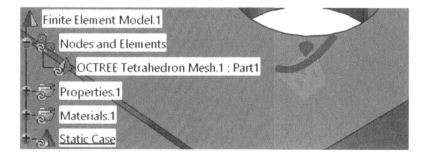

FIGURE 4.46 Showing the mesh symbol.

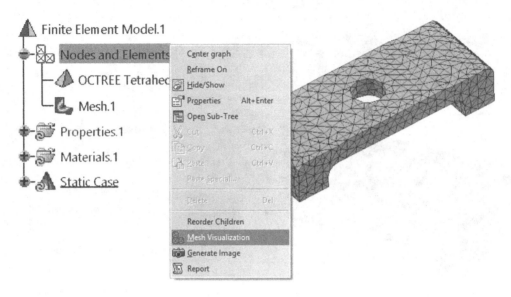

FIGURE 4.47   Mesh visualization.

For a first display of the created mesh, the user may right-click the *Nodes and Elements* feature in the specification tree and select the *Mesh Visualization* option (Figure 4.47) from the context menu. After a small amount of time required to compute the mesh, the program displays a message, then adds a *Mesh.1* feature in tree.

The *OCTREE Tetrahedron Mesh.1* feature and symbol should be hidden, to keep only the mesh displayed on screen. This does not mean that the *OCTREE Tetrahedron Mesh.1* feature and, implicitly, the representation of the part model have been removed. The user may open its dialog box (double-click) to check and specify, for example, the *Proportional sag* value (0.05 mm).

Figure 4.48 presents both *Sag* options checked, to be taken into account by the part's mesh. Once an option has been checked or a value changed, the *Mesh.1* feature in the specification tree is automatically deactivated and needs to be reactivated and updated. For this, the user applies the *Activate/Deactivate* option from the context menu.

Figure 4.49 shows two partial representations of the part without and with the option *Proportional sag* checked. On the representation on the right, the user can see the larger number of finite elements and their smaller size. The user should note that smaller mesh/better discretization has been

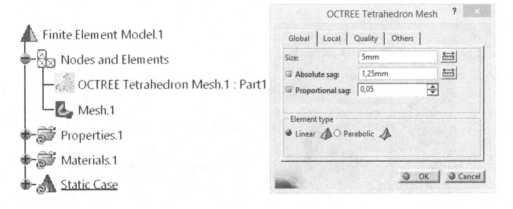

FIGURE 4.48   Activation of the *Proportional sag.*

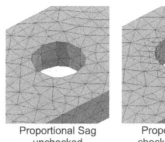
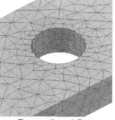

Proportional Sag
unchecked

Proportional Sag
checked, 0.05 mm

FIGURE 4.49 Different partial representations of the part depending on the checking of the option *Proportional sag.*

generated automatically by the program in and around the hole. Also, other fillet surfaces were adapted in the area of the stiffening elements (Figure 4.50).

The program considers that these curved surfaces are more complex than flat ones, thus requiring a better discretization to obtain correct and precise results. However, the computation time will increase significantly as the curved surfaces are more numerous and more complicated.

In the specification tree, *Links Manager.1* must become visible to present the part's geometry on which the loading force and restraints will be applied. To make correct selections, the *Mesh.1* feature should be hidden.

The user applies a *Distributed Force* of 2940 N, oriented perpendicular to the surface (Figure 4.51), in the opposite direction to the Z axis (the minus sign in the Z field).

A restraint is set on the two bottom surfaces. Restraining the Z directional displacement determines whether the part remains static against the distributed force. This restraint can be applied with the *User-defined Restraint* icon in the *Restraints* toolbar (Figure 4.52).

A dialog box is opened and the user selects the two bottom faces, then checks the *Restrain Translation 3* option, as shown in the figure. When a mesh is defined on the part's body, it was used the 3D tetrahedron element.

It is important to know that 3D elements do not have rotational degrees of freedom, so the *Restrain Rotation* options in the dialog box may remain unchecked. The numbers 1, 2 and 3 refer to global X, Y and Z axes/directions.

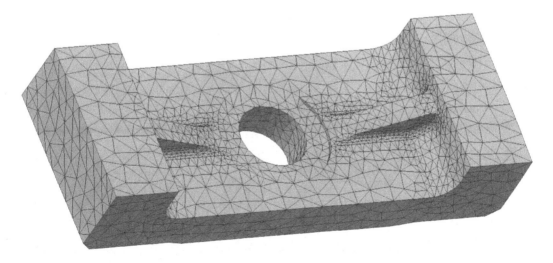

FIGURE 4.50 Automatic reconstruction of the part's mesh in the area of stiffening elements and filleted surfaces.

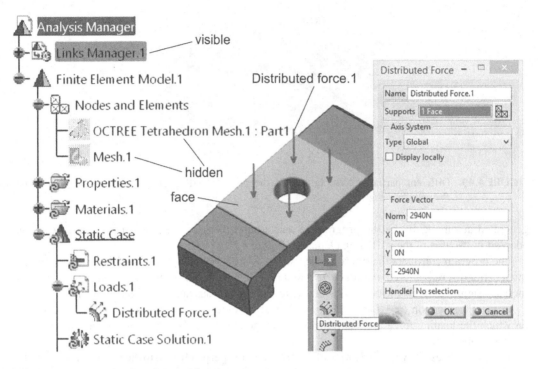

FIGURE 4.51 Application of the loading *Distributed Force*.

Once the material, restraint and load are applied, the user should check the finite element model for basic requirements. From the *Model Manager* toolbar, the *Model Checker* icon is clicked and the dialog box in Figure 4.53 opens. If the *OK* statement is present in the *Status* column, the created problem is correct and *The whole model is consistent*. This important message is displayed in the dialog box, which means that the model can move on to the computation phase.

By pressing the *Compute* icon in the toolbar with the same name, the user opens a selection box to choose what to analyse to solve the problem. After pressing the *OK* button with the *All* option,

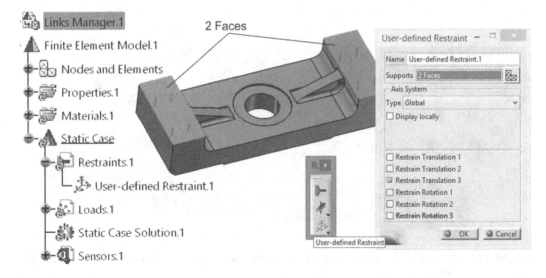

FIGURE 4.52 Application of the *User-defined Restraint*.

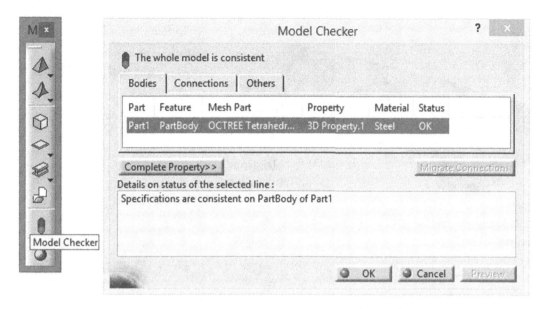

FIGURE 4.53 Checking the 3D model.

the *Computation Resources Estimation* information box becomes available on screen (Figure 4.54) showing the resources needed to complete the analysis.

Confirming the start of the analysis, the user expects a smooth solving of the problem and good results. An error message is displayed instead. It means that some degree of freedom is not correctly set. This type of message is quite frequent and often appears in the case of complex models of

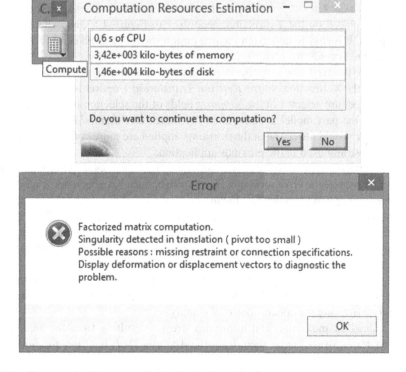

FIGURE 4.54 *Computation Resources Estimation* information box.

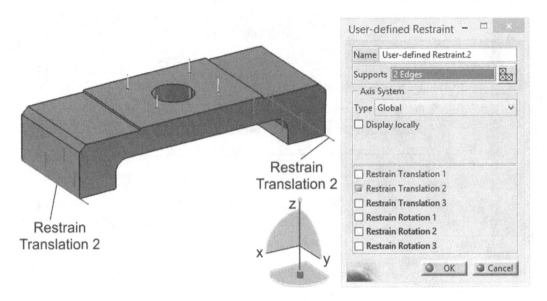

FIGURE 4.55   Adding *Restrain Translation 2*.

parts and assemblies, or when the user does not have enough experience to establish the restraints. Understanding and knowing how to deal with this kind of error is a very important issue of the linear static analysis.

The part was restrained only for the *Z* directional translation. This restraint seems quite logical because no loads were applied to the part on the *X* and *Y* axes. Even so, this is not correct in numerical analysis and the user should restrain the part on *X* and *Y* too, so no motion appears in these directions. The idea is that the part's motion induces an acceleration, correctly solved by dynamic analysis.

The user opens a new *User-defined Restraint* selection box and selects two long edges at the bottom of the part, placed on the *Y* direction, according to Figure 4.55, and checks the *Restrain Translation 2* option. It is easy to understand which option should be checked (2 for the direction of the *Y* axis) taking into account the compass.

Similarly, the user inserts a new restraint selecting the four short edges at the bottom of the part. These edges are on the *X* direction, so the *Restrain Translation 1* option is checked (Figure 4.56). Flat surfaces and edges are selected in the *Supports* fields of the selection box.

At this moment, the part model is constrained by three *User-defined Restraints*, presented in Figure 4.57. The user should observe that the restraints applied are more complex and versatile than the simple *Clamp* restraint used in the previous application.

Having these restraints applied to the part's model, the user resumes the computation phase and he should no longer receive the error message. If so, the user must save his work (Figure 4.58) using the *Save Management* option from the *File* menu.

After the computation is finished, the tools in the *Image* toolbar become available to the user to view the results. The specification tree is completed according to the inserted images, by default one (the last inserted) is active by deactivating the previous ones. Figure 4.59 shows a list of three images/results and their icons.

Figures 4.60–4.63 show four results/images, corresponding to the computation of the considered part model. The user should note that the deformations are presented graphically in an exaggerated manner to facilitate drawing the conclusions of the analysis.

To find the values for maximum and minimum stresses resulting from the analysis, the user activates the *Von Mises Stress* image, then, from the *Analysis Tools* bar, uses the *Information* tool to display the box with the same name.

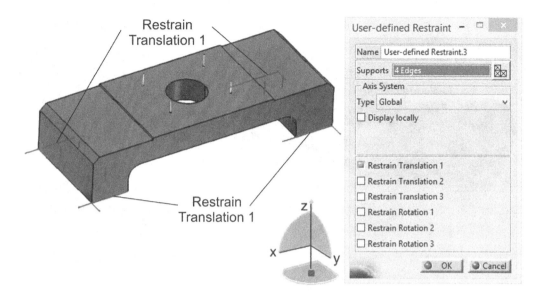

FIGURE 4.56 Adding *Restrain Translation 1.*

FIGURE 4.57 List of three restraints in the specification tree applied at the bottom of the part.

State	Name	Location	Action	Access
Modified	Analysis2.CATAnalysis	C:\Users\Ionut\Dr...	Save	Read Write
Open	App2.CATPart	C:\Users\Ionut\Dr...		Read Write
Open	Analysis2.CATAnalysisResults	C:\Users\Ionut\A...		Read Write
Open	Analysis2.CATAnalysisComputations	C:\Users\Ionut\A...		Read Write

FIGURE 4.58 Saving the files.

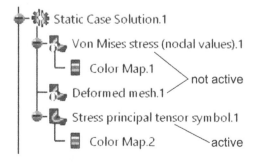

FIGURE 4.59 Active and inactive images in the specification tree.

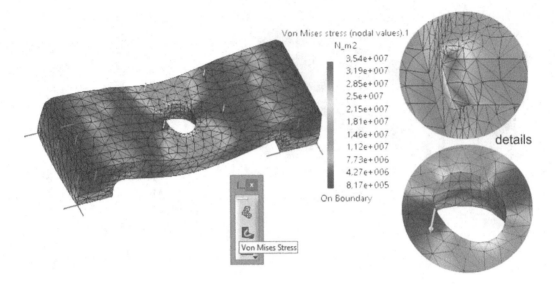

FIGURE 4.60   *Von Mises Stress* result.

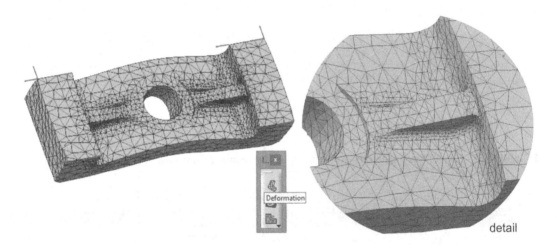

FIGURE 4.61   *Deformed Mesh* result.

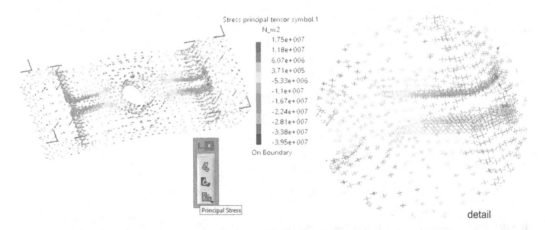

FIGURE 4.62   *Stress principal tensor symbol* result.

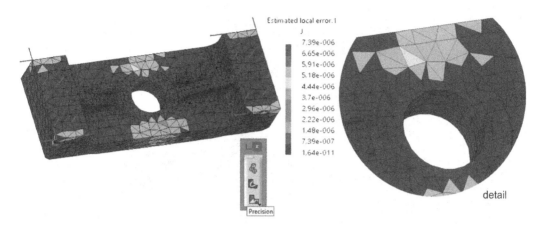

FIGURE 4.63 *Estimated local error* result.

In Figure 4.64, next to this information box, the colour palette that accompanies the results for the *Von Mises* image is presented. The lowest values of the stresses are at the bottom of the palette, the maximum at the top, but the box also contains these explicit values, in the *Extrema Values* area, as follows: Min: 8.17×10^5 N/m² and Max: 3.54×10^7 N/m².

Considering that the yield strength of the steel material is 2.5×10^8 N/m², it can be concluded that the part model will withstand the applied distributed force in the conditions of material and

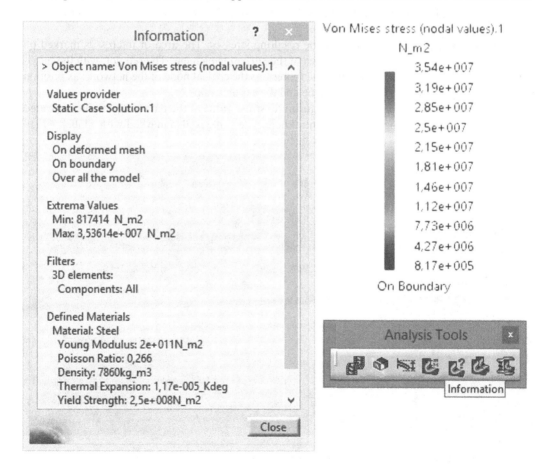

FIGURE 4.64 Displaying the maximum and minimum stress values.

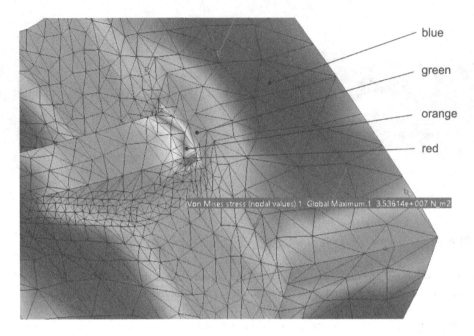

FIGURE 4.65 Marking the position of the maximum stress value.

restraints applied. This means that the part will only deform elastically and no cracks or breaks will appear under the action of the loading force.

Figure 4.65 shows the distribution of the resulting stresses. The area of interest is marked on screen by orange and red colours, in the area of the stiffening element. The red dot, with the maximum stress, value, also shown in the palette, represents the critical node of the network, its position being marked by a label containing the name and the explicit value.

Figure 4.66 shows the area where the minimum stress value of the part is highlighted, located between the stiffening element and the central hole. The value is also marked with a label, which

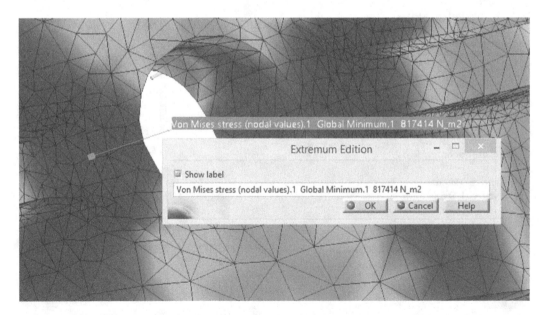

FIGURE 4.66 Marking the position of the minimum stress value.

FIGURE 4.67   Creation of a *Reaction Sensor*.

can be shown or hidden by double-clicking on the value and checking/unchecking the *Show label* option.

To obtain the reaction force for the considered restraints, the user has to define a *Reaction Sensor*. This reaction force is important because it is required to put the part's model in a static equilibrium.

At the bottom of the specification tree, the user finds the *Sensors* feature. By default, once the computation is done, it contains two sensors: *Energy* and *Global Error Rate*. The user is able to add more sensors by right-clicking on this feature and from the context menu he should select *Create Resultant Sensor* → *Reaction Sensor*, according to Figure 4.67.

The dialog box with the same name is displayed and a new feature is created in the specification tree: *Reaction Sensor.1*. The user should select in the *Entity* field (Figure 4.68) the restraint along the Z direction within the *Restraints* feature. As seen in Figure 4.52, the name of this restraint is *User-defined Restraint.1*. The *Axis System* is set on *Global* and the user clicks the *Update Results* button.

The force values in the fields below are displayed according to each direction X, Y and Z. Note that the reaction force along the Z direction is 2940 N, equal with the *Distributed Force*.

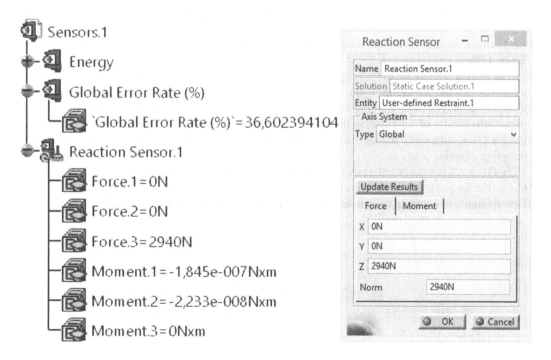

FIGURE 4.68   Calculating the reaction force for the first restraint.

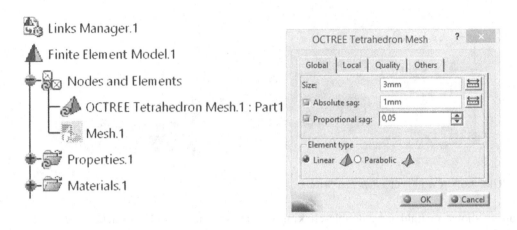

FIGURE 4.69   Mesh refinement to increase the number of finite elements and nodes.

It is recommended that the user determine the reaction force for the other two restrictions (*User-defined Restraint.2* and *User-defined Restraint.3*) as defined on the X and Y directions. Having no loads on these directions, the reaction forces should be zero.

Figure 4.68 shows a *Global Error Rate* sensor and its value is above 36.6%. Such a value is very high to be accepted in a finite element analysis. The error is, in fact, a possible deviation of the virtual, simulated model from the real part in real conditions of loads and restraints. A high value captured by this sensor means an imprecise FEM analysis that cannot be used to correctly simulate an important problem from a real project. Most often, decreasing the value of the error rate is done by refining the mesh and resuming the analysis. A convenient and usual value is between 10% and 15%, but there are cases in which a rate of 20% is also accepted.

To refine the mesh, the user has to double-click on the *OCTREE Tetrahedron Mesh* feature in the specification tree and the dialog box with the same name opens. According to Figure 4.69, the new values are set for *Size*: 3 mm and *Absolute Sag*: 1 mm. If the part had curved and/or complex surfaces, a good and recommended idea would have been to select the *Parabolic* option for *Element type*. In this moment, just by updating the *Global Error Rate* sensor, the new value becomes 30.5% (Figure 4.70).

After refining the mesh, the user applies the *New Adaptivity Entity* tool. Generally, this adaptivity consists in selectively refining the mesh in such a way as to obtain a desired results accuracy in a specified region. The mesh refining criteria are based on a technique called predictive error estimation, which consists of determining the distribution of a local error estimate field for a given *Static Analysis Case*. Adaptivity management consists of setting global adaptivity specifications and then computing adaptive solutions.

The *Global Adaptivity* dialog box opens (Figure 4.71), the user selects the *OCTREE Tetrahedron Mesh* in the *Supports* field and enters an *Objective Error*: 10%. Thus, the *CATIA v5* program has to adapt the mesh, to compute the whole problem and to try to reach this much smaller error rate. In a

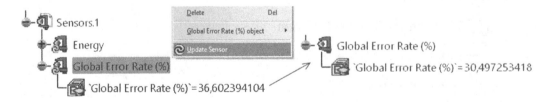

FIGURE 4.70   Updating the *Global Error Rate* sensor.

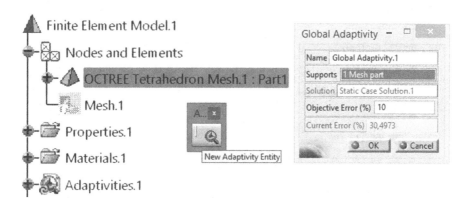

FIGURE 4.71   Applying the *New Adaptivity Entity* tool and setting a desired error rate.

non-editable field, the current value of the error rate is displayed after refining the mesh and updating the sensor. A new feature, *Adaptivities,* is added to the specification tree.

Once the new condition is imposed in the problem, the user applies the *Compute with Adaptivity* tool so that the computations take this condition into account. Figure 4.72 presents the *Adaptivity Process Parameters* dialog box and the user enters the *Iterations Number*: 2 and a *Minimum Size*: 2 mm for a finite element. Thus, the program is restricted to use finite elements smaller than 2 mm on any side of the tetrahedron. A greater number of iterations may lead to a smaller error rate and better results, but the time required for computations increases considerably, even for simple 3D models. Each new iteration takes results from the previous iteration(s) and tries to improve the mesh in order to obtain the required error rate.

Note that if the user continues to use the *Compute* tool, the conditions introduced by adaptivity are not taken into account.

If the desired/requested error rate in the dialog box of Figure 4.71 is not reached, the user receives a warning message (Figure 4.73) and the new value of the error rate is displayed (18.68%). The user may try a new mesh refinement and/or computing the analysis with more iterations.

Following these imposed conditions (adaptivity), the stress values (for example) change, as well their maximum and minimum locations. Figure 4.74 presents the palette with values for the *Von Mises* results, various details of the refined mesh and the new position of the minimum stress value. The colours are visible on screen from dark blue to red, according to the resulted values.

It is also observed in details that the maximum stress value no longer appears in the area of the stiffening element, but on the edges at the bottom of the part. Additionally, the values of the stresses

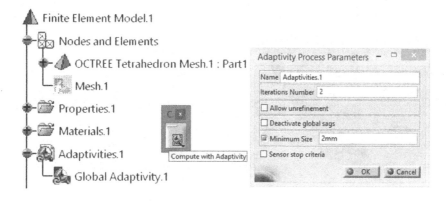

FIGURE 4.72   Computing the model with adaptivity in two iterations.

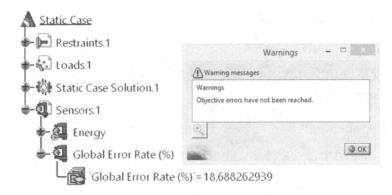

FIGURE 4.73 A warning message and the new value of the error rate.

around the central hole decrease, and the entire distribution of the stresses in the solid part looks smoother from one area to another.

Accepting the error rate value of 18.68%, the user clicks the *Displacement* icon in the *Image* toolbar and the *Translational Displacement vector* palette is displayed (Figure 4.75). Two other details are presented to mark the maximum displacement vector and to show vectors and colours in the stiffening element area. As seen in the colour palette (on screen), the maximum value is 0.0094 mm, too small to affect the part or its assembly.

To display the Z directional displacement, the user double-clicks the *Translational displacement component* feature in the specification tree and the *Image Edition* selection box (Figure 4.76) opens.

In the *Visu* tab, *Types* field, the user selects *Average iso* to change the representation of the part's mesh with displacements. Note that for the image results in Figure 4.75, *Symbol* type was used.

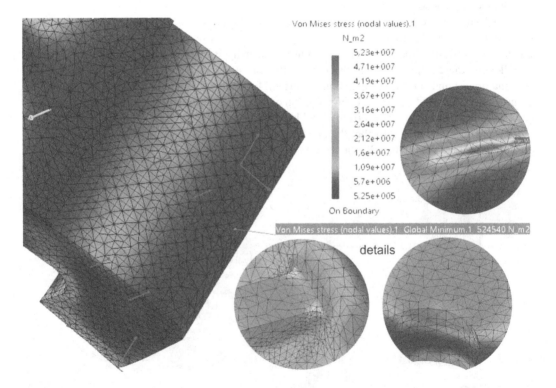

FIGURE 4.74 New colours palette for *Von Mises* results and various details of the new mesh.

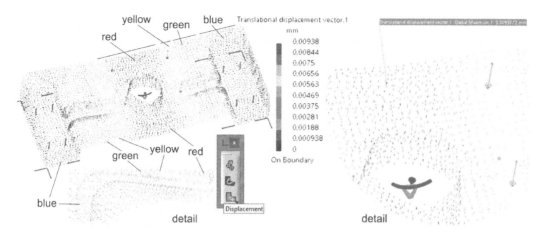

FIGURE 4.75 Palette for *translational displacement vector* results.

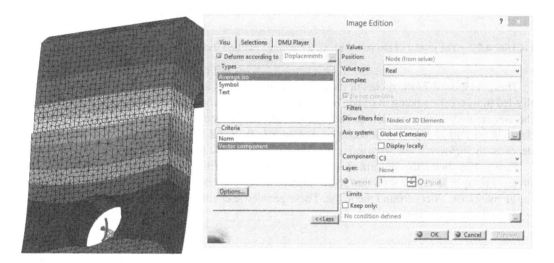

FIGURE 4.76 Showing the Z component of displacement.

Also, in the *Criteria* field, *Vector component* should be selected. Pressing the *More*≫ button the user extends the *Image Edition* box to the right to select *C3* (corresponding to the *Z* direction) in the *Component* list. This selection, *C1* (for *X*), *C2* (for *Y*) or *C3* (for *Z*) is possible only after the selection of the *Vector component* option. Figure 4.76 shows on screen a distribution of colours according to the displacement values and, obviously, the most affected area is the central area of the part, around the hole.

4.3 INFLUENCE OF FILLET RADIUS ON STRESS VALUES

The application uses the part analysed and presented by its 2D drawing in the previous application, but some changes are necessary due to the project update. The project manager asks that several constructive variants of the model should be taken into account. Thus, to obtain each new model of the part, updated according to these changes, the 3D solid should be created again and again. The user has an option to initiate a parametrization method, adapted to this model, that can be applied and by which certain features of the part are disabled, depending on his choice.

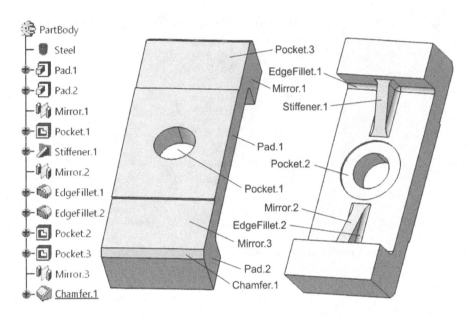

FIGURE 4.77 The model of the part with all its features marked and active in the specification tree.

Thus, in Figure 4.77 the model of the part from the previous application is presented, the features that led to its creation being marked. They are positioned successively in the specification tree, and the user created, during modelling, some dependencies between these features (examples: *Mirror.1* depends on *Pad.2, Mirror.2* depends on *Stiffener.1, Pocket.2* depends on *Pocket.1, EdgeFillet.2* depends on *Stiffener.1* and *Mirror.2,* etc.). The part is available to download.

Figure 4.78 shows the simplified model of the part, according to the changes imposed by the project manager. It can be seen that some surfaces have been removed: the central hole, the stiffening elements, the fillets around them, etc. These geometric elements, however, were not deleted, but

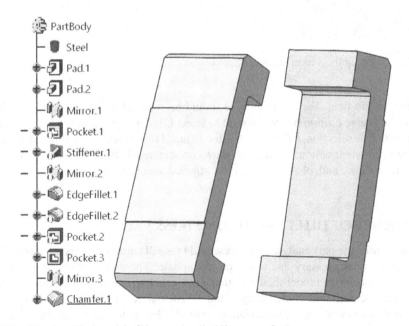

FIGURE 4.78 The simplified model of the part by disabling some features.

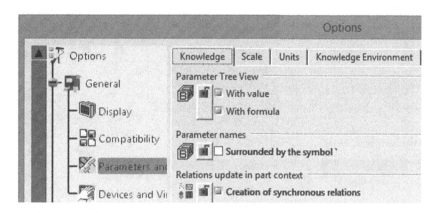

FIGURE 4.79 Displaying the value and formula of each parameter in the specification tree.

only deactivated (as features in the specification tree), so that the user can easily switch between the two models.

Thus, through a correct parametric modelling and the creation of a rule, the user can successfully fulfil this challenge imposed by the project changes. Deactivating the features by which the part was created can also be done manually by the user, but it is a difficult process if the part has a more complex geometry.

In the *CATIA v5 Part Design* workbench, a full set of options is opened from the *Tools →Options* menu for customizing the program, displaying the work environment, the presence of features in the specification tree, etc. In order for the user to be able to display his own parameters in the specification tree, he must check the *With value* and *With formula* options from the *General →Parameters and Measure* category, in the *Knowledge* tab (Figure 4.79).

In the same menu *Tools → Options → Infrastructure → Part Infrastructure → Display* tab (Figure 4.80), the user will check the *Constraints*, *Parameters* and *Relations* options so that they become also visible in the specification tree.

From the *Knowledge* toolbar, the user clicks the *Formula f(x)* icon and the *Formulas* dialog box opens (Figure 4.81). In order to display existing and newly inserted parameters, some filters can be established through the options in the *Filter Type* list, then the type of the

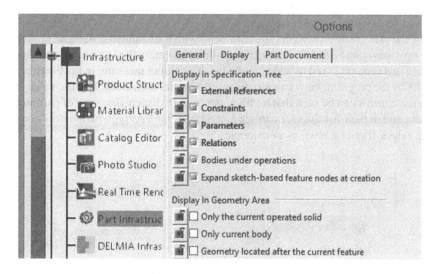

FIGURE 4.80 Display of constraints, parameters and relations in the specification tree.

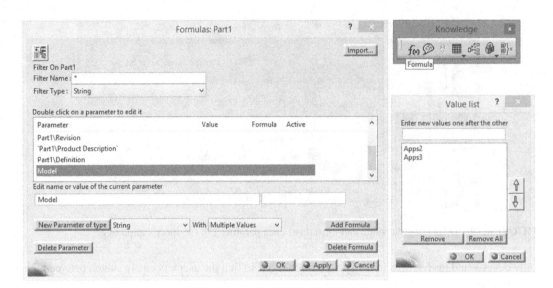

FIGURE 4.81   Creating a user parameter of type *String* with two values.

parameter to be created is chosen, using the *String* option in the list to the right of the *New Parameter of type* button. Also, so that the parameter can take two values, the user chooses the *Multiple Values* option. The user presses the button to create the new parameter and a dialog box becomes available for entering two names for the initial part and for the modified part (*Apps2*, respectively, *Apps3*).

In the *Value list* dialog box, the user types the first name, presses the *Enter* key, the parameter is created and goes down in the list, then the user types the second name. By pressing the *OK* button, the dialog box closes, and the user is asked to set a name for the parameter. Initially, a simple name is proposed (example: *String.1*), but to be suggestive and easy to follow throughout the code of the (future) rule, the *Model* name is chosen. The names of the parameters are case sensitive, so *Model* is different from *model*, and the values that can be assigned (*Apps2* and *Apps3*) must be used exactly in this syntax (for example, *apps2* or *Apps 2* cannot be used in the rule code).

Figure 4.82 contains the *Model* parameter with the value *Apps2*, the user double-clicks on it and opens an *Edit Parameter* selection box. The program displays a short list with the two possible values that can be assigned to the parameter.

Similarly, the user creates a *Length* type parameter with four values (2, 4, 6 and 8, in mm). In the field *Enter new values one after the other* (Figure 4.83) he enters the values, presses the *Enter* key after each one and they are saved in/added to the list. The unit of measure, in millimetres, is entered automatically by the program, but it can also be written by the user, along with the value (example: 8 mm). From the figure it can be seen that no filter was used to display the types of parameters (*Filter Type* list: *All*), and in their list the user can see *Length* and *Boolean* type parameters (the latter with the possible values *True* or *False*), as an example.

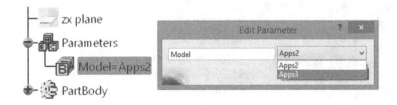

FIGURE 4.82   Displaying the *String* parameter *(Model)* and selecting one of the two values.

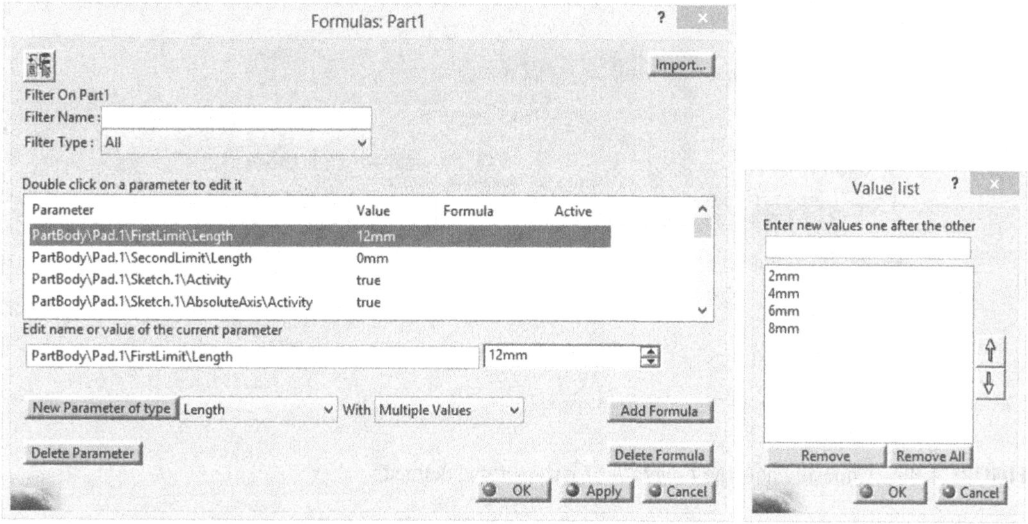

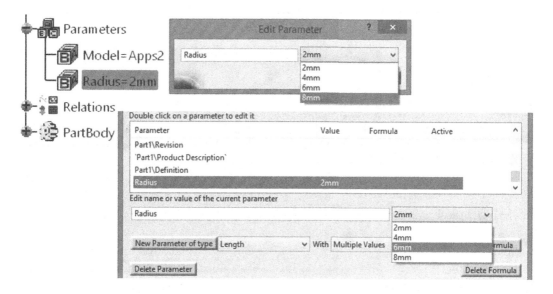

FIGURE 4.83 Creating a user parameter of type *Length* with four values.

When creating the parameter, *CATIA v5* proposes a name (example: *Length.1*), but the user will change it to *Radius,* according to Figure 4.84, in the field *Edit name or value of the current parameter.* To the right of the field is available the list with the four values in the order in which they were entered by the user.

After confirming the creation of the *Radius* parameter, it is added to the specification tree, along with one of the values it can take. By double-clicking on the parameter name, the *Edit Parameter* selection box and the list of values become available. The user has to remember that the unit of measurement is particularly important in *CATIA v5*: formulas, rules, checks, tables with parameters contain values together with these units of measurement.

The user must assign the four possible values to an intrinsic parameter, created automatically by the program during the modelling of the part. Thus, in the specification tree, the user double-clicks on the

FIGURE 4.84 Display of the *Radius* parameter of *Length* type and the way to select one of the four values.

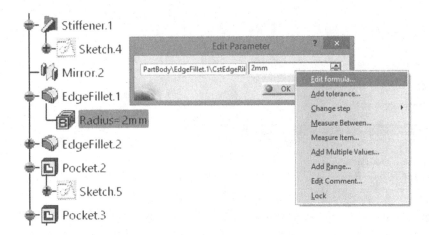

FIGURE 4.85 Choosing how the *EdgeFillet.1* parameter is defined.

Radius parameter within the *EdgeFillet.1* feature (Figure 4.85). The *Edit Parameter* box opens, and the user right-clicks in the field where the default value is written and a context menu becomes available.

From this menu, the user chooses the *Edit formula* option and opens the editing box in Figure 4.86. At the top of the *Formula Editor* box there are two fields: the first is not editable and contains the intrinsic parameter, and in the second the user selects the *Radius* parameter, created with the four values. Between the two fields is placed the equal symbol (=).

Figure 4.86 shows the complex name of the *Radius* intrinsic parameter, which contains the feature it belongs to, *EdgeFillet.1*. In fact, the user must refer to this parameter by its full name, *PartBody\EdgeFillet.1\CstEdgeRibbon.2\Radius,* as determined by the *CATIA v5* program.

Through the equality established by the formula, the intrinsic parameter is assigned four values, easy to select from a list, instead of a single value. There are many cases where design engineers create such user parameters with multiple values and equalize them with intrinsic parameters to highlight the respective parameters and simplify their editing. Also, through this formula, the intrinsic parameter will be able to take only the values from the established list, and this fact is visible in Figure 4.87 by the presence of the symbol $f(x)$. The figure also explicitly shows *Formula.1* in the specification tree within the *Relations* feature.

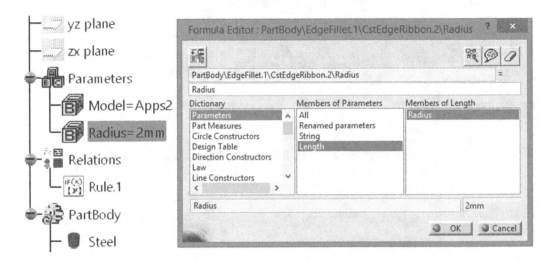

FIGURE 4.86 Establishing an equality relation between the two *Radius* parameters.

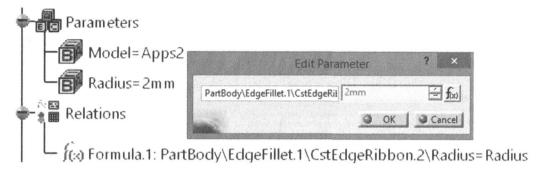

FIGURE 4.87   Displaying the symbol *f(x)* next to the first value of the parameter.

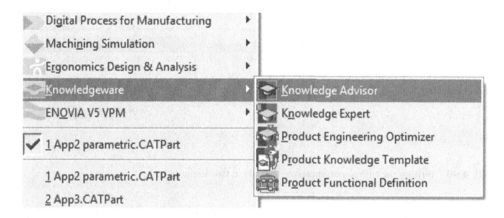

FIGURE 4.88   Accessing the *CATIA v5 Knowledge Advisor* workbench.

To create a rule to manage the model of the proposed part (between the initial and the modified variant), the user must access the *CATIA v5 Knowledge Advisor* workbench from the *Start →　Knowledgeware* menu (Figure 4.88).

From the *Reactive Features* toolbar, the user clicks the *Rule* icon and opens the *Rule Editor* information box (Figure 4.89). The name of this rule, a short description (who wrote it and when it was created) and its position in the specification tree are established.

By pressing the *OK* button, this information is confirmed, the rule is added within the *Relations* feature and its respective box opens for editing and entering a simple to understand *Visual Basic*

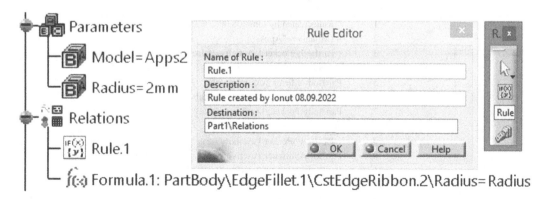

FIGURE 4.89   Establishing the name and position of the rule in the specification tree.

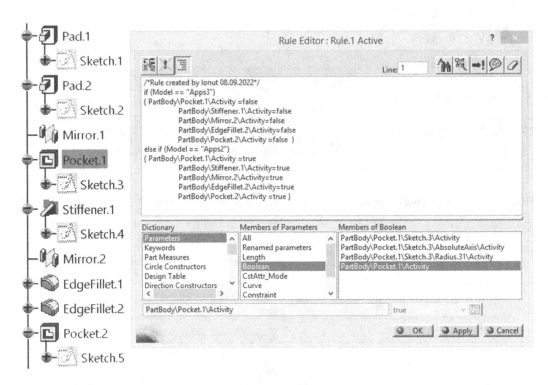

FIGURE 4.90 Writing the rule to parameterize the part in two variants.

code. Figure 4.90 shows the *Rule Editor* box which contains two main areas: the upper one where the rule code is written and the lower area divided into three fields (*Dictionary*, *Members of Parameters* and *Members of Boolean* in this case). The fields help the user in selecting the parameters to be inserted with a double-click in the rule code.

This code is presented in Table 4.1 and explained later. The numbers in front of the code lines are not taken into account, and are used only for easy identification of each explained code line.

In line 1 there is a comment automatically inserted by the program regarding who wrote the rule and when, on what date. Any other comment can be placed there by the user as long as it is between

TABLE 4.1

The Rule Code

1	/*Rule created by Ionut 08.09.2022*/
2	if (Model == "Apps3")
3	{ PartBody\Pocket.1\Activity =false
4	PartBody\Stiffener.1\Activity=false
5	PartBody\Mirror.2\Activity=false
6	PartBody\EdgeFillet.2\Activity=false
7	PartBody\Pocket.2\Activity =false }
8	else if (Model == "Apps2")
9	{ PartBody\Pocket.1\Activity =true
10	PartBody\Stiffener.1\Activity=true
11	PartBody\Mirror.2\Activity=true
12	PartBody\EdgeFillet.2\Activity=true
13	PartBody\Pocket.2\Activity =true }

/* and */. Also, such comments can be added anywhere in the code of the rule to provide certain explanations, mark the beginning of a logical loop, etc.

In the second line, the content of the *Model* parameter is compared with the value of *Apps3*. Between lines 3 and 7 the user placed the code of the rule that is executed if the comparison is fulfilled. The code is contained between curly brackets and disables the features in the specification tree that determine the removal of certain surfaces from the original model.

For this purpose, *Activity* parameters of type *Boolean* are used, one for each feature to be deactivated. The insertion of these parameters in the rule code is done by identifying them in the *Members of...* field. By double-clicking on each parameter, it is uploaded/inserted to/in the rule code, or the user may write it manually, but this requires a lot of attention and knowledge of the complex name of parameters. The *false* value is written by the user.

In Figure 4.90, at the bottom of the editing box, it can be seen that, by default, the value of the first *Activity* parameter (for *Pocket.1*) is set on *true,* and the respective field cannot be edited. Assigning the value of a parameter is done with the simple equal symbol (=). In lines 2 and 8, it was used the double equal symbol (==) because the current value of a parameter is compared with a value written by the user.

In line 8, the content of the *Model* parameter is compared with the *Apps2* value. If the condition is met, then it means that the user wants to restore the initial part, as it was created in the second application using the 2D drawing presented. Of course, the geometric elements are activated in the specification tree and the surfaces become visible on the part. The activation code is also enclosed in curly brackets.

The rule is simple to write if the syntax imposed by the program is respected, the correct names of the parameters are entered, the units of measurement are present after the values, there are no missing brackets, conditions, etc. Everything, however, is based on a clear algorithm, which must be previously established by the user, the written code of the rule being only the implementation of the algorithm in the *CATIA v5* program.

Without the presence of this rule, the user would have had to manually deactivate all the geometrical elements/features in the specification tree, then, if necessary, reactivate them in the same way.

If the deactivation is done manually, the order in which the features of the part are deactivated/activated is very important. For example, it is not possible/recommended to disable *Pocket.1* before *Pocket.2* because *Pocket.2* depends on *Pocket.1* by the way it was defined (concentricity between the two holes, and *Pocket.1* is the first feature).

Figure 4.91 shows an attempt to disable the *Pocket.1* feature and the message received from the program. The manual deactivation of a feature in the specification tree has nothing to do with the

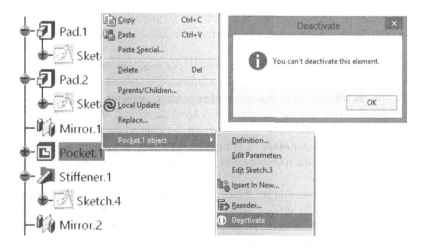

FIGURE 4.91 Attempting to disable a feature on which another is dependent.

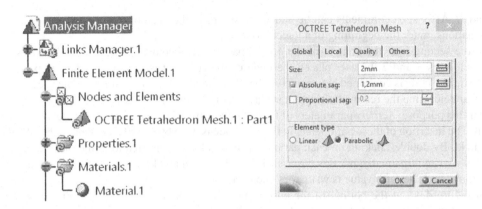

FIGURE 4.92 Refinement of the part mesh and definition of finite element type.

created rule, but it acts at once and quickly on all the features that must be deactivated/activated. Not included in the rule, the *Radius* parameter is not influenced by it, it can be modified by the user during the FEM analysis, acting both on the initial model of the part and on the modified one.

With these explanations, the parameterization phase of the considered part ends, and the user will apply a force and a restriction on the modified, simplified variant *(Model = Apps3)*. The first value considered for the *Radius* parameter is 2 mm. Moreover, the user can save this 3D model (together with the created parameters and formula).

From the *Start → Analysis & Simulation* menu, the user accesses the *Generative Structural Analysis* workbench for a static analysis, then applies the *Steel* material to the part (yield strength of 2.5×10^8 N/m^2). On the part model, the user notices the mesh symbol, but to define it, he double-clicks on this symbol or on the *OCTREE Tetrahedron Mesh.1* feature in the specification tree. This mesh symbol placed on the geometry means that the mesh has already been defined. The values for *Size* and *Absolute Sag* are set to be 2 mm and 1.2 mm, respectively, and for *Element Type*, *Parabolic* is chosen (Figure 4.92). These settings, being in the *Global* tab, are considered to be applied to the entire part.

There are, however, areas on the part that can be further refined, at a local level, in order to provide, after the analysis, better and more accurate results. For the analysed part, the user considers that these surfaces, which must be refined locally, are the fillet radii of the *EdgeFillet.1* feature (Figure 4.93). These radii are defined by the *Radius* user parameter, with four values, but local refinement is not required for each value that the parameter can take. Thus, on the *Model Manager* toolbar, the user clicks the *Local Mesh Size* icon and opens the selection box with the same name. In the *Supports* field, he selects the two fillets on the 3D model of the part or by choosing the *EdgeFillet.1* feature in the specification tree. For these surfaces, the size of the finite element is set as equal to 1.2 mm.

To visualize the new refinement of the part model, especially in the fillet area, the user right-clicks to open the context menu of the *Nodes and Elements* feature and chooses *Mesh Visualization* from the list of options (Figure 4.94). A high number of finite elements can be observed in the fillet area, and in the specification tree the features *Local Mesh Size.1* and *Mesh.1* appear.

To apply the restraint and the loading force, from the context menu, through the *Hide/Show* option, the feature *Finite Element Model.1* is hidden and *Links Manager.1* is displayed (it can be seen that at this moment it is hidden (Figure 4.94). Also, a *Shading with Edges* visualization can be used, according to Figure 4.95.

The user defines a *Clamp* restraint on the two flat surfaces at the base of the part model and a distributed force on its upper flat surface. Thus, the force is defined on the three directions of the *X*, *Y* and *Z* axes: 1000 N on the *X* axis and 2000 N on the *Z* axis, in its negative direction, therefore the value is negative. From the composition of the two components, applying the theorem of Pythagoras, the resulting value is automatically computed and verified, *Norm*: 2236.068 N.

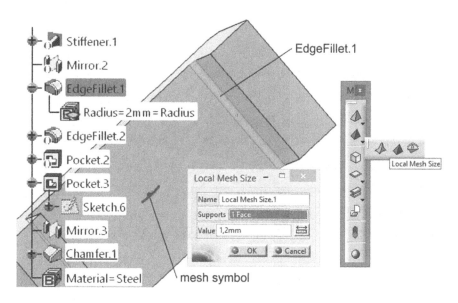

FIGURE 4.93 Application of a local refinement on the radius surfaces.

Figure 4.95 shows the application of the *Clamp* restraint and the distributed force on the simplified model of the part (parameter *Model=Apps3*).

The user is recommended to check the model to confirm its validity and if the restraint and loading conditions were correctly established. If *OK* appears on the last column *(Status)* of the *Model Checker* information box (Figure 4.96), the user has successfully completed all the steps preceding the finite element analysis.

The next step consists of successive computations of *Von Mises Stresses* for the four values of the fillet radii (imposed by the *Radius* parameter). The user changes the visualization mode to *Shading*

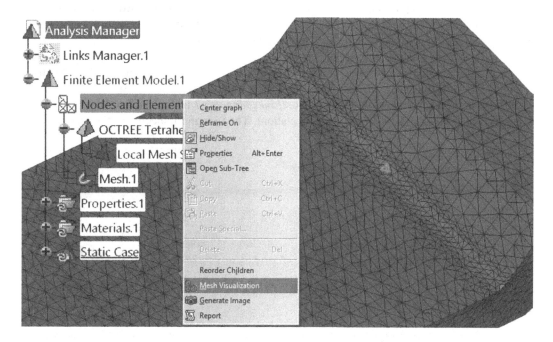

FIGURE 4.94 Displaying the mesh in the fillet area.

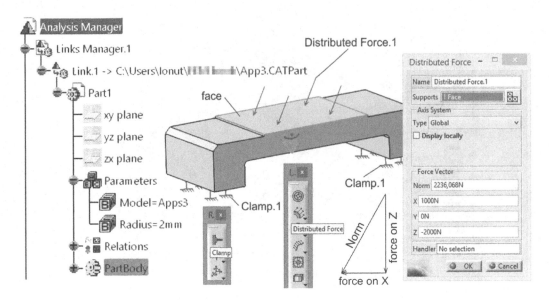

FIGURE 4.95 Establishing the *Clamp.1* restraint and the distributed force.

with Material, clicks the *Compute* icon and waits for the computation of the analysis. Knowing that the finite element type has been established as *Parabolic,* the calculus takes longer than if *Linear* elements were used.

Figure 4.97 shows the results of the *Von Mises Stress* analysis, the maximum stress value being 3.49×10^7 N/m² (34.9 MPa), lower than the yield strength of the material established for this part (2.5×10^8 N/m²). To display the minimum and maximum stress values, the *Image Extrema* tool is used.

As expected, the maximum value of the stress is located in one of the fillet areas, so this requires a more careful analysis. For this purpose, the user will create a group of the two fillet surfaces. He should hide the *Von Mises stress* result and makes visible the *Links Manager.1* feature in the specification tree, then changes the visualization mode to *Shading with Edges*.

From the *Groups* toolbar, the user clicks the *Surface Group by Neighborhood* icon, the selection box with the same name opens, and the user selects the fillet surfaces in the *Supports* field (Figure 4.98). Note the appearance of the *Groups.1* feature in the specification tree.

The user returns to the previous display state (*Links Manager.1* feature hidden, *Von Mises stress* feature visible, *Shading with Material* visualization). The user must specify to the analysis that

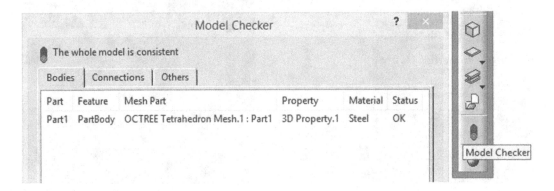

FIGURE 4.96 Checking the model.

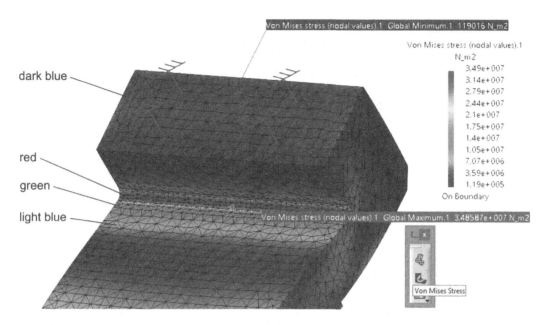

FIGURE 4.97 Displaying and locating the *Von Mises Stress* results.

he wants to display only the results from the area of the surfaces placed in the previously defined group. Practically, the transition is made from the display of stresses at the global level to a display of the stresses in a certain area of interest, at the local level.

The *Image Edition* selection box (Figure 4.99) is opened by the user by double-clicking on the feature *Von Mises stress (nodal values).1*. Then, he selects the group *Surface Group by Neighborhood.1* in the upper list *(Available Groups)* to move it to the bottom list *(Activated Groups)* using the simple arrow. Thus, the group becomes active and only the finite element analysis results for the two fillet surfaces are displayed on the screen.

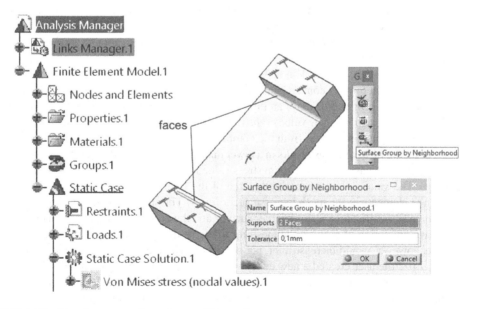

FIGURE 4.98 Creating a group from the two fillet surfaces.

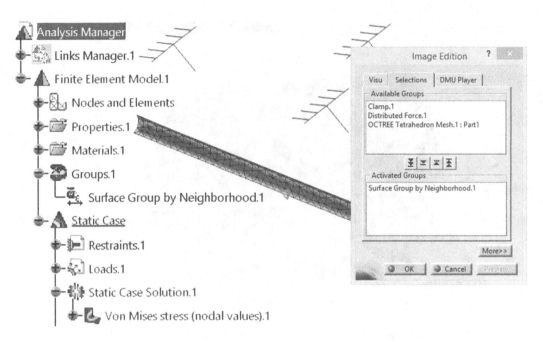

FIGURE 4.99   Displaying the *Von Mises stress* results on one of the fillet surfaces, *Radius* = 2 mm.

After selecting the group, it is not necessary to restart the computation by using the *Compute* tool, nor to update the analysis results, which is very important because this stage may require significant time resources. The maximum stress for the *Radius* = 2 mm case is the one found previously, of 3.49×10^7 N/m^2.

To change the fillet radii, the user double-clicks on the *Radius* parameter in the *Links Manager.1* → *Part1* → *Parameters*. From the list of available values, the user chooses the next value of the radius, 4 mm, and notice that the program has automatically switched to the *Part Design* workbench. The *Links Manager.1* feature becomes visible, and the 3D model of the part is displayed in the *Shading with Material* visualization. After modifying the radius, in order for the finite element analysis to take into account the new geometry and compute the stresses for the modified model, the user double-clicks on the *Finite Element Model.1* feature in the specification tree (Figure 4.100) to return to the *Generative Structural Analysis* workbench.

The user notices that the feature *Von Mises stress (nodal values).1* has become inactive following the radius change. From its context menu (right-click), the *Activate/Deactivate* option is used (Figure 4.101), and the program displays an information box because it needs to update the solution of the problem in the case of the new value of the radius. Moreover, the user is informed that this activation and update requires resources from the computer system and takes a certain period of time.

Re-computing the FEM analysis results in a lower maximum stress value of 2.69×10^7 N/m^2, compared with the previous case. This maximum stress value is also placed in the area of the fillet radius (Figure 4.100). The value is lower because the area of the fillet surface has been increased, so the number of finite elements in the mesh increase accordingly. The stress is better distributed between the finite elements, the fillet area forms a larger surface and the computation time also increases. The user could have the option of refining the mesh for fillets so that the results are more accurate. The calculus time, however, would increase a lot, but not necessarily the accuracy of the analysis.

In Figure 4.102 the user created a new entity with the help of the *New Adaptivity Entity* tool in order to selectively refine the mesh in such a way as to obtain a desired results accuracy in a specified region. It is observed, however, that the percentage/rate of error obtained is very small (<2%), which is due to the correct previous refinement with *Parabolic* elements.

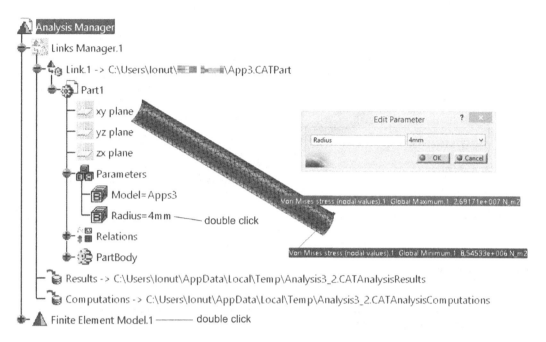

FIGURE 4.100 Displaying the *Von Mises stress* results on one of the fillet surfaces, *Radius* = 4 mm.

Of course, for such a percentage it is no longer necessary to repeat the computation, as the results of the analysis are considered to be very good. Figure 4.102 also shows the arrangement of the finite elements and, above all, their density in the analysed fillet area.

Similarly, the user can determine the values and positions of the maximum stresses in the other two cases, *Radius* = 6 mm, respectively, *Radius* = 8 mm, according to the parameterization made at the beginning of the application.

Some important information regarding the four cases is presented next: the values of the maximum stresses, the error rates, the number of finite elements and the number of nodes for the entire

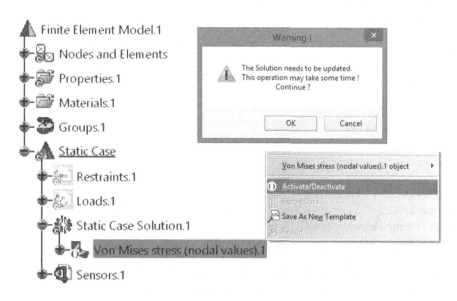

FIGURE 4.101 Activating and updating the FEM analysis result.

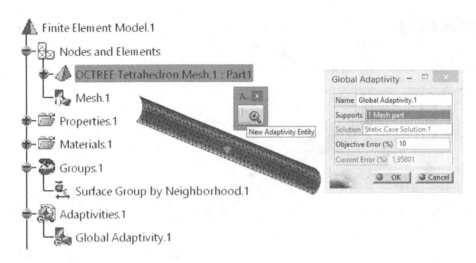

FIGURE 4.102   Display of the error percentage.

mesh of the 3D model. To determine some of this information, as well as others, the *Report* tool on the *Analysis Results* toolbar will be used.

Case 1. Radius parameter = 2 mm

Von Mises stress: 3.49×10^7 N/m^2
Error rate: 2.178%
Number of finite elements: 52,190
Number of nodes: 81639

Case 2. Radius parameter = 4 mm

Von Mises stress: 2.69×10^7 N/m^2
Error rate: 1.958%
Number of finite elements: 53865
Number of nodes: 84227

Case 3. Radius parameter = 6 mm

Von Mises stress: 2.26×10^7 N/m^2
Error rate: 1.824%
Number of finite elements: 58905
Number of nodes: 91527

Case 4. Radius parameter = 8 mm

Von Mises stress: 1.99×10^7 N/m^2
Error rate: 1.802%
Number of finite elements: 60847
Number of nodes: 94684

From the data presented synthetically in the previous list, the maximum stress decreases with the increase of the fillet radius and stays lower than the yield strength of the material from which the analysed part will be manufactured. Also, the error rate decreases and remains at very low values, below 2%. The number of finite elements and the number of nodes also increases, the time required for the FEM analysis increases because the mesh is more refined and uses *Parabolic* type elements.

The FEM analysis can continue if the user changes the initial conditions of restraints and loading forces or if the geometry of the part model is modified. Generally, it is recommended that the user save the files generated by the analysis before making any changes to the restraints, applied forces, part geometry, etc.

At the end of this application, as an additional topic, the user is challenged to establish a new variant, called *Apps_Stiff*, by modifying the rule code and introducing a new parameter. Thus, the stiffener is taken into account (parameter *PartBody\Stiffener.1\Activity = true*), and its width (parameter *PartBody\Stiffener.1\Thickness*) should take three values: 6, 7 and 8 mm. All the parameterization and computation steps will be repeated and the results will be compared highlighting the role of the stiffener in changing the value and position of the maximum stress.

4.4 ANALYSIS OF A CONNECTING ROD LOADED BY A VIRTUAL PART

In the application, the user will perform a finite element analysis on a part of type connecting rod, presented by its 2D drawing and isometric view (Figure 4.103).

The part has two cylindrical surfaces at both ends, provided with through holes, and at one end there is a keyway. At the other end, it can be seen that the hole has a conical shape. The cylindrical ends of the rod are connected by an arm, both of its flat faces being contour milled in the manufacturing phase to reduce the mass of the part.

The part has the role of transmission of movement/connection between two components of an assembly. One component is actuating, the other is actuated. For this purpose, the rod is provided with assembly, support and tightening surfaces, having a double T-shaped profile in section.

Modelling the part is simple and done easily in the *CATIA v5 Part Design* workbench. Thus, in the *XY Plane*, two circles of diameters Ø42 and Ø60 are drawn, being symmetrical to the *H* axis of the sketch's coordinate system. According to Figure 4.104, two tangent lines are drawn between the circles (applying the *Bi-Tangent Line* tool). The distance between the ends of the rod is 210 mm. Two circular arcs are removed from inside the profile using the *Quick Trim* tool. The user can check the correctness of this profile (if it is closed) using the *Sketch Analysis* option from the *Tools* menu. The profile is extruded with a value of 13 mm on either side of the plane containing the sketch by applying the *Mirrored extend* option in the *Pad Definition* dialog box.

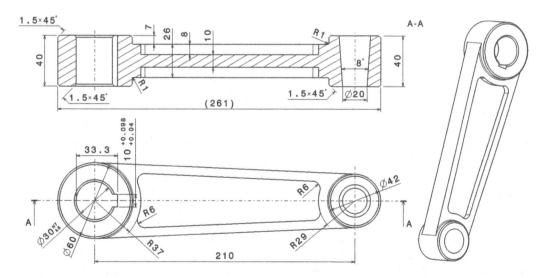

FIGURE 4.103   Rod type part.

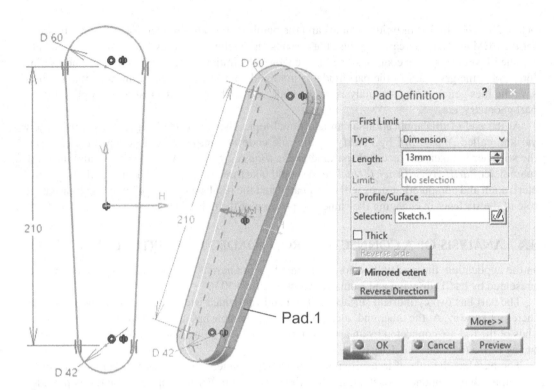

FIGURE 4.104   The first sketch and the creation of the *Pad.1* feature.

At the ends of the rod, in the same *XY Plane,* the user draws two circles coinciding with the circular edges, then extrudes them (with the *Pad* tool) on distances of 20 mm. Thus, two sketches (*Sketch.2* and *Sketch.3*) are used to obtain by extrusion the *Pad.2* and *Pad.3* features and not a single sketch and a single *Pad* feature. This is done because during the development of the project, the two *Pad* extrusions can become different values, and in this manner they are easier to edit.

In the *Pad Definition* dialog box from Figure 4.105 the user can see what the *Mirrored extent* option means: the same extrusion value (of 20 mm) of the selected sketch is present in the *Length* fields of the *First Limit* and *Second Limit* areas. The *Mirrored extent* option is available for selection only if the user has also chosen the *Dimension* option from the *Type* list. It is also observed that the extrusion is done in a direction perpendicular to the *XY Plane* in which the sketch was drawn. The *Pad Definition* dialog box is expanded to the right by pressing the *More≫* button.

Figure 4.106 shows the *Sketch.4,* drawn on one of the flat faces of the connecting rod. The sketch is used in a *Pocket.1* cut, at the depth of 8 mm.

This cut is also copied on the other side of the part, resulting in the *Mirror.1* feature. The user adds several fillets (*EdgeFillet* features) of radii R6 mm and R1 mm, a detail shown in Figure 4.107. It can be seen that each of these fillets contains several edges of the part.

At the end of the part with a diameter of Ø60 mm, a cylindrical through hole, *Hole.1*, is created, with a diameter of Ø30 mm. This hole is concentric with the circular edge of the *Pad.3* feature. On one of its flat sides, the user draws a rectangle with the dimensions and in the position of Figure 4.108. By cutting it out of the part, the *Pocket.3* feature is obtained, representing the keyway.

Similarly, the second hole is obtained at the other end of the part. *Hole.2* is conical on an angle of 8°. In the *Hole Definition* dialog box, the type of the hole, *Tapered,* can be observed in the list of options from the *Type* tab (Figure 4.109). The value Ø20 mm is entered in the *Diameter* field of the *Extension* tab.

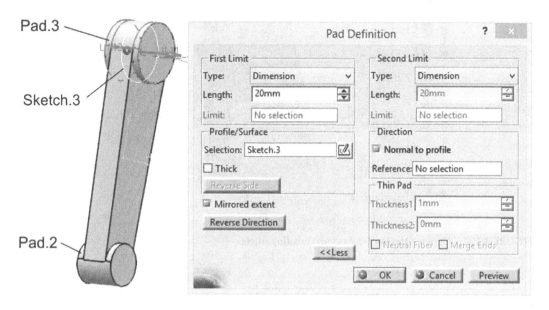

FIGURE 4.105 *Sketch.3* extrusion to obtain the *Pad.3* feature.

At the end of the part modelling phase, the user adds some $1.5 \times 45°$ chamfers, according to the 2D drawing, then applies a material (steel) to its solid, having the physical properties listed in the previous applications.

Figure 4.110 shows the *External Storage* dialog box to establish the files destination chosen by the user, as well as the structure of *Links Manager* features. Note that the analysis results and computations will be saved in the same folder as the part model file.

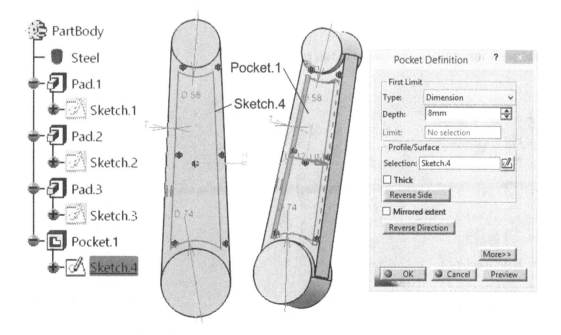

FIGURE 4.106 *Pocket.1* cut on one of the flat faces of the connecting rod.

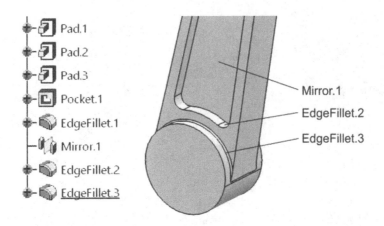

FIGURE 4.107 Copying the *Pocket.1* cut and creating fillets.

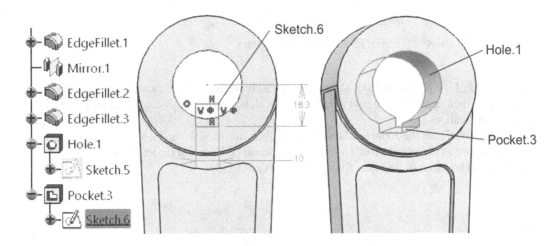

FIGURE 4.108 Creation of the *Hole.1* and the keyway.

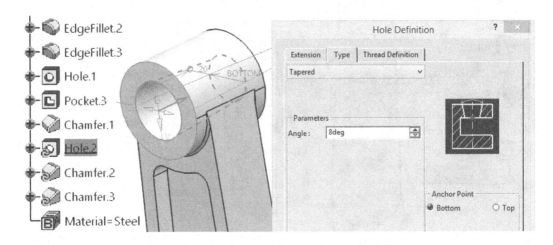

FIGURE 4.109 Creating the conical *Hole.2*.

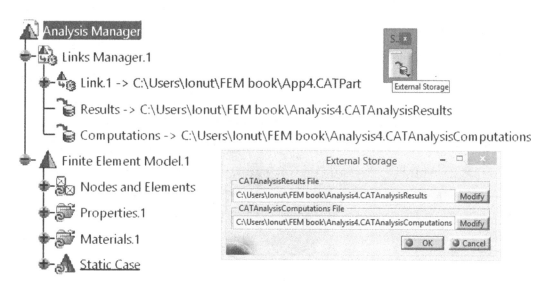

FIGURE 4.110   Setting up the saving destination of the analysis files.

The discretization of the network of nodes and elements of the model is done by double-clicking on the *OCTREE Tetrahedron Mesh* feature in the specification tree, and in the displayed dialog box the following are set: the finite element *Size* (3 mm), the minimum tolerance (0.5 mm) and the element type as *Linear* (Figure 4.111).

To apply the restraints, the *Clamp* tool is used on the active lateral surfaces of the keyway and on the hole surface (Figure 4.112). The restraints were thus established because there is an assembly between the considered part and another shaft-type connecting part (not shown). Setting the *Clamp.1* restraint ensures that there is no relative movement between the hole surface and the mating surface of the shaft, and also between the active side surfaces of the keyway and the key.

Considering that the analysed rod is detached from its assembly, the component with an actuation role is missing and its place must be taken by a virtual part. Thus, an assembly with conical surfaces is provided between the respective part and the rod, at the free end of the latter.

The application point of the force acting on it through a virtual part is positioned on the axis of the conical hole at a distance of 100 mm from the longitudinal plane of symmetry of the rod.

From the *Window* menu (Figure 4.113) the user chooses the part model *(.CATPart)* to create that force application point. The model can also be accessed by double-clicking on the *PartBody* feature in the specification tree.

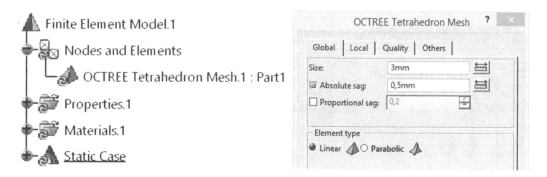

FIGURE 4.111   Discretization of the part model.

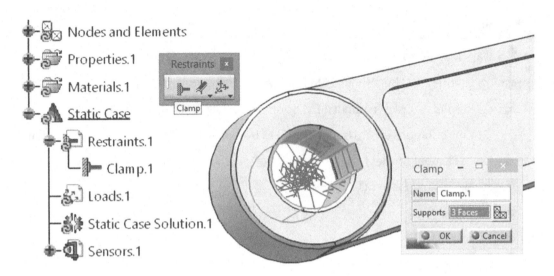

FIGURE 4.112 Applying the *Clamp.1* restraint.

In the *CATIA v5 Part Design* workbench, using the *Plane* icon from the *Reference Elements* toolbar, a new plane is created to be parallel to the *ZX Plane*, at a distance of 105 mm from it, so that the axis of the conical hole is contained in this new plane.

In the dialog box in Figure 4.114, in the *Plane type* field the user chooses the *Offset from plane* option, takes the *ZX Plane* as a reference in the *Reference* field, the distance between the two planes being given by the value entered in the *Offset* field. This value, as well as the reference plane, may differ depending on how the user built the model of the rod part.

In the newly created plane, called *Plane.1,* the user creates a point using the *Point* icon on the same *Reference Elements* toolbar. In the *Point type* field, the user chooses the *On plane* option, the *Plane* field contains the plane, and in the *H* and *V* fields, the user enters the coordinates of the point (Figure 4.115). In the considered case, the value H: 0 mm means that the point is on the axis of the conical hole, and the value V: 100 mm is the distance from the longitudinal plane of symmetry of the rod.

Next, in the *CATIA v5 Generative Structural Analysis* workbench, the user clicks the *Smooth Virtual Part* icon to display the dialog box with the same name (Figure 4.116).

In the *Supports* field, the user chooses the surface of the conical hole, and in the *Handler* field, the previously created point on the axis of the same hole. Figure 4.116 shows the selected point and

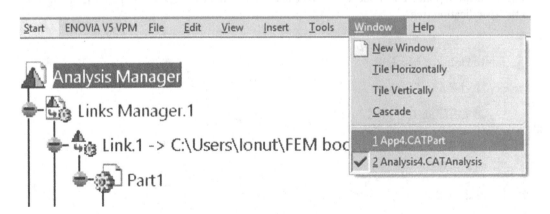

FIGURE 4.113 Accessing the rod model.

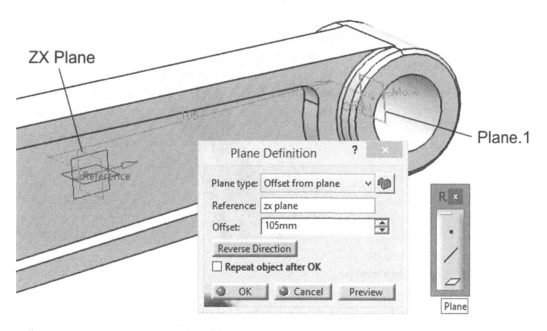

FIGURE 4.114 Creating a new plane parallel to the *ZX Plane*.

the symbol of the virtual part. This part has no geometry nor volume, it's virtual, but has only the role to transmit a load to the rod, without adding rigidity.

The virtual part was used because the analysis (and the user) must focus on the rod model only.

Figure 4.117 shows the phase of applying a distributed force at this point. The value and direction of the load are in the *Force Vector* fields. The support of the force is, in this case, only the previously created point, as a result, the symbolization consists only of an arrow, as presented in the adjacent detail.

The direction of the arrow corresponds to the negative direction of the *X* axis (given by the minus sign in front of the value). The rod model is now ready for computation.

Launching the computation phase is done using the *Compute* tool. Considering the parameters used in the discretization of the network of nodes and elements of the model, the required calculation time is significantly higher. Analysis results can be displayed on screen using the tools on the *Image* toolbar.

Figure 4.118 shows the minimum stress values (nodes of the network without the touch of the loading force, 0 N/m²), respectively, the maximum stress values (8.64×10^7 N/m²). The most stressed

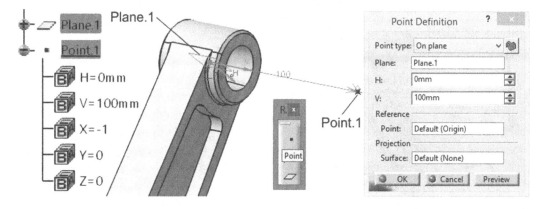

FIGURE 4.115 Defining the point to apply the loading force.

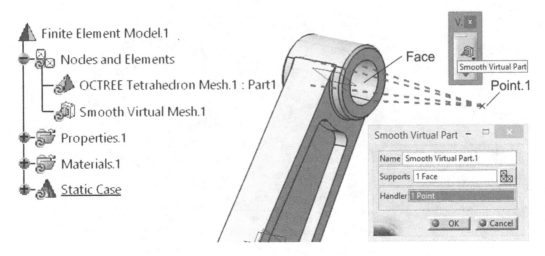

FIGURE 4.116   Creating the virtual part.

areas on the analysed model are highlighted using the *Image Extrema* icon on the *Analysis Tools* toolbar. The *Global Maximum/Minimum* and *Local Maximum/Minimum* features appear in the specification tree (Figure 4.119). Double-clicking on any of them displays an indicator containing the extreme value type, along with its value. The indicator is connected to the area where the respective extreme value was reached.

The elastic deformation of the model can be visualized in real time using the *Animate* tool located on the same *Analysis Tools* toolbar. A continuous sequence of successive frames is obtained, each frame (Figure 4.120) presenting the obtained result, with a certain amplitude of the deformation. The frames follow each other quickly, giving the impression of movement, the deformations are displayed in an exaggerated way, but they do not represent the real behaviour of the model.

Applying a loading force at the point of the virtual part leads to a bending and twisting deformation of the rod model, a fact clearly highlighted during the visualization of the animation.

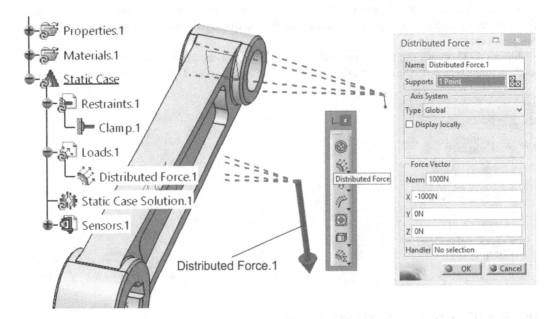

FIGURE 4.117   Applying the distributed force at the considered point.

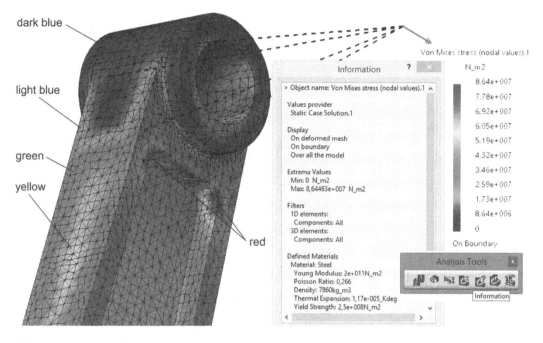

FIGURE 4.118 *Von Mises stress* analysis results.

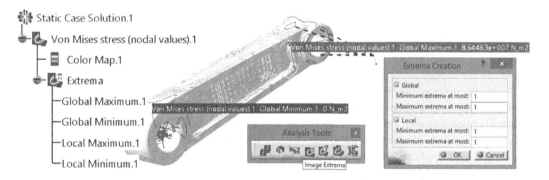

FIGURE 4.119 Marking of extreme stress values.

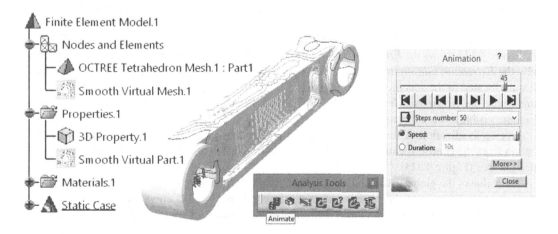

FIGURE 4.120 Animated visualization of elastic deformation.

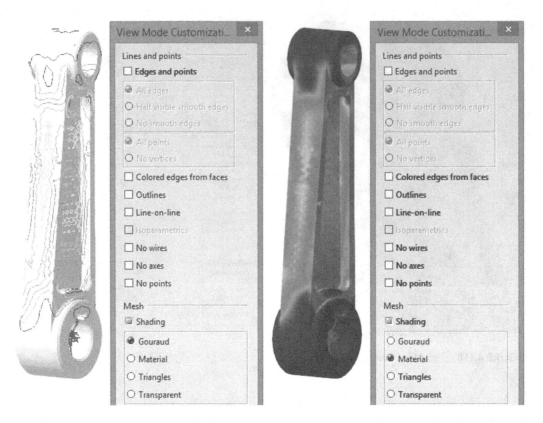

FIGURE 4.121 Comparative display of *Von Mises stress* results.

To display the part as in Figures 4.119 and 4.120, the *Gouraud* option was used in the *View Mode Customization* selection box (Figure 4.121, left). For the representation in Figure 4.121, right, the *Material* option was used and the distribution of stresses on the part is very visible. If the *Edges and points* option had been checked, the finite element network would have been displayed, as in the case of previous applications. Figure 4.121 compares the two display modes in the case of the *Von Mises stresses* determined after the analysis.

The maximum stress that appears is compared with the yield strength of the material proposed for the part and the result is that the rod model resists the applied stress (load and restraint). In the next phase, the user wants to determine the percentage of correctness of the performed computations. For this, the *Precision* tool on the *Image* toolbar is used, together with the *Image Extrema* tool. The areas containing the extreme values of global and local estimated errors are identified.

Figure 4.122 shows the *Precision* view of the model, two extreme values of estimated errors, but also the *Extremum Edition* information box obtained by double-clicking on the *Global Maximum.1* feature located in the specification tree (within *Estimated local error.1* → *Extreme*) and on the figure. Some *Image Extrema* features are not displayed, but by checking the *Show label* option they become visible as an indicator.

Precise localization of the area containing a certain error is done by choosing the *Focus On* option from the context menu (pressing the right mouse button) of the feature selected in the specification tree (Figure 4.123).

The value of the global estimated error can be found with the help of the *Information* tool. Thus, the information box with the same name is displayed (Figure 4.124), in which the user notes the *Global estimated error rate*: 36.5046%. The same value appears, of course, in the *Sensors.1* feature in the specification tree.

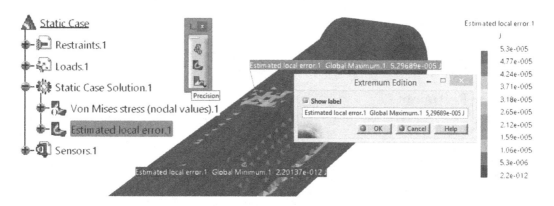

FIGURE 4.122   Identification of extreme values of global and local errors.

The percentage/rate of error is high because it basically represents the difference between the analysed model of the rod and this one, as a real part subject to real restraints and loads. The reduction of this value is possible by applying the *New Adaptivity Entity* tool. Thus, in the *Global Adaptivity* dialog box, in the *Supports* field, the user selects the *OCTREE Tetrahedron Mesh* feature (Figure 4.125), then, in the *Objective Error (%)* field, he enters the desired percentage: 15%. A possible cause of this error high value can be the simple discretization of the model mesh (*Size* = 3 mm and *Element type* = *Linear*, Figure 4.111).

Resuming the computation phase is necessary for the program to re-refine the rod model in an attempt to reach the error target set by the user.

Thus, the *Compute with Adaptivity* icon is clicked to display the *Adaptivity Process Parameters* dialog (Figure 4.126).

In the dialog box, in the *Iterations Number* field, the user sets the number of computation iterations to which the model will be subjected for analysis. The decrease in the size of the finite elements is also observed, from 3 mm to 2 mm in the *Minimum Size* field. Increasing the number of iterations and imposing a smaller error rate leads to a considerably longer calculation time, during which the *Computation Status* information box is displayed. During the analysis the user cannot work on the

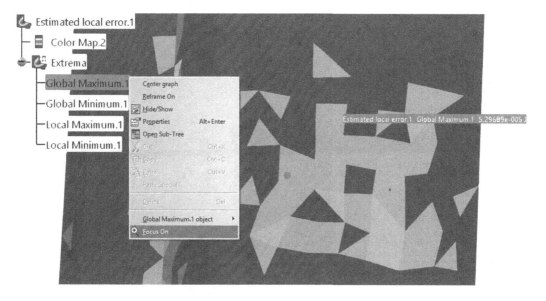

FIGURE 4.123   Locating a certain error on the analysed model.

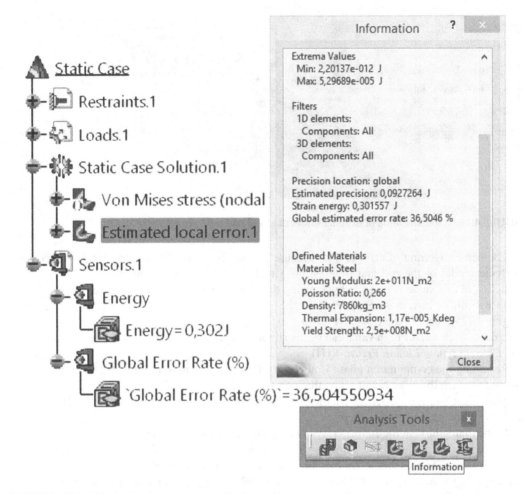

FIGURE 4.124 Displaying the percentage of estimated error.

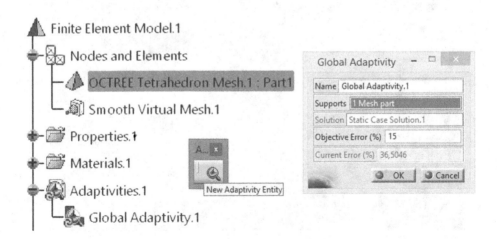

FIGURE 4.125 Establishing the desired percentage/rate of global error.

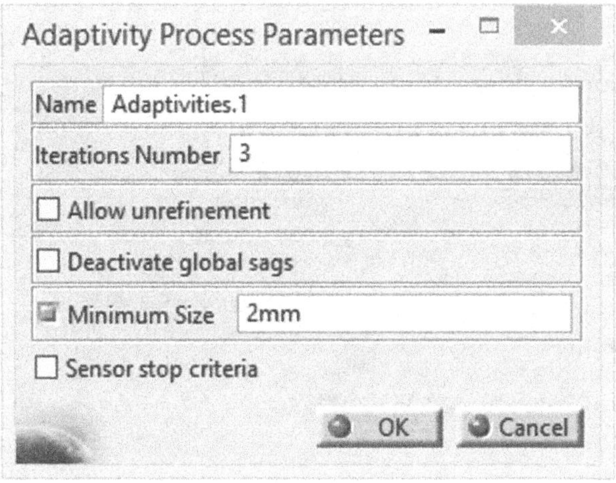

FIGURE 4.126 Resuming the computation phase.

3D model, cannot establish new conditions for the analysis, cannot view the resulting images, etc. At the end of the computation, the information/warning box from Figure 4.127 is displayed.

Following the computation based on the new settings, the value of 15% of the global error imposed by the user was not reached, but only 22.3435% (Figure 4.128), a much more convenient percentage than the previous one of 36.5046%.

Also, for the *Von Mises* result, there is a change in the value of the maximum stress, of 1.19×10^8 N/m^2, compared with 8.64×10^7 N/m^2, as resulted from the first computation, before applying the *Adaptivity* tools.

For better accuracy of the results, the user has the possibility to continue the analysis process, applying another refinement on the model, with the reduction of the minimum size of the finite elements and, possibly, by switching to *Parabolic* elements (and not *Linear* elements).

However, according to the current results, the user should note that the value of the maximum stress is only two times lower than the value of the material's yield strength and the error rate is still high. Even if the user's experience in interpreting the results of the FEM analysis says that the part will only deform in the elastic domain, a new analysis should take place under similar conditions, but with a different load.

Thus, the user can perform another analysis by creating a new case of static analysis with reference to the previous/actual case. For this new static analysis case it is possible to define a new load and/or a new restraint.

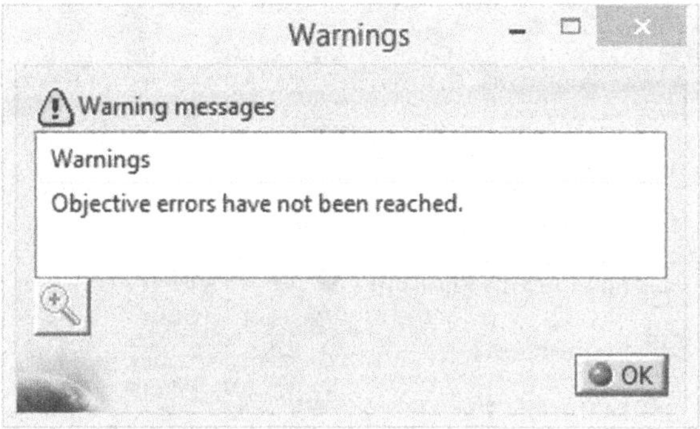

FIGURE 4.127 The user is informed that the desired error rate has not been reached.

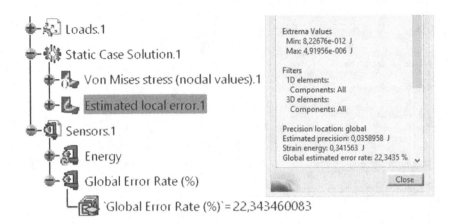

FIGURE 4.128   The new global error value.

Following the steps presented earlier, from the *Insert* menu the user selects the *Static Case* option (Figure 4.129).

Of course, the previous analysis and results are kept in the specification tree, but they will be hidden automatically.

By default, the name of the new analysis is *Static Case* (as for the first analysis), but the user will change it to *Static Case 2* (from the context menu → *Feature Properties* tab). The selection box from Figure 4.130 is displayed and the user chooses the *Reference* option for *Restraints* and a field becomes available in which the *Restraints.1* feature from the previous analysis will be selected. For *Loads*, the *New* option is kept because the user will establish a new load. Also, if the *Masses* option remains checked, the feature with the same name is added to the specification tree.

In Figure 4.130 it can be seen that the *Restraints.1* feature is contained by both *Static Case* analyses, as it is about keeping the conditions imposed by *Clamp.1*.

The new load is chosen by the user as being the *Bearing* type. The previous virtual part and the distributed force used for the present analysis are not taken into account.

In the conical hole, therefore, *Bearing Load.1* is applied, in two directions: on *X* with a value of 1000 N and on *Z* with a value of –1500 N. This load is distributed over an angle of 120° using a radial orientation and a sinusoidal type profile. In the *Supports* field, the user selected the conical hole, and in the *Distribution* field, an outward orientation of the load. All these options and *Bearing Load.1*, added to the specification tree, are presented in Figure 4.131.

Solving the newly created problem is done by pressing the *Compute* icon on the toolbar with the same name. The program displays the selection box from Figure 4.132 and from the dropdown list

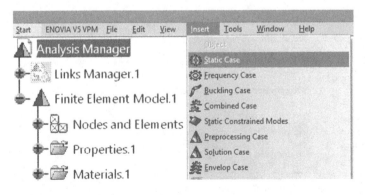

FIGURE 4.129   Inserting a new static analysis case.

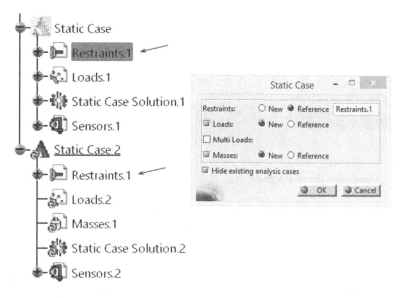

FIGURE 4.130   Settings to use the previous restraints.

the user chooses the *Analysis Case Solution Selection* option. In the *Solution(s) to Be Computed* field, the user selects the *Static Case Solution.2* feature from the context menu.

The program quickly checks the validity of the model, the load and the applied restraint, determines the resources needed for processing, then displays the information box *Computation Resources Estimation*. If the estimation values are zero or if the static case is not fully defined, the

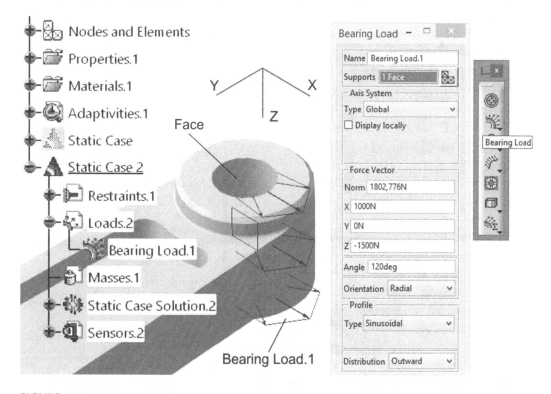

FIGURE 4.131   Applying the *Bearing* load.

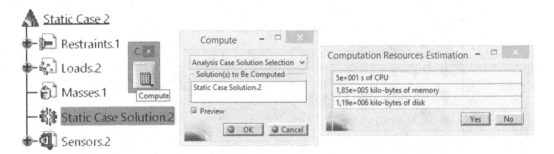

FIGURE 4.132 Computing the new *Static Case*.

user becomes aware that the result is impossible to be obtained. Otherwise, like in Figure 4.132, the user should press the *Yes* button and proceed with the computation step.

The user should note that this computation does not apply to both situations (*Static Case* and *Static Case 2*), as defined and presented previously. In the *Compute* selection box, the user has clearly specified that the calculation step applies only to *Static Case 2*. However, the computation can be applied to both solutions if the user has made some changes compared with the initial situation, but the calculation/computation time will increase significantly.

Next, the user wants to compare the results obtained by analysing the two cases: *Static Case* and *Static Case 2* by displaying them side by side.

From the specification tree, the restraints and loads are hidden (Figure 4.133) and a *Shading with Material* view mode is chosen using the *Customize View Parameters* icon on the *View* toolbar. The figure also shows that the user decided to make visible the *Static Case* feature.

To display on screen the analysis results for the stresses of the *Static Case 2,* the user clicks the *Von Mises stress* icon. Immediately, these stresses generated in the part by the restraint *Clamp.1*

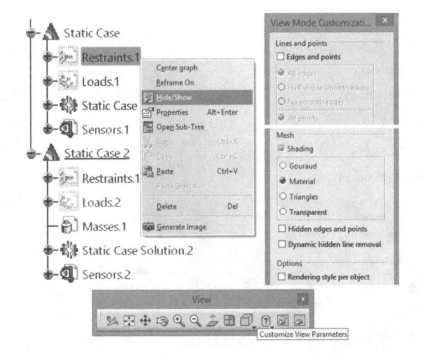

FIGURE 4.133 Hiding loads/restraints in the specification tree and choosing the *Shading with Material* view mode.

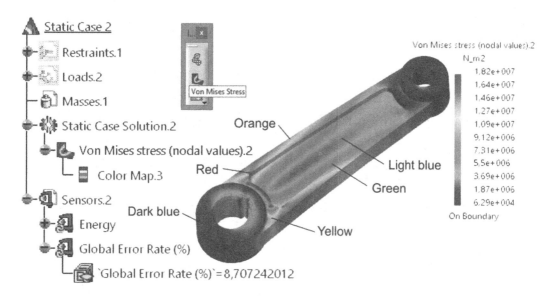

FIGURE 4.134   Showing the *Von Mises* results for *Static Case 2*.

and *Bearing Load.1* are displayed by colours on the part together with the corresponding values in the palette. First of all, it is observed that the stress value in the second case is lower than in the first case, and the error rate has a very convenient value, of only 8.707% (Figure 4.134).

In order to compare the stress values with each other, but also with the yield strength value of the part material (2.5×10^8 N/m²), the user must establish a scale to display the maximum and minimum values for each case, taking into account the maximum values calculated.

Thus, within the *Von Mises stress (nodal values)* features, the user double-clicks on the *Color Map*. The *Color Map Edition* box opens in which the user chooses the maximum number of colours to represent the stresses, imposes the minimum value (dark blue colour on screen) in the *Imposed min* field and has the possibility to set their maximum stress value in the *Imposed max* field.

It should be noted that these values entered by the user in the respective fields only influence the way the stresses are represented and, of course, do not change the values resulting from the computation.

Thus, in the two cases in Figures 4.135 and 4.136, the user can observe the palettes with the resultant values, while they kept the default values for the maximum stresses. The two *Von Mises stress* results can be displayed side by side using the *Images Layout* icon on the *Analysis Tools* toolbar.

Figure 4.137 shows the results and, apparently, the part is more stressed, with higher values, in the second case. This is, however, false, the maximum value for *Static Case Solution.2* is 1.82×10^7 N/m². The representation can easily mislead an inexperienced user because it uses different scales for displaying the maximum and minimum values, as presented earlier (the same value was not chosen in the *Imposed max* fields).

By pressing the *Images Layout* icon, the selection box with the same name from Figure 4.137 opens, the user ticks the *Explode* option, chooses the *X* axis along which to position the two *Von Mises stress* results and enters a distance of 150 mm. By checking the *Default* option, the results are displayed overlapped. Also, the different deformation of the part in the two cases can be observed.

This is why a common scale is important. To set up such a scale for displaying the maximum values, the user reopens the *Color Map Edition* boxes of each case, ticks the *Imposed max* option and enters the same value: 1.5×10^8 N/m². The distribution of colours and, implicitly, of the stress values calculated for each case changes, according to the representations in Figure 4.138.

By pressing the *Image Extrema* icon and selecting the *Von Mises stress* results in the specification tree for both cases, the places where the maximum stresses appear are accurately identified. It

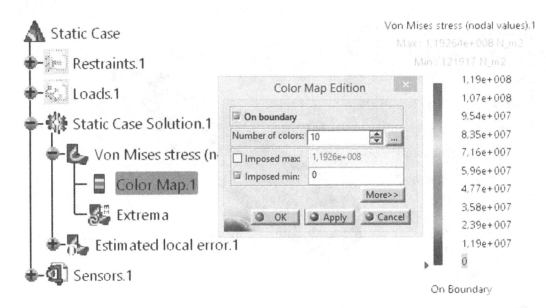

FIGURE 4.135 Adjusting the range of the *Color Map* for *Static Case Solution.1*.

can be seen that these maximum stresses are placed approximately in the same area of the part. The representations in Figure 4.138 are obviously different from those in Figure 4.137 even though the calculated stresses are the same. The difference is imposed by the common display scale of these stresses.

The maximum values of the stresses determined in this application are reasonably lower than the yield strength value of the chosen material (steel). Also, the error rates are convenient, but by refining the part's mesh it is possible to obtain a better precision of the computation.

The user should note that the application presents a static analysis for both cases. Thus, the load is applied very slowly until the part reaches an equilibrium state. When the load is applied very fast or repeatedly, the part would resist or break under a very similar loading condition.

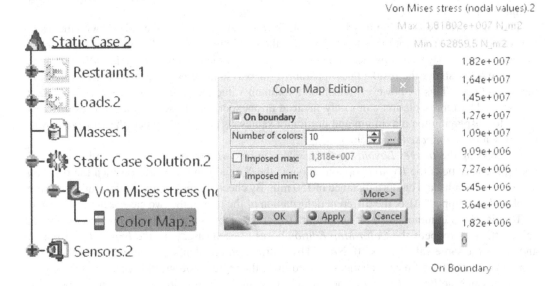

FIGURE 4.136 Adjusting the range of the *Color Map* for *Static Case Solution.2*.

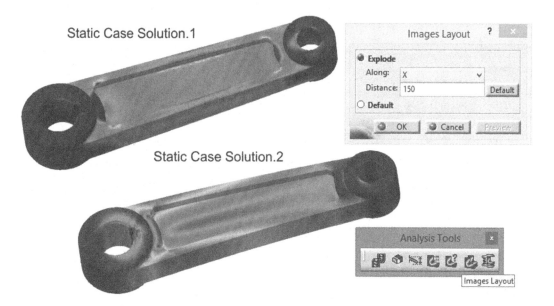

FIGURE 4.137 Separated result images.

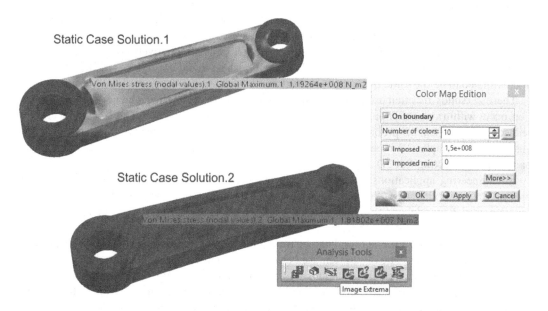

FIGURE 4.138 Positions and maximum values of *Von Mises stresses*.

Fatigue influences should also be analysed and computed with linear results based on fatigue theory.

4.5 ANALYSIS OF A 2D TWISTED PART

In this application, the finite element analysis of a connecting twisted part will be performed, the 2D drawing and the isometric projection are presented in Figure 4.139. The part has two connection surfaces at its ends, provided with holes for screws. Thus, the role of the part is to connect two components of an assembly, using two screws. The part's holes have the axes in perpendicular planes. This requirement also determines the twisted shape of the analysed part.

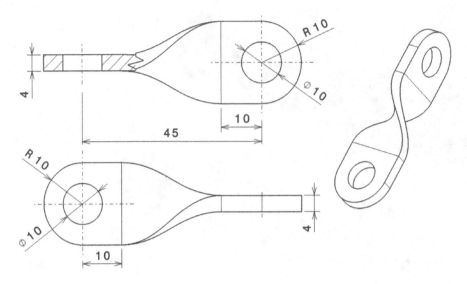

FIGURE 4.139 Two-dimensional drawing of the connecting twisted part.

In the practice of finite element analysis, if a certain part is relatively thin and has a specific shape, it is considered to be a 2D solid. This solid has no volume, contains fewer finite elements, is easier to discretize and its thickness is defined as a 2D property.

The 3D solid can be reduced to a 2D solid if its dimensions are relatively large compared with its thickness, if the part has the shape of a plate, shell or sheet metal cover, etc. The deformation is approximated to be the same throughout its thickness.

These 2D solids are often used for structural problems where the loading, and hence the deformation, occur within a plane. Though no real part or structure can be completely 2D, FEM engineers can reduce many practical 3D problems to 2D problems and to obtain satisfactory results. They perform analyses using simpler 2D models, which can be very much more efficient and cost-effective compared with conducting full 3D analyses.

In engineering applications, there are ample practical problems that can be modelled as 2D problems for simplification. Practice has shown that the differences between FEM calculations for a complex 3D model and those for a simplified 2D model are very small, but the resources and computation time required are much smaller.

The finite elements for the 2D solid can be triangular, rectangular or quadrilateral in shape with straight or curved edges. The most often used elements in engineering practice are linear, but quadratic elements are also used for situations that require high accuracy in stress. A linear element uses linear shape functions, and therefore the edges of the element are straight. A quadratic element uses quadratic shape functions, and their edges can be linear and/or curved.

The part considered in the application can be easily modelled in the *CATIA v5 Part Design* and *CATIA v5 Generative Shape Design* workbenches, therefore, it is used a hybrid modelling.

In the *XY* plane, in *Sketch.1*, the user draws the profile from Figure 4.140, positioned at the origin of the coordinate system. The sketch contains a circle of diameter Ø10 mm, an arc of a circle of radius R10 mm and three lines, two of which are horizontal and one vertical. The position of the horizontal lines is symmetrical with respect to the *H* axis of the coordinate system.

The profile is extruded on either side of the *XY* plane using the *Pad* tool and its *Mirrored extent* option. In the *Length* field, the user enters the value of 2 mm to obtain the 4 mm thickness of the part (Figure 4.141).

In the *ZX* plane, a second profile is drawn, identical to the first and placed at a distance of 45 mm, according to Figure 4.142. The distance is considered between the centres of the two holes,

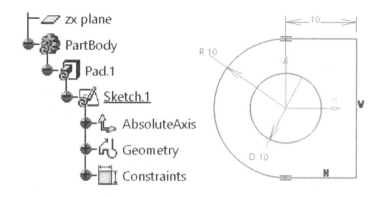

FIGURE 4.140 Profile from one end of the twisted part.

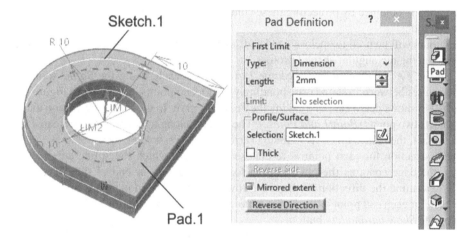

FIGURE 4.141 *Pad* extrusion of the profile.

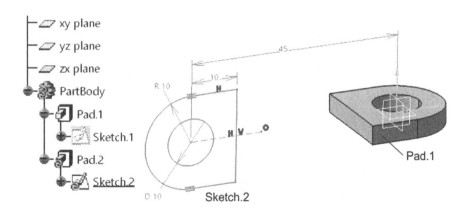

FIGURE 4.142 Drawing the second profile in the *ZX* plane.

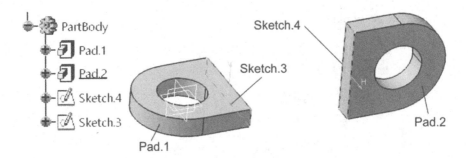

FIGURE 4.143   Drawing two rectangles in *Sketch.3* and *Sketch.4*.

this being a requirement of the assembly of which the considered part belongs. Similarly, *Sketch.2* is *Pad* extruded by 2 mm in both directions of the *ZX* plane.

Having both features *Pad.1* and *Pad.2* in the specification tree, the user draws two sketches (*Sketch.3* and *Sketch.4*) of rectangular shape, each one located on a flat face towards the interior of the part (Figure 4.143). The two profiles are exactly on the contours of the flat faces and can be easily and quickly obtained by projecting these contours in the respective sketches using the *Project 3D Elements* tool.

The contours will be used to enclose a twisted surface bounded and guided by two spline curves drawn between the rectangular shaped profiles.

Thus, in the *CATIA v5 Generative Shape Design* workbench, using the *Spline* tool, the user draws two curves between certain correspondingly positioned points, according to Figure 4.144.

Figure 4.145 shows details about drawing the *Spline.1* curve. In the *Spline Definition* dialog box, the user selects two points *(Point1 and Point2)*. The points belong to the previously created sketches and have, in fact, the names taken from them (*Vertex.22* and *Vertex.23*).

The next selection for each point is to choose an edge to which the spline curve is tangent. Obviously, the edge contains the point chosen for the beginning or end of the curve. Also, the user must determine the direction of the red arrow tip. The selected elements are highlighted in Figure 4.145. For each end point of the spline curve, the user sets the value 1 in the *Tensions* column by pressing the *Show parameters* button.

By using the *Multi-Sections Surface* icon, the selection box in Figure 4.146 opens. To create the twisted surface, the user first selects *Sketch.3* and *Sketch.4,* then two points with the role of closing the surface. It is possible that by default, *Closing Point1* and *Closing Point2* are not the ones in the figure. In this case, the user must change at least one of the points: right-click in the *Closing Point* column and choose the *Replace Closing Point* option from the context menu that opens.

Below, in the *Guides* tab, the two spline curves must be specified in the *Guide* column. In this way, the rectangular profile on the left (*Sketch.3* in this case) will move towards the rectangular

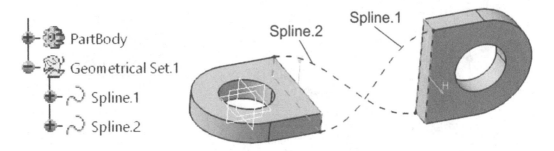

FIGURE 4.144   Drawing the two spline curves.

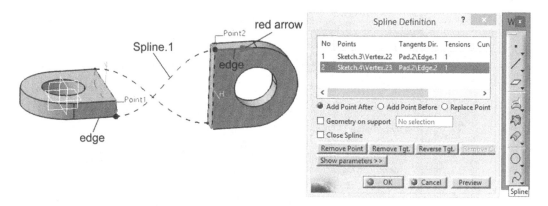

FIGURE 4.145 Details on drawing the *Spline.1* curve.

profile on the right *(Sketch.4)* along the spline curves and, thus, the twisted surface is obtained. If the *Closing Points* are correctly chosen, the generated surface is also correct, as in Figure 4.146.

Multi-sections Surface.1 is open at its ends, it has no volume, only area; it's a surface. To transform the surface into a solid element, the user must apply the *Close Surface* tool from the *CATIA v5 Part Design* workbench. Once obtained, it connects the ends created in the beginning of the application. The user can hide (*Hide/Show* option from the context menu) the surface. For verification, the volume of the part at the end of modelling is 3752.398 mm³, and the area is 2865.182 mm².

Video solution for a correct modelling of the part is found at: https://youtu.be/zAWMkHAsCVo.

To prepare the model for FEM analysis, the user creates a surface in the middle of the part's thickness. Thus, the user returns to the *Generative Shape Design* workbench, clicks the *Offset* tool icon and the *Offset Surface Definition* dialog box opens (Figure 4.147). In the *Surface* field, certain faces of the solid must be selected, but for them to become a single element, the user clicks the right mouse button and chooses *Create Join* option from the context menu.

The dialog box closes so that the selection can be made in another selection box: *Join Definition* (Figure 4.148). The three selected surfaces are marked on the part's solid. At the bottom of the figure, it can be seen that the *Join* tool was launched within the *Offset* tool (*Running Commands* information box). The specification tree contains the *Geometrical Set.1* feature in which the *Join.1*

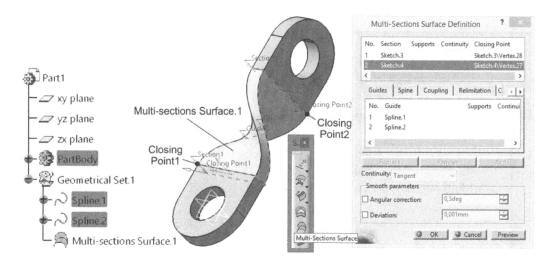

FIGURE 4.146 Creating the surface between the solid ends of the part.

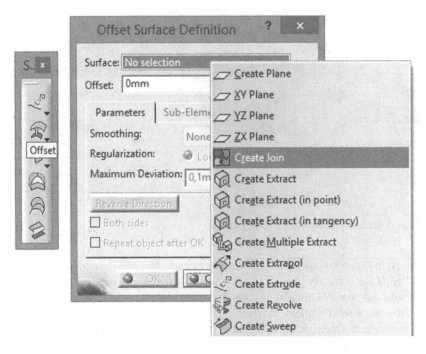

FIGURE 4.147   Accessing the *Offset* tool.

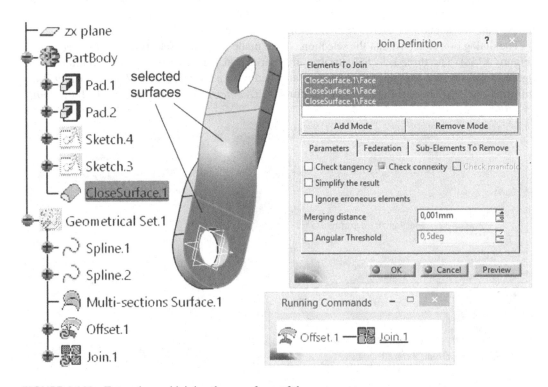

FIGURE 4.148   Extracting and joining three surfaces of the part.

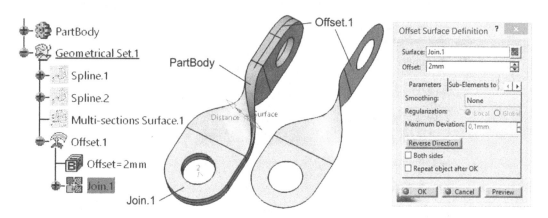

FIGURE 4.149   Obtaining the *Offset.1* surface.

and *Offset.1* features will be added. Once the selection is finished, the names of the surfaces appear in the *Elements To Join* field, then by pressing the *OK* button they are saved and joined to be part of the *Offset.1* feature.

Figure 4.149 shows the reopened *Offset Surface Definition* box and the created surfaces. Obviously, the direction in which the *Offset.1* surface is created is towards the inside of the part, but if by default the direction is towards the outside, the user must press the *Reverse Direction* button. In the *Surface* field, the *Join.1* surface can be observed, and in the *Offset* field, the user has entered half the value of the part's thickness: 2 mm. In the representation on the left, it can be seen the *Join.1* surface, but also the edge of the *Offset.1* surface, this one being hidden by the *PartBody* solid. In the representation on the right, the solid has been hidden and the *Offset.1* surface is completely visible.

The obtained and displayed surface has no thickness or volume, but material properties can be assigned to it. Thus, the user presses the *Apply Material* icon, then in the *Library* selection box they choose *Aluminium* (density 2710 kg/m^3, yield strength 9.5×10^7 N/m^2) from the *Metal* tab. If the visualization type chosen is *Shading with Material,* the representation of the surface is done according to Figure 4.150. It can be seen that the chosen material is added to the specification tree within the *Offset.1* feature.

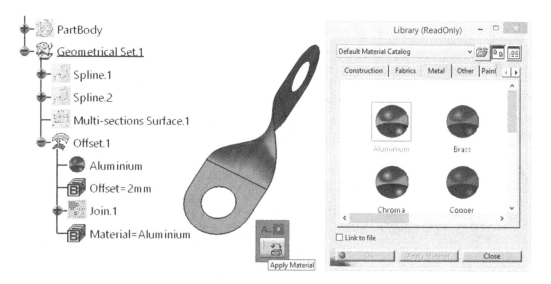

FIGURE 4.150   Applying the *Aluminium* material on the *Offset.1* surface.

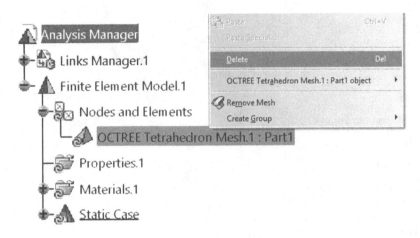

FIGURE 4.151 Deleting *OCTREE Tetrahedron Mesh.1* and *3D Property.1* features.

To start the FEM analysis, the user accesses the *Generative Structural Analysis* workbench and chooses a static analysis. A warning message is displayed: *Material is not properly defined on Part1* because the user defined the material on the surface and not on the 3D solid of the part.

In the specification tree there are two features that must be removed: *OCTREE Tetrahedron Mesh.1* from *Nodes and Elements* and *3D Property.1* from *Properties.1*. From the context menu of each, the user chooses the *Delete* option (Figure 4.151), and other features will be added later in their place.

Thus, the *CATIA v5 Advanced Meshing Tools (AMT)* workbench is launched from the *Start →* *Analysis & Simulation* menu. This workbench offers a lot of advanced meshing tools. The user may define *Octree Triangle Mesh* in the *Generative Structural Analysis* workbench, but for this application he must define a quadrangle element.

From the *Meshing Methods* toolbar, the user clicks the *Surface Mesher* icon to open the *Global Parameters* dialog box (Figure 4.152) and selects the *Offset.1* feature from the specification tree. In the *Mesh* tab, the icon *Set frontal quadrangle method* is selected, then the user selects the *Parabolic* option, enters the value of 2 mm in the *Mesh Size* field and also checks the *Quads only* option.

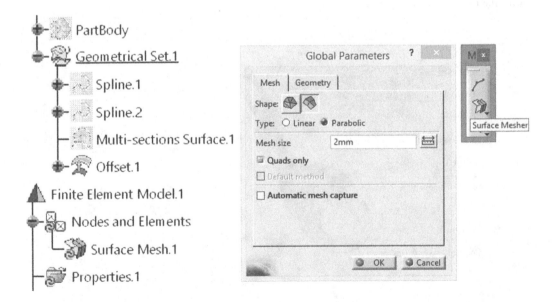

FIGURE 4.152 Defining the new mesh.

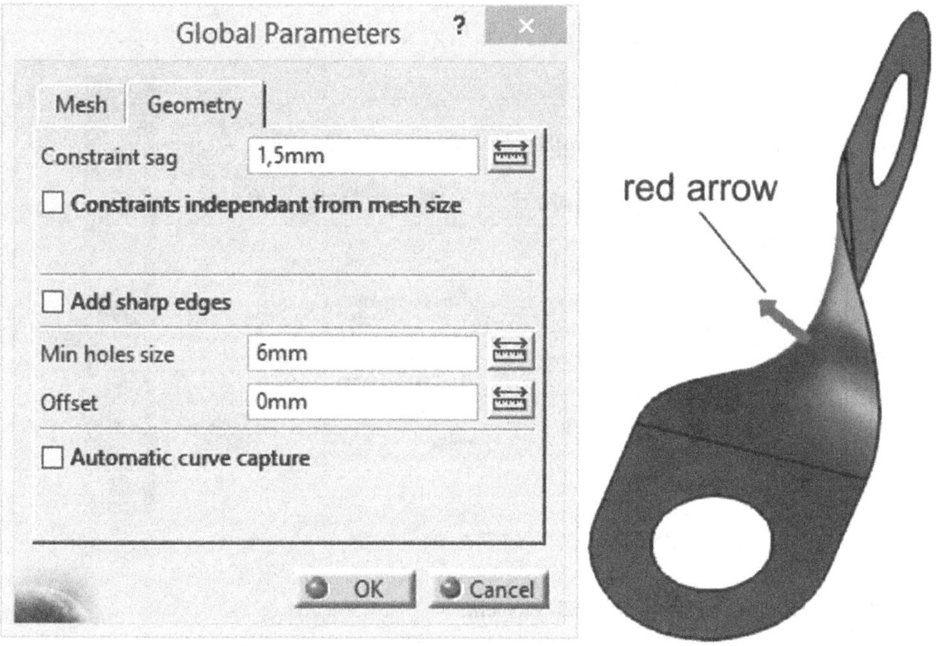

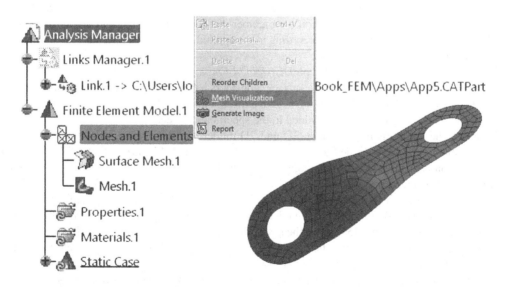

FIGURE 4.153 Setting the minimum hole size.

An arrow (Figure 4.153) is displayed on the model and specifies the upper direction of the 2D element. In the *Geometry* tab, the user enters the value of 1.5 mm in the *Constraint sag* field and, because the *Offset.1* surface has two holes, the user sets a value of 6 mm in the *Min holes size* field.

The user returns to the *Generative Shape Design* workbench and displays the mesh of elements and nodes for the *Surface Mesh.1* feature added to the specification tree. Thus, by right-clicking on *Nodes and Elements*, the contextual menu from Figure 4.154 opens, and the *Mesh Visualization* option is chosen. At this moment, the network is displayed on the geometry of the *Offset.1* surface. It can be seen that the elements of the network are two-dimensional.

FIGURE 4.154 Mesh visualization.

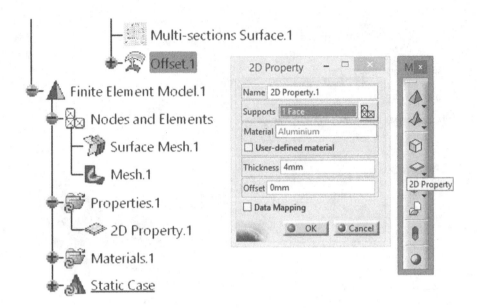

FIGURE 4.155 Defining thickness of the 2D element.

The user must define a thickness for the mesh. From the *Model Manager* toolbar, the user clicks the *2D Property* icon and the dialog box with the same name from Figure 4.155 opens.

In the *Supports* field the user selects the *Offset.1* surface and notices that *Aluminium* is automatically added in the *Material* field. Also, he enters a value of 4 mm in the *Thickness* field to specify the thickness of the part.

To apply the restraints and loads to which the part must be subjected, the user hides the network of nodes and elements, but also the *2D Property.1* feature and makes the *Links Manager.1* feature visible in the specification tree.

The restraint applied to one end of the part is complex, of type *User-defined Restraint*. The user presses the icon with the same name on the *Restraints* toolbar and the selection box in Figure 4.156 becomes available. The circular edge at one end of the part is selected, the *Supports* field is completed with *1 Edge,* then the user ticks the options *Restrain Translation 1, 2* and *3*, as well as *Restrain Rotation 1* and *3*. The user should remember that the numbers *1, 2* and *3* refer to the *X, Y* and *Z* axes of the global axis system. According to the applied restraint, the part has only one degree of freedom: rotation around the *Y* axis.

The load applied to the part is of type *Distributed Force,* placed to its other end. In the dialog box in Figure 4.157, the user selects the surface, considers a *Global* axis system, then in the *Force Vector* fields, sets a value of 200 N on the negative direction of the Z axis.

Restraint and distributed force are added to the specification tree.

To check the entire created problem (3D and 2D model, mesh, restraint, distributed force) the user must apply a check using the *Model Checker* icon on the *Model Manager* toolbar. The result of the verification is positive (*Status OK*, Figure 4.158), and the user can go to the next stages, saving the project and then computing.

Before performing the calculations necessary to solve the problem initiated in this application, the user is recommended to save the part model and the analysis using the *Save Management* option from the *File* menu. Figure 4.159 shows the saved files, the *Action* column of the selection box and the available options. According to the figure, the part model should not be saved because it has the status *Open* (but was saved in a previous iteration).

The calculation stage is launched by pressing the *Compute* icon. In the selection box with the same name, the user chooses the *All* option, presses the *OK* button and the program estimates the

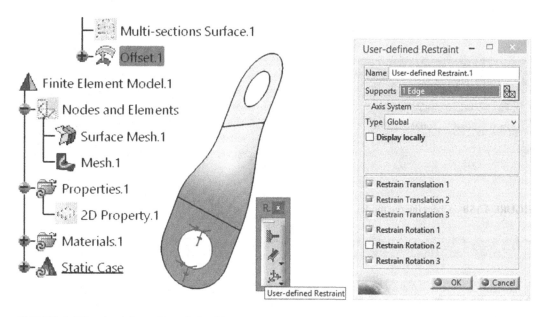

FIGURE 4.156 Applying a *User-defined Restraint*.

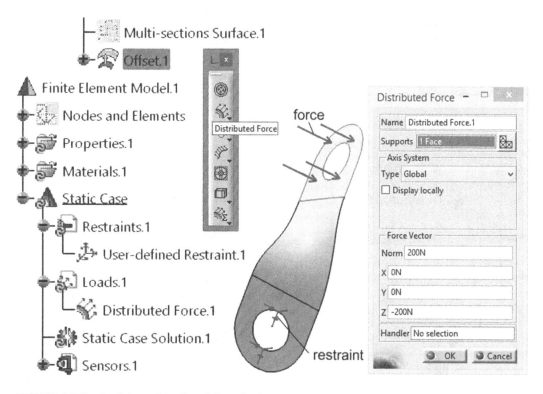

FIGURE 4.157 Applying a *Distributed Force* load.

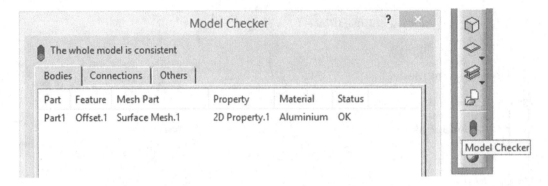

FIGURE 4.158 Checking the model.

FIGURE 4.159 Saving the model and the analysis files.

amount of required resources. If the estimations are zero or if *CATIA v5* cannot determine the estimation, it means that the user won't be able to obtain the results. The computation phase is pretty fast due to the reduced number of nodes and elements of the mesh.

The first result identified by the user should be related to the stresses that appear in the part. Figure 4.160 shows the *Von Mises stress* result. The stress distribution on the *Offset.1* surface can be observed, their correspondence in the accompanying palette, but also the very low value (1.043%)

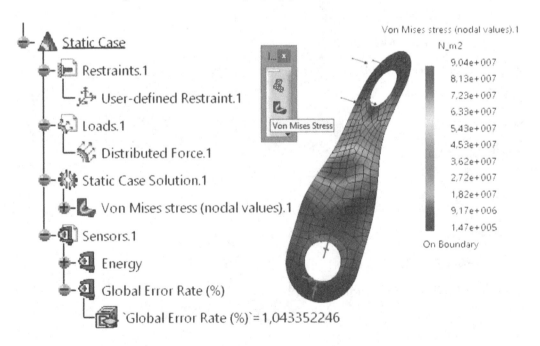

FIGURE 4.160 *Von Mises stress* result.

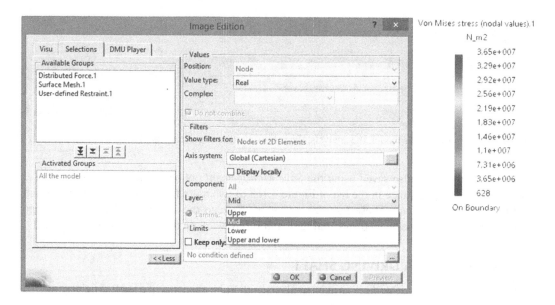

FIGURE 4.161 *Von Mises stress Mid* result.

of the *Global Error Rate*. The maximum stress resulting from the computation phase is close to the yield strength of the material applied to the part, but lower.

The user is able to display the results on the *Upper, Lower* and *Middle* areas of the part's thickness. This is important because in practice a fracture in the part starts from the outer surface and then goes through the whole volume.

The user double-clicks on the *Von Mises stress (nodal values).1* feature in the specification tree to open the *Image Edition* dialog box (Figure 4.161). By pressing the *More≫* button, the right side of the dialog box opens, and from the *Layer* drop-down list, the user chooses one of the *Upper, Mid, Lower* and *Upper and lower* options.

The user should also note that the *Upper* area is coincident with the direction of the arrow (Figure 4.153) that appears when the *2D Property* was defined.

Figure 4.162 presents the *Von Mises stress* values for *Upper, Mid* and *Lower* areas.

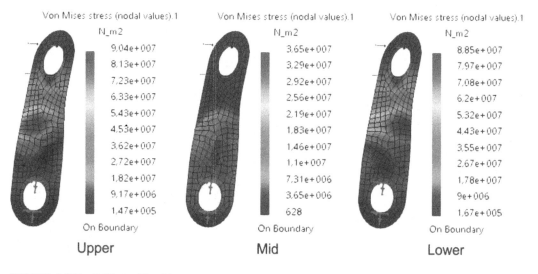

FIGURE 4.162 Different *Von Mises stress* results.

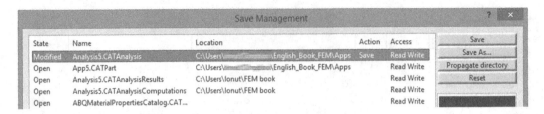

FIGURE 4.163 Saving the model, the analysis, the results and the computations files.

Analysis results are also saved using the *Save Management* option in the *File* menu. The user should notice the appearance in the list of Figure 4.163 of two more files: *CATAnalysisResults* and *CATAnalysisComputations*. These files are necessary to keep the obtained results on a storage device, which can later be opened and analysed, and future work on the project can be continued.

4.6 ANALYSIS OF A DRIVING SHAFT

In this application, the user will perform an analysis with finite elements of a shaft type part, with a driving role, having the 2D drawing in Figure 4.164.

Its functional role consists in transmitting a rotational movement, with a certain moment, between a crank or a wheel, mounted on the end of a cuboid shape (it is a square with rounded corners in section) to another coaxial shaft with the analysed part. The connection between them is made by joining a key and its $4 \times 10 = $ mm keyway, provided on the front surface of the part. It is also guided and supported in a bearing on the cylindrical surface with a diameter of Ø30 mm.

The application will analyse the shaft with a constant circular section between the drive element and the driven element. The analysis takes into account the fulfilment of the resistance condition $\tau_{max} \leq \tau_a$, as well as of stiffness $\theta_a = 5$ degrees/m $= 8.73 \times 10^{-5}$ rad/mm.

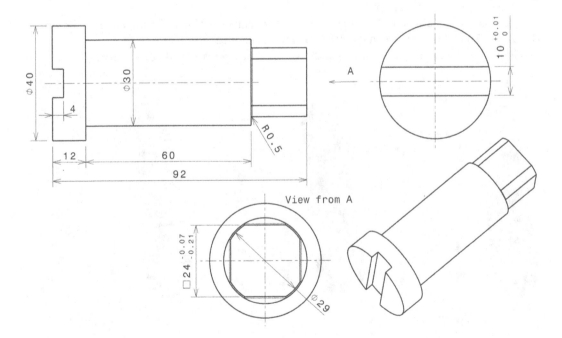

FIGURE 4.164 Two-dimensional drawing of a simplified driving shaft.

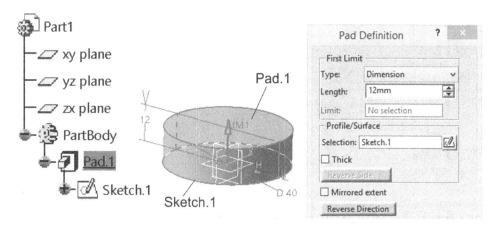

FIGURE 4.165   Creating the first solid feature of the part, *Pad.1*.

Modelling the part is simple and is done in the *CATIA v5 Part Design* workbench. Thus, in the *XY Plane,* the first sketch is created, it contains a circle of diameter Ø40 mm, having the centre at the origin of the coordinate system. The circle is extruded *Pad* over a distance of 12 mm, according to Figure 4.165. It can be seen that the extrusion direction of the circle is directed upwards (see the arrow), in the direction of the Z axis.

On the top face of *Pad.1,* the user draws a second circle in *Sketch.2,* diameter Ø30 mm, concentric with the circular edge of the face. The circle is also extruded in the direction of the Z axis, over a distance of 60 mm, to obtain the element *Pad.2* (Figure 4.166).

Similarly, on the upper flat surface of *Pad.2,* a circle is drawn with a diameter of Ø29 mm *(Sketch.3),* concentric with the circular edge of the face. Thus, by extrusion over a distance of 20 mm, the user obtains the third element, *Pad.3* (Figure 4.167). All three *Pad* elements have a cylindrical shape and are placed on top of each other, according to the 2D drawing and the previous figures.

At the *Pad.1* end of the part, it has a cutout, necessary for assembly with a 4 × 10-mm key. A 10-mm wide rectangle is drawn in the *XY Plane.* The user applies the *Pocket* tool to extract a

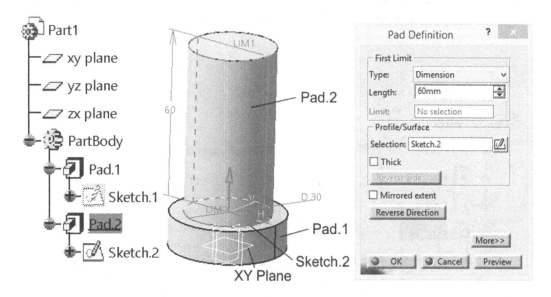

FIGURE 4.166   Creating the second solid feature of the part, *Pad.2*.

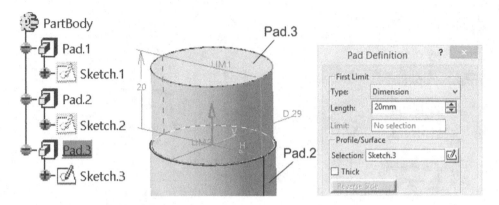

FIGURE 4.167 Creating the third solid feature of the part, *Pad.3*.

volume of rectangular shape, on a depth of 4 mm. Figure 4.168 shows the current shape of the part and the specification tree that contains the four features: three *Pad*(s) and *Pocket.1*.

At the other end of the part, on the upper flat surface of the *Pad.3* feature, a square profile is drawn, with a side of 24 mm. The profile is involved in a *Pocket.2* cutout (Figure 4.169) and results in the profiled end of the part, represented in the 2D drawing, *View from A*. The direction of the two arrows can be observed: the one in the centre is downwards, being the cutting direction of *Pocket.2*. The other arrow is directed towards the outside of the *Sketch.5* profile, the part is cut with the outside of the square. The directions of the two arrows can be changed by clicking directly on each or by pressing the *Reverse Direction* and *Reverse Side* buttons in the *Pocket Definition* dialog box.

The last feature added to the part is *EdgeFillet.1* of radius R0.5 mm at the base of *Pad.3/Pocket.2*. For this fillet, the user selected the eight existing edges (four straight and four curved). The options and the shape of the part at this end can be seen in Figure 4.170.

Once the solid modelling is finished, the user assigns a polyamide material to the part, having a yield strength of 9.5×10^7 N/m^2 (95 MPa) and a tensile modulus of elasticity of 3450 MPa. This material, Ertalon 66 SA, has a high mechanical strength, stiffness, heat and wear resistance. It also has a good creep resistance; its impact strength and mechanical damping ability are reduced. Ertalon 66 SA is well suited for machining on automatic lathes, sprockets of high modules, wheels and rollers, bushings, separators, shafts, supports, parts subjected to high loads, etc.

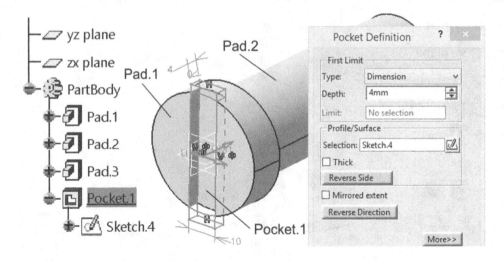

FIGURE 4.168 Creating the keyway *Pocket.1*.

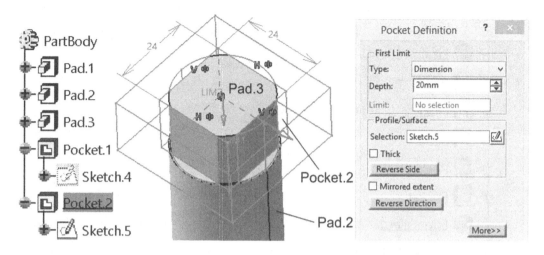

FIGURE 4.169 Cutting *Pocket.2* at the *Pad.3* end of the part.

From the list of materials available by pressing the *Apply Material* icon, the *Library (Read Only)* selection box opens. The user may select any material and apply it to the part, then edit its properties (*Properties* option is used from the context menu). In the editing box with the same name, in the *Feature Properties* tab, the name of the material to Ertalon 66 SA is changed, then in the adjacent tab, *Analysis,* the user edits its properties, according to Figure 4.171.

Once the material is applied to the studied part, it is saved together with it, but the default material library of the *CATIA v5* program is not updated. In cases when the user needs to use defined materials often, it is recommended to create a new material library. Thus, the *Material Library* workbench is accessed from the *Start → Infrastructure* menu (Figure 4.172).

The user notices the opening of a dialog box in which there is a default material *(New Material)* and the *Material Library* toolbar, its icons being marked next to it (Figure 4.173).

By right-clicking on the default material, a context menu opens from which the user chooses the *Properties* option. Thus, the editing box with the same name, similar to the one in Figure 4.171, becomes available, and the user edits the name and properties of the material.

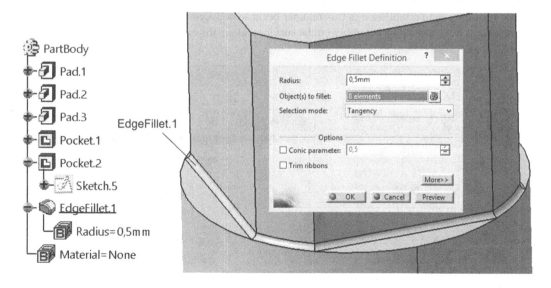

FIGURE 4.170 Creating the *EdgeFillet.1.*

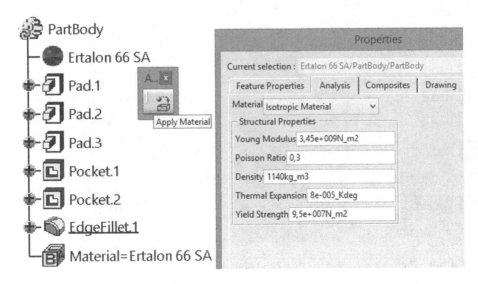

FIGURE 4.171   Editing material properties.

In most cases, the material is part of a family of materials with similar properties (plastic materials of a certain type, iron alloys, aluminium alloys, copper alloys, etc.). Thus, the user must specify the name of each family of materials, add these materials and define their properties (name, characteristics, etc.).

Assuming that the user has completed the definition of the materials, the library can be saved with the *CATMaterial* extension (Figure 4.174, left), and then it will be loaded in the *Library (Read Only)* selection box, according to Figure 4.174, right, so that the user has access to the created materials.

Next, in the *CATIA v5 Generative Structural Analysis* workbench, in a *Static Analysis,* a discretization of the part model is applied, by double-clicking on the *OCTREE Tetrahedron Mesh* feature in the specification tree, then, in the dialog box displayed, the user chooses the size of the finite element (3 mm), the minimum tolerance/absolute sag (1 mm) and its type as *Linear* (Figure 4.175).

When applying the restraints, the *Clamp* tool is used on the active side surfaces (three) of the keyway (Figure 4.176).

The restraints were established in this manner because between this part and its connecting part (not represented) is applied an assembly on flat surfaces, along the entire length of the part's keyway.

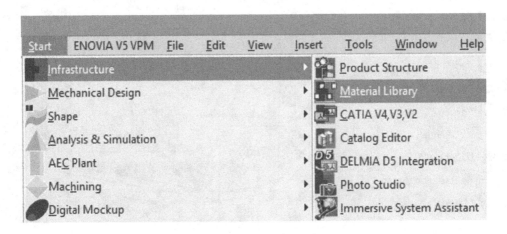

FIGURE 4.172   Creating a new catalogue with user-defined materials.

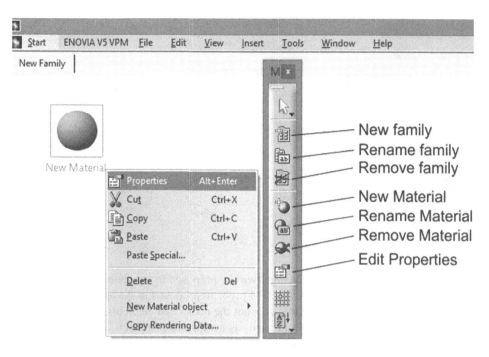

FIGURE 4.173   Options for creating a family of materials and editing properties.

The application of *Clamp* restraints ensures that there is no relative movement between the active surfaces (of the part and of the key) in contact. The part is loaded with a torque M_t of 120 N × m, which is applied to the other end of the part (*Pad.3/Pocket.2*), on each of its eight surfaces, considered to be in contact with the driving part (not represented) in assembly.

Figure 4.177 shows the *Moment* dialog box which contains these eight surfaces in the *Supports* field, and in the *Moment Vector* area where the torque values are placed around the three coordinate axes *X*, *Y* and *Z*.

The specification tree is completed with the features *Clamp.1* and *Moment.1*. It is also observed that the *Restraints.1* and *Loads.1* features require update, which can be achieved by launching the computation. However, before this, the user is recommended to save the part file and the analysis file

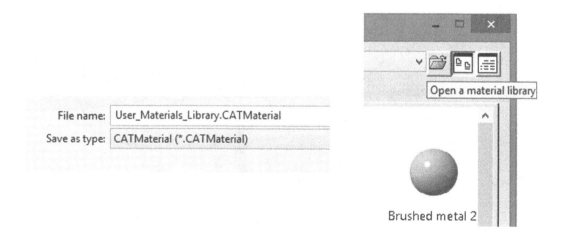

FIGURE 4.174   Saving and loading the material library created by the user.

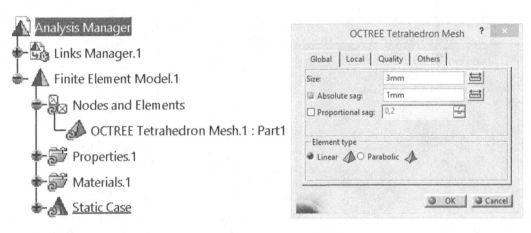

FIGURE 4.175 Discretization of the part model.

(Figure 4.178). In the next steps, some changes will be proposed to the geometry of the part, thus it is important to save this first version of the FEM problem proposed in the application.

Using the *Compute* tool and taking into account the nodes and elements network discretization parameters, its constraints and loading, the analysis results are computed, which can then be displayed using the tools on the *Image* bar.

Thus, the stresses appearing in the part model are highlighted graphically and numerically with the help of the *Von Mises Stress* tool. A *Shading with Material* visualization is chosen by the user to see the distribution of areas with maximum and minimum stresses. The numerical values of these stresses are displayed by the program alongside the *Von Mises stress (nodal values).1* colour palette, Figure 4.179.

The user applies the *Image Extrema* tool, of the *Analysis Tools* toolbar, to locate the minimum and maximum stress values. This is possible both globally and locally.

The *Global Maximum/Minimum* and/or *Local Maximum/Minimum* features appear in the specification tree depending on the user's option. These features are also part of the graphic representation with the meshed 3D model as indicators that contain the type of extreme stress value, together with its value (Figure 4.179). The indicator is connected to a certain node of the network, in the area where the respective extreme value has been reached.

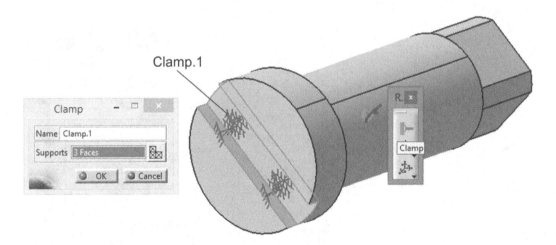

FIGURE 4.176 Applying the *Clamp* restraints.

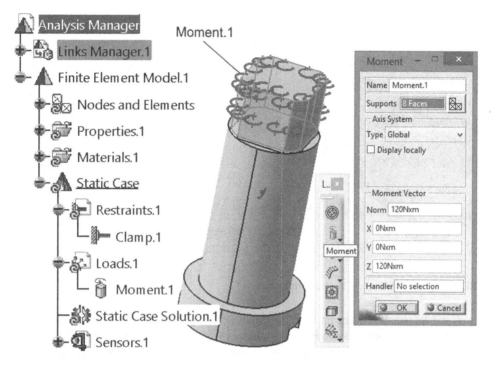

FIGURE 4.177 Applying the *Moment* type load.

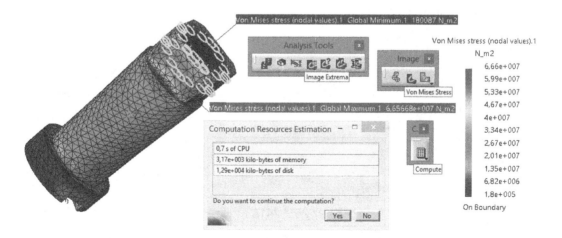

FIGURE 4.178 Saving the part and the FEM analysis.

FIGURE 4.179 *Von Mises Stress* results and indication of extreme stresses.

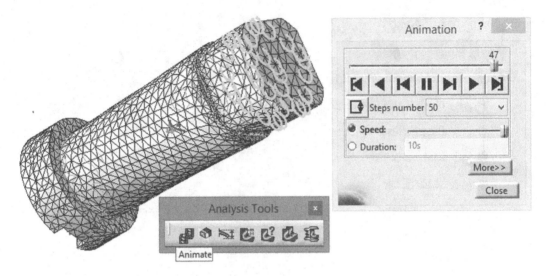

FIGURE 4.180 Animated view of elastic deformation.

The elastic deformation of the part can be visualized using the *Animate* tool on the same *Analysis Tools* toolbar. In this manner, a continuous sequence of successive frames is obtained, which deliberately exaggerates the computed result, with a certain amplitude of the deformation. The application of the moment load has the effect of a torsional deformation of the part model. This fact is highlighted during the running of the animation (Figure 4.180).

In the figure, a *Shading with Edges* representation was used. The animation shows how the nodes and finite elements of the network move/deform. It is also observed that the maximum stress appears in the area between *Pad.2* and *Pad.3* on the fillet radius, it seems that is a stress concentrator due to the small value (R0.5 mm).

In the next step, to find out the correctness percentage of the results of the computations performed during the FEM analysis, it is possible to use the *Precision* tool on the *Image* bar, together with the *Information* tool on the *Analysis Tools* toolbar.

Figure 4.181 shows a fragment of the *Information* box, displayed after selecting the second tool and choosing the *Estimated local error.1* feature in the specification tree.

Thus, for the finite element analysis process presented earlier, the *CATIA v5* program provides the results with an estimated global error rate of 27.265%, which is considered to be unacceptable for this type of loading and part model.

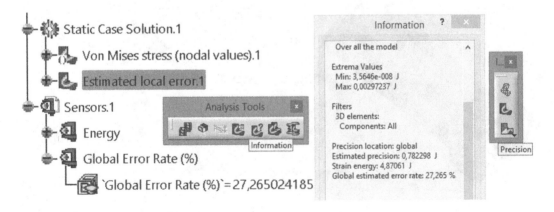

FIGURE 4.181 The estimated global error rate.

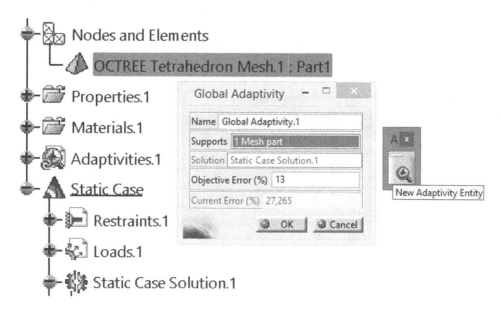

FIGURE 4.182   Establishing the percentage of target global error rate.

Resuming the analysis process in order to reduce the value of the global error rate is done by applying the *New Adaptivity Entity* tool. Thus, in the *Global Adaptivity* dialog box, in the *Supports* field (Figure 4.182), the user selects the *OCTREE Tetrahedron Mesh.1* feature, then, in the *Objective Error (%)* field, enters the objective percentage/rate value of 13%. The user also notices the current error rate value in the non-editable field *Current Error (%)*.

This restart of the computation process is necessary for the program to re-refine the part's model trying to achieve this previously specified error objective. Thus, the user clicks the *Compute with Adaptivity* icon, the *Adaptivity Process Parameters* dialog box is displayed, then, in the *Iterations Number* field, two calculation iterations are set to which the part model will be subjected. In the *Minimum Size* field, the user enters the minimum value of the finite element, equal to 0.5 mm (Figure 4.183).

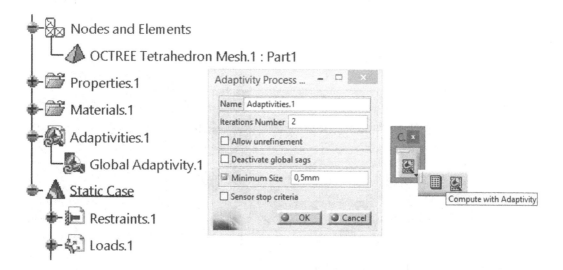

FIGURE 4.183   Resuming the computation step in two iterations using adaptivity.

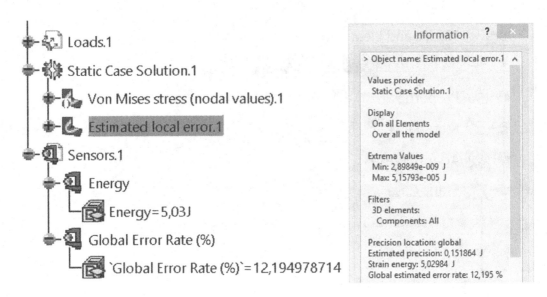

FIGURE 4.184 Obtaining the imposed global error rate.

Comparatively, in the first iteration, this value was considered to be 3 mm (Figure 4.175), the modification performed having, mainly, the role of increasing the number of finite elements that discretize the part's model.

As a result, on a better defined model, the results are precise and closer to reality, the global error rate becoming equal to 12.195%. This value is slightly lower than the one sought and imposed by the user in the *Global Adaptivity* dialog box.

Increasing the number of iterations and imposing a new error rate leads to a considerably longer computation time, but, in this case, the proposed objective is achieved, the *Information* box presents the new values (Figure 4.184).

Precise localization of the area where the estimated error rate reaches an extreme value (maximum or minimum), globally or locally, is achieved using the *Image Extrema* tool, then choosing the *Focus On* option from the context menu of each value. Figure 4.185 indicates, for example, the position of the area where the maximum value of the global error is located. The area with the maximum error is, however, of small extent.

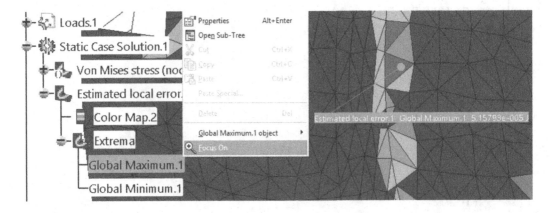

FIGURE 4.185 Locating the maximum global error.

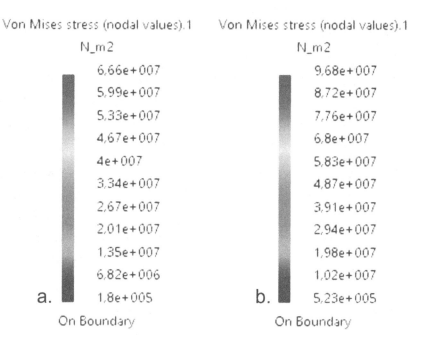

Von Mises stress (nodal values).1   Von Mises stress (nodal values).1

N_m2   N_m2

6,66e+007   9,68e+007

5,99e+007   8,72e+007

5,33e+007   7,76e+007

4,67e+007   6,8e+007

4e+007   5,83e+007

3,34e+007   4,87e+007

2,67e+007   3,91e+007

2,01e+007   2,94e+007

1,35e+007   1,98e+007

6,82e+006   1,02e+007

a.   1,8e+005   b.   5,23e+005

On Boundary   On Boundary

FIGURE 4.186   Comparison of values resulting from FEM analyses (before and after refinement).

Figure 4.186 shows the *Von Mises stress* values at the end of each analysis process. On the colour palette visible on screen (Figure 4.186a) are the values obtained before the refining operation, the value of the maximum stress is 6.66×10^7 N/m^2 (see also Figure 4.179). Through refinement, following the new process, the value of the estimated error rate decreased, but the value of the maximum stress increased, up to 9.68×10^7 N/m^2 in Figure 4.186b. The location of the stressed areas coincides.

In the initial data of this application, a polyamide material was chosen, having a yield strength of 9.5×10^7 N/m^2 (95 MPa) and a tensile modulus of elasticity of 3450 MPa. This value of the yield strength was, however, reached and even exceeded after the refinement, for an error rate of 12.195%. This is a convenient value, but the stress value tends to increase with the decrease of the error rate.

In this situation, there are mainly two possibilities for the user to choose: change the constructive shape of the part in the area of maximum stress or select another material with a higher yield strength. In this way, the initial load applied the analysed part is preserved.

Thus, the part modelled according to the 2D drawing from the beginning of this application changes as follows: in *Sketch.3* the user edits the value of the diameter of the circle from Ø29 mm to Ø30 mm (see Figure 4.167). Also, the fillet between *Pad.2* and *Pad.3* features, called *EdgeFillet.1* (Figure 4.170), is disabled by the user from the context menu. The dimensions of the *Pocket.2* feature remain unchanged so that the user does not have to change the conjugate part of the assembly.

For changes, the user will double-click on the geometric features of the part located within *Links Manager.1* in the specification tree. The shape of the part at its drive/assembly end is visible in Figure 4.187.

By this editing of the part model, the user removed the sensitive area between *Pad.2* and *Pad.3*, including the fillet radius, i.e. exactly the area where the maximum stress appeared.

After resuming the computation stage that also includes the refinement of the part's finite elements network, a new maximum stress value of 8.56×10^7 N/m^2 (85.6 MPa) results, located at the

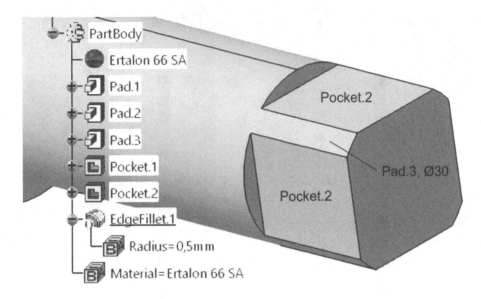

FIGURE 4.187   Changes applied to the part model.

other end of the part, between the *Pad.1* and *Pad.2* features, according to Figure 4.188. That edge is a stress concentrator too, but its diameter is larger and it is at some distance from the drive end (*Pad.3*). The low value of the *Global Error Rate* of only 7.745% and the fact that the stress is lower than the value of the yield strength can be observed.

The FEM analysis can continue, in order to make other changes to the geometry of the part, thus, the edges of the parallelepiped *Pad.3* are to be chamfered and filleted to the feature *Pad.2* (Figure 4.189), replacing the square section with a profile of circular arcs (Figure 4.190). The dimensions of these elements are left to the discretion of the user to experiment with different modelling solutions, which will also lead to different and interesting results for the FEM analysis.

Also, for a better discretization of the part model, parabolic finite elements can be used, but they require important computing resources (time, RAM memory and storage space). Different metal materials can be applied to the part while keeping the same load to compare how the areas with maximum stress are formed and its value.

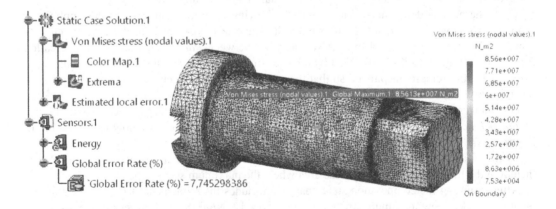

FIGURE 4.188   The results of the FEM analysis after applying the geometric changes.

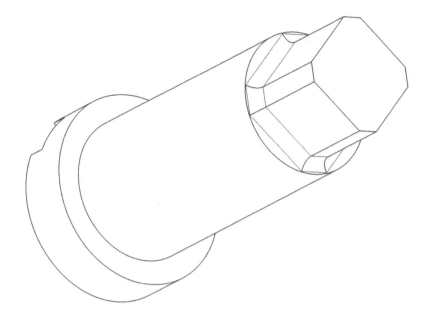

FIGURE 4.189 Example 1 of a 3D model proposed for FEM analysis.

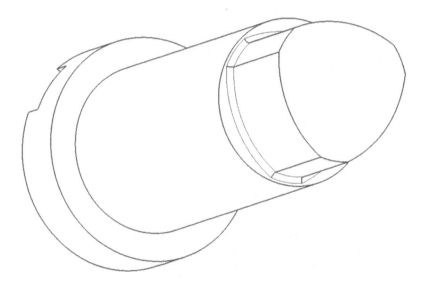

FIGURE 4.190 Example 2 of a 3D model proposed for FEM analysis.

4.7 ANALYSIS OF AN ARM TYPE PART

In this application, the user will perform a finite element analysis of an arm-type part, which has the 2D drawing represented in Figure 4.191. The part belongs to a mechanical assembly having the role of support and driving. The lower area of the part (7.8 × 6 mm) is assembled with another part in the mechanism. In the upper area, where the elongated hole (R1.6 × 5 mm) is present, the user applies a driving force.

The 3D modelling of the part is relatively simple and can be done in less than 20 minutes by the user following the video tutorial showed at https://youtu.be/P-nd4wOG5OE. For modelling, he may use only the *CATIA v5 Part Design* and *CATIA v5 Sketcher* workbenches.

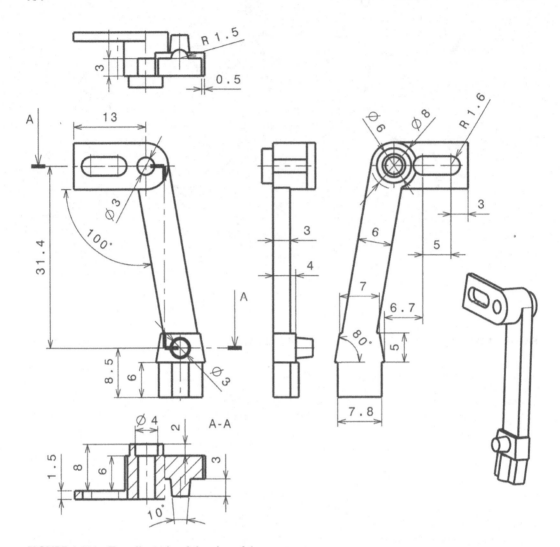

FIGURE 4.191 Two-dimensional drawing of the arm type part.

Once the modelling is finished, the part is represented according to Figure 4.192 together with its specification tree. All geometric features of the part are marked on its 3D model. Any modelling option is possible if the correct model of the part is obtained, with a volume of 1239.178 mm^3.

To fulfil the functional role, the part model is provided with assembly surfaces with other components, but also with a connection surface, on which a loading force will be applied.

The part is considered to be manufactured from a material (steel), having the following physical and mechanical properties (Figure 4.193), important during the analysis: Young's modulus (2×10^{11} N/m^2), Poisson's ratio (0.266), density (7860 kg/m^3), the coefficient of thermal expansion (1.17×10^{-5} ^{0}K) and the yield strength (2.5×10^8 N/m^2).

After applying the material (Figure 4.194) to the part model, the specification tree is completed with the *Material = Steel* feature. To display some analysis results, the user requires another way of viewing the model. Thus, from the *View* toolbar the user expands the group of *Render Style* icons (Figure 4.195) and chooses *Customize View Parameters,* then, from the *View Mode Customization* dialog box that appears, he checks the *Shading* and *Material* options. As a result, the model acquires a dark grey colour, with metallic reflections, specific to this display mode.

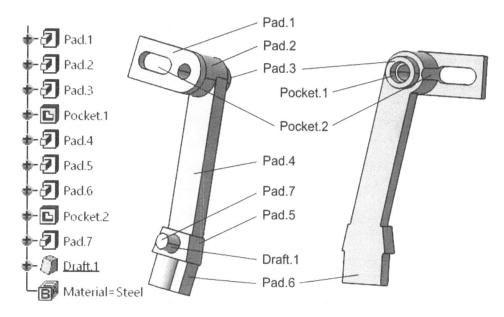

FIGURE 4.192 The final model of the part and its specification tree.

The user accesses the *CATIA v5 Generative Structural Analysis* workbench from the *Start* → *Analysis & Simulation* menu and sets the analysis type to *Static Case*, the specification tree displays the feature with the same name.

Although the *CATIA v5* program implicitly defines the network of nodes and elements, it is recommended to edit it and set the parameters by the user according to the dimensions of the part and its role. Thus, the user double-clicks on the *OCTREE Tetrahedron Mesh* feature located in the specification tree (Figure 4.196).

Figure 4.196 also shows the dialog box with the same name. Next, the user sets the finite element *Size* (2 mm), *Absolute sag* (1 mm) and the finite element type as *Linear*. A *Clamp* type restraint is applied to each surface at the base of the part, as can be seen in Figure 4.197.

The specification tree is completed with the *Clamp.1* feature, the dialogue box contains in the *Supports* field six selected surfaces, highlighted by the restraint symbols. The *Clamp* tool is available on the *Restraints* toolbar.

A distributed force of 150 N is applied to the surface of the bore in the upper area of the part. This force is directed outwards, in the direction opposite to the *X* axis. This is the initial value of

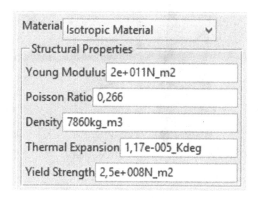

FIGURE 4.193 Properties of the steel material applied to the arm type part.

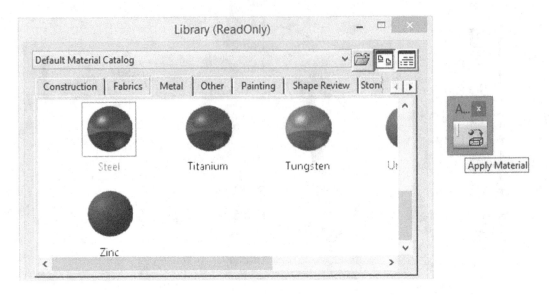

FIGURE 4.194 Selecting the part material from the material library of the *CATIA v5* program.

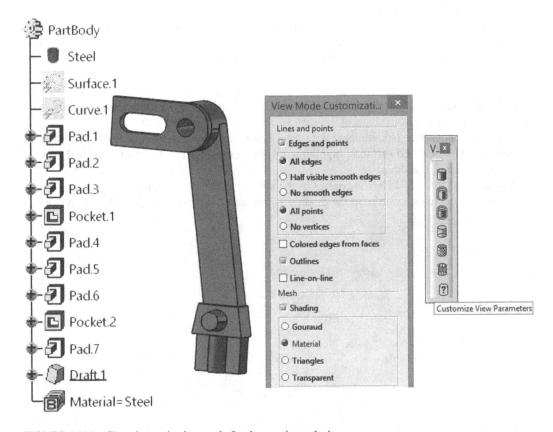

FIGURE 4.195 Choosing a viewing mode for the part in analysis.

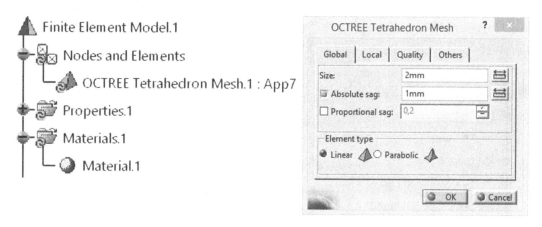

FIGURE 4.196 Refinement of the part mesh.

the force, later in the application the user will create a method to change the value based on a code sequence and some conditions.

In the specification tree, the *Distributed Force.1* feature becomes available, the force is symbolized by four arrows on the surface, its value and orientation, but also the coordinate system in which it was created. All these parameters can be entered in the corresponding fields of the dialog box in Figure 4.198. The force value, of 150 N, is, in fact, negative in the *X* field of the *Distributed Force* dialog box because the direction of the four arrows must be in the negative direction of the *X* axis. Simplified, the tips of the arrows must touch the surface on which the force is applied. No components of this force are applied on the *Y* and *Z* axes, so their value is 0 N, and the *Norm* field contains only the value of 150 N.

After the restraints and loading force setup comes the actual stage of the analysis computation. Pressing the *Compute* icon on the toolbar of the same name leads to the opening of the dialog box in Figure 4.199. For this case, the user selects the *All* option, the first effect of the action being the update of the *Static Case Solution.1* feature.

Unchecking the *Preview* option allows shortening the analysis process by not displaying the *Computation Resources Estimation* information box. This is important, however, in the case of very complex analyses, because it provides information on the computation time and the required disk space for stored files. If the values presented in this box are equal to 0 (zero), the computation will not generate the expected results, most likely due to an error made by the user in the setup stage of the FEM analysis.

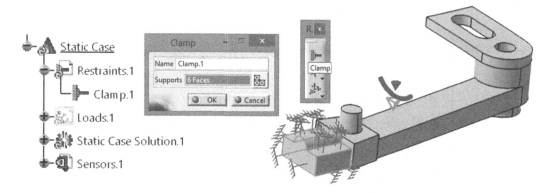

FIGURE 4.197 Setting the *Clamp.1* constraint.

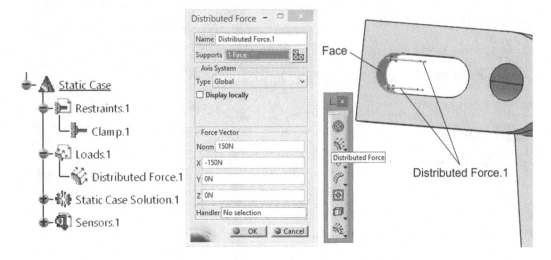

FIGURE 4.198 Setting the *Distributed Force.1* load.

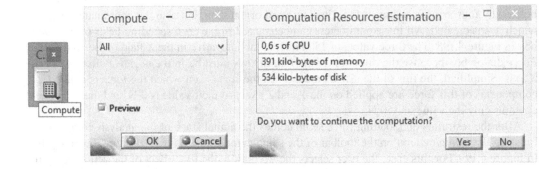

FIGURE 4.199 Launching the computation stage of the FEM analysis.

After the computation is finished, the *Image* toolbar icons/tools are available to the user to view the results. These are known from previous applications. The specification tree is completed according to the inserted images. By default, the most recent one becomes active by deactivating the others/previous ones. In Figure 4.200 the specification tree is exemplified, containing a list of four images and their icons, three of which are disabled/deactivated and one active.

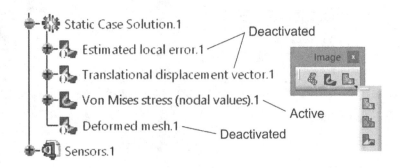

FIGURE 4.200 List of possible results to be obtained through FEM analysis.

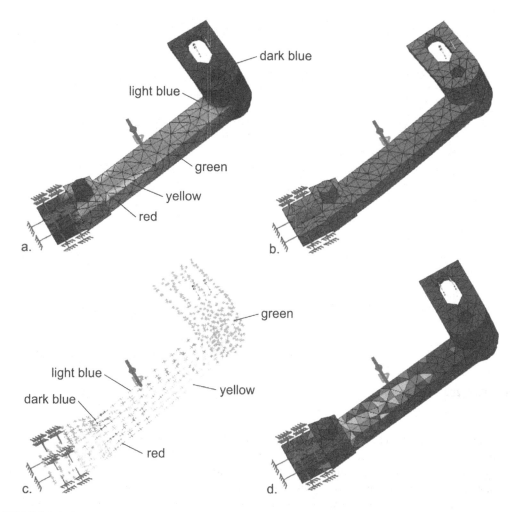

FIGURE 4.201 Four graphical results of the finite element analysis.

In Figure 4.201, four results/images are displayed using the tools, (a) *Von Mises stress,* (b) *Deformation,* (c) *Principal Stress* and (d) *Precision,* corresponding to the computation of the part model and the considered loading force.

The user should be aware that the deformations/displacements are presented graphically exaggerated to facilitate the analysis conclusions.

To find the maximum and minimum stress values computed by the FEM analysis, the user should activate the *Von Mises stress* result, then, from the *Analysis Tools* toolbar, he must use the *Information* icon to display the information box with the same name.

In Figure 4.202, next to this information box, *CATIA v5* presents the colour palette accompanying the *Von Mises stress* image result. The lowest stress values can be found at the bottom of the palette, and the highest at the top. The box also contains the explicit values, in the *Extrema Values* area, as follows: Min: 0 N/m² and Max: 2.13×10^8 N/m². The user can understand how the stresses are distributed on the part, also by the colours displayed.

Displayed on screen, dark blue and light blue colours indicate low stress values (e.g. 4.26×10^7 N/m²), and yellow to red colours high stress values (e.g. 1.49×10^8 N/m² or 1.7×10^8 N/m²). These are close, as is the case in this application, to the admissible yield strength value of the chosen material applied to the part model.

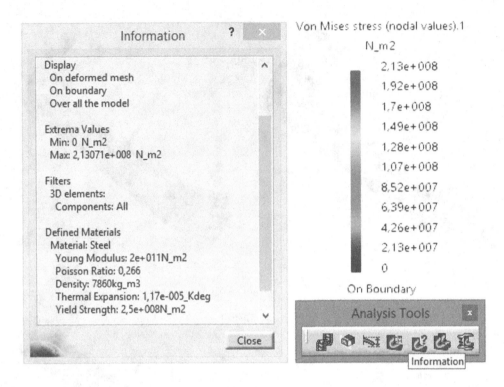

FIGURE 4.202   Display of minimum and maximum stress values.

Considering that the yield strength of the material is 2.5×10^8 N/m^2, an initial conclusion can be drawn, namely that the part model will withstand the applied distributed force of 150 N, but the safety in use is at the limit (the values of 2.13×10^8 N/m^2 and 2.5×10^8 N/m^2 are very close) because it is not known precisely how the part will behave in real working conditions.

Thus, in the next step, the user wants to find out the correctness percentage of the computations of the performed analysis. For this, the *Precision* tool on the *Image* toolbar is used, together with the *Image Extrema* tool on the *Analysis Tools* toolbar. The areas containing the extreme values of estimated global and local errors are identified (Figure 4.203).

Precise localization of the area containing a certain error is done by choosing the *Focus On* option from the context menu for each indicator. In Figure 4.204, some important values resulting from the FEM analysis can also be observed very well: maximum stress 2.13×10^8 N/m^2, maximum displacement 0.197 mm, global error rate: 43.78%.

Among these values, the *Global estimated error rate* value can be observed: 43.78%. The user appreciates that the error rate is very high because it basically represents the difference between the analysed virtual model and the real part in a real assembly. Reducing the error rate value is possible by applying the *New Adaptivity Entity* tool on the *Adaptivity* toolbar. In the *Global Adaptivity* dialog box, in the *Supports* field, the user selects the *OCTREE Tetrahedron Mesh.1* feature, then, in the *Objective Error (%)* field, he enters a smaller desired percentage/rate: 25% (Figure 4.205).

Basically, the user tries to impose a new error value. The current value is displayed in a non-editable field. By the new value and then by refining the mesh, the user forces the program to create a new mesh for a new computation through several iterations. The computation stage will take longer.

Resuming the computation is mandatory for the program to refine the part model in an attempt to reach the error rate imposed by the user. Thus, the *Compute with Adaptivity* icon on the *Compute* toolbar is pressed and the *Adaptivity Process Parameters* dialog box is displayed (Figure 4.206).

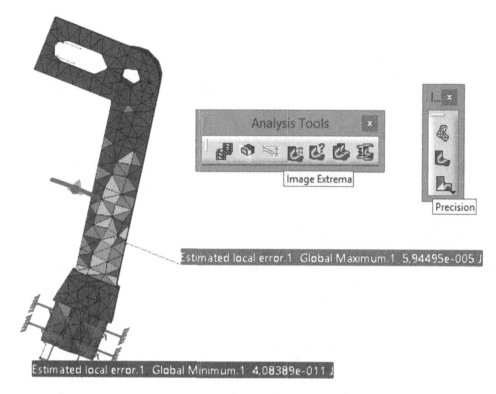

FIGURE 4.203   Display of extreme values of global errors.

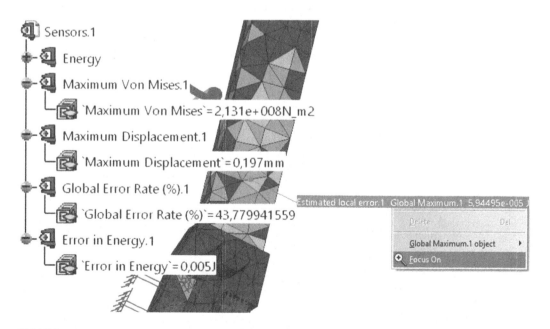

FIGURE 4.204   Display of the estimated global error value.

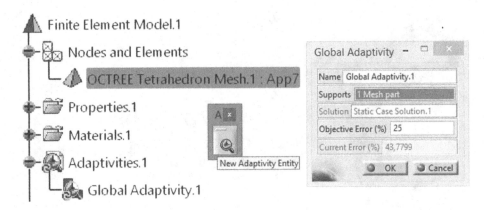

FIGURE 4.205   Establishing/imposing a new error rate.

In the *Iterations Number* field of the dialog box, the user sets the number of computation iterations to 5. The part model will face different analysis conditions. A refinement of the mesh was
also applied, by changing the parameter in the *Minimum Size* field, thus, 0.5 mm instead of 2 mm,
which is the value initially established according to Figure 4.196. Increasing the number of iterations and imposing a smaller error rate leads to a considerably longer computation time, meanwhile
a *Computation Status* information box is displayed.

Following the computation based on the new settings, the user notices that the value of 24.37% of
the global error has been reached (Figure 4.207), a much more convenient rate than the previous one
of 43.78%. Also, for the *Von Mises stress* result, in the new conditions, a change in the maximum
stress value is noted of 3.09×10^8 N/m^2, compared with 2.13×10^8 N/m^2, which resulted from the
first computation stage, before the application of the *Adaptivity* tools.

Also, the maximum displacement of a node of the part model is 0.264 mm compared with 0.197
mm before refinement. The increases (maximum stress and displacement) are considerable for a fairly
high error rate. For a better accuracy of the results, the user has the possibility to continue the analysis
process, applying another refinement to the model, also imposing a global error rate of less than 25%.
It is possible that after several iterations such a percentage can be obtained, simultaneously, however,
with the increase of the maximum stress value. This will far exceed the default yield strength of
2.5×10^8 N/m^2 of the standard steel provided by the *CATIA v5* program in its library of materials.

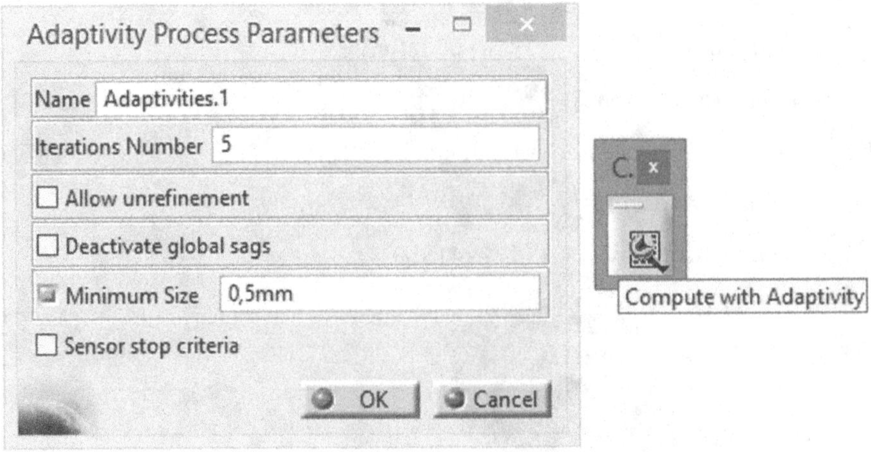

FIGURE 4.206   Resuming the computations through five iterations and refining the mesh.

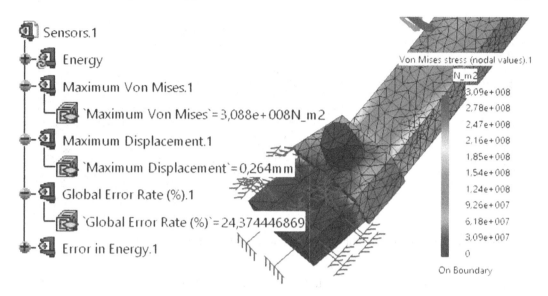

FIGURE 4.207 Achieving the imposed error rate and displaying the new values for maximum stress and displacement.

In such cases, the user must be informed about the exceedance and be offered a solution to change the part material, considering that its geometric shape and dimensions cannot be edited, as the part has a predetermined role in a certain assembly.

The values of stresses, displacements, global and local estimated errors, etc. computed during the application, represent important parameters that can be involved in formulas, rules and reactions, but they must be identified with the help of sensors. By using the *Sensors* feature in the specification tree (Figure 4.208), the user can find out various concise information about the results of the analysis process. Thus, the user right-clicks on this feature *Sensors* to choose the option *Create Global Sensor* from its context menu that appears. In the *Create Sensor* dialog box, in the available list, the user selects the sensors needed in the part analysis. Some of these sensors are already present in the specification tree, according to the figure.

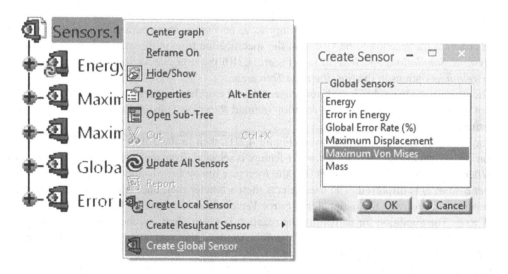

FIGURE 4.208 Creating global sensors.

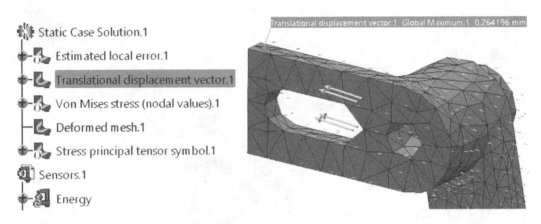

FIGURE 4.209 Identification of the area with maximum displacements of the mesh nodes.

In this application, the user is interested in the value of the maximum stress (*Von Mises stress*) and the maximum displacement appearing in the part due to its loading with a set consisting of a *Clamp* restraint and a distributed force of 150 N.

As previously observed in the previous explanations, the specification tree is completed with the features *Energy, Maximum Von Mises* and *Maximum Displacement* (Figure 4.207). From the values displayed by the displacement sensor, it follows that certain surfaces of the part (from the area where the force is applied) move on a maximum distance of 0.264 mm. This deformation is accepted by the user, but it must not exceed an imposed value of 0.37 mm, established considering the functional role of the part in the assembly.

Figure 4.209 shows the *Deformed Mesh* and *Translational displacement vector* representations superimposed. The user observes the displacement vectors that have the maximum values to the left of the surface on which the distributed force was applied.

The previous analysis also shows that the part deforms plastically when a force of 150 N is applied, having a material with a yield strength of 2.5×10^8 N/m^2. Increasing the force value will produce a maximum stress even higher than this value, and the part will deform more, with the possibility of cracks and breakage.

Considering the deformation and stress restrictions that appear, the user may write a reaction using the *CATIA v5 Knowledge Advisor* workbench. The reaction will monitor the change in the value of the applied distributed force and impose, as necessary, some parameters changes.

In order for the reaction to be visible in the specification tree, in the menu *Tools → Options → Analysis & Simulations → General* tab (Figure 4.210) the user checks the *Show parameters* and *Show relations* options in the *Specification Tree* area.

The user opens the *CATIA v5 Knowledge Advisor* workbench, then, within the *Relations* feature of the specification tree, adds a reaction (named *Reaction.1*) using the *Reactions* icon on the *Reactive Features* toolbar.

The *Reaction* dialog box is displayed (Figure 4.211), then in the *Source Type* field the *Selection* option is chosen and the *Distributed Force.1* feature is selected (from the expanded specification tree).

When *Distributed Force.1* is selected, the *Reaction* dialog box disappears, but a selection box *(Select a source)* is displayed. The user selects the parameter that, once modified, will trigger/run the reaction. In this case, the parameter is *Force Vector.1\Force.1* because it represents the force on the *X* axis. The condition for activating the reaction is given, then, by choosing the *ValueChange* option from the *Available Events* field of the *Reaction* box, that reappears after selecting the mentioned parameter and pressing the *OK* button.

In the non-editable field below *ValueChange,* the type of this parameter can be observed as a force (*parameter: Force*). In the *Action* area, the user checks the *Knowledgeware Action* option to

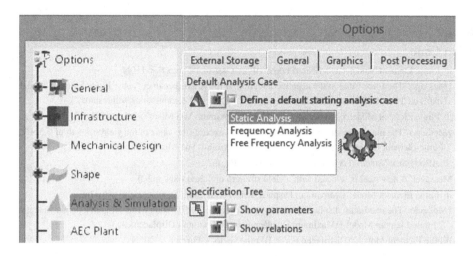

FIGURE 4.210 Checking the options for displaying parameters and relations in the specification tree.

write the code sequence using the identified parameters and syntax peculiarities of the *CATIA v5 Knowledge Advisor* workbench.

The writing of code lines takes place directly in the respective box, which is small in size, but which can be enlarged by pressing the *Edit action* button.

As can be seen in Table 4.2, the reaction contains a complex code sequence, composed of several lines, where they are displayed and fully presented in the table. Each line of code was subsequently numbered to facilitate comments and observations. Of course, the numbers on the left column do not belong to the reaction code. The user has to enter this code in the application exactly as it is presented in Table 4.2.

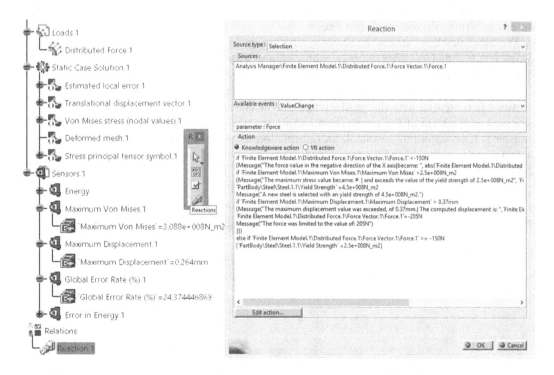

FIGURE 4.211 Complex dialog box for writing the reaction.

TABLE 4.2

Reaction Code

1   if `Finite Element Model.1\Distributed Force.1\Force Vector.1\Force.1` <-150N

2   {Message("The force value in the negative direction of the X axis|became: ", abs(`Finite Element Model.1\
     Distributed Force.1\Force Vector.1\Force.1`),"N.", " The FEM computations will resume.")

3   if `Finite Element Model.1\Maximum Von Mises.1\Maximum Von Mises` >2.5e+008N_m2

4   {Message("The maximum stress value became: # | and exceeds the value of the yield strength of 2.5e+008N_m2",
     `Finite Element Model.1\Maximum Von Mises.1\Maximum Von Mises`)

5   `PartBody\Steel\Steel.1.1\Yield Strength` =4.5e+008N_m2

6   Message("A new steel is selected with a yield strength of 4.5e+008N_m2.")

7   if `Finite Element Model.1\Maximum Displacement.1\Maximum Displacement` > 0.37mm

8   {Message("The maximum displacement value was exceeded, of 0.37mm.| The computed displacement is:
     ",`Finite Element Model.1\Maximum Displacement.1\Maximum Displacement`)

9   `Finite Element Model.1\Distributed Force.1\Force Vector.1\Force.1`= -205N

10   Message("The force was limited to the value of: 205N")

11   }}}

12   else if `Finite Element Model.1\Distributed Force.1\Force Vector.1\Force.1` >= -150N

13   {`PartBody\Steel\Steel.1.1\Yield Strength` =2.5e+008N_m2}

In line 1, the value of the applied force is compared with −150 N, initially established in the analysis and for which the results were presented above (Figure 4.202). The minus sign indicates the negative direction of the X axis, as the force value was entered in the dialog box in Figure 4.198.

By double-clicking on the *Distributed Force.1* feature in the specification tree, the user modifies the force value in the dialog box with the same name. The change can be made to increase or decrease the force value. If the force is greater than −150 N (example: −160 N, −190 N, etc.), according to line 2, an information message is displayed regarding the change made. The character "|" has the role of writing the text on several lines.

The "*abs*" function extracts the absolute value (example: the value −212 becomes 212). The displayed text is completed after the value with "N." and with an information regarding the resuming of the FEM computations (Figure 4.212). The text passages are separated by commas.

The program performs the computations, and in line 3 the maximum *Von Mises stress* obtained is compared with the yield strength of the chosen steel. Its value of 2.5×10^8 N/m^2 is observed as the term of the comparison and its special syntax (2.5e+008N_m2). If that comparison is true, according to line 4, an information message from Figure 4.213a. is displayed. It contains the new value of the stress, and the user is warned that the initial yield strength, of 2.5×10^8 N/m^2, was exceeded.

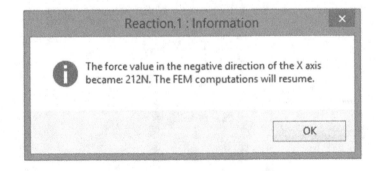

FIGURE 4.212   Displaying the information box on the change of the force value.

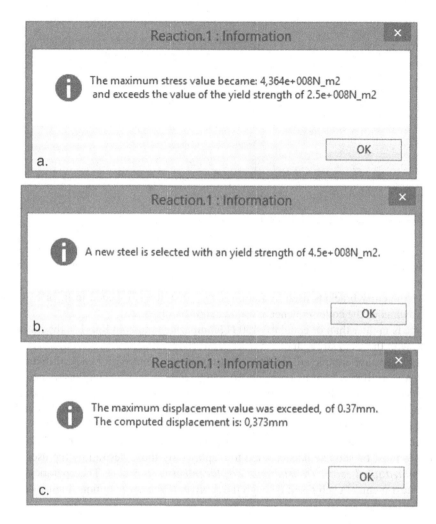

FIGURE 4.213   (a–c) Displaying information boxes with different resulting values and regarding the choice of a new material.

In line 4 there is also the character "#". It takes the place of the parameter *Finite Element Model.1\Maximum Von Mises.1\Maximum Von Mises*; the user preferred to enter all the information text, to mark the position of the parameter, it being added at the end, but displayed the "#" character instead.

Analysing the maximum stress value and if the comparison in line 3 is fulfilled, in line 5 it is necessary to choose another steel (e.g.: C22C38/14) with the yield strength of 4.5×10^8 N/m². The user is informed by a message in line 6 about the change of the initially chosen steel (Figure 4.213, b.).

For the new conditions, the program computes the maximum value of the displacement, and in line 7 compares it with an imposed value, of 0.37 mm. Line 8 displays an information message (Figure 4.213c.) if the respective comparison is met.

A value higher than 0.37 mm is not accepted by the user and, therefore, not by the program according to the reaction code, the solution consisting in the automatic reduction of the force value, which becomes –205 N, through line 9. That value basically represents the maximum force that can be supported by the part in its assembly, considering that it will be manufactured of the newly chosen material, a steel with better mechanical characteristics.

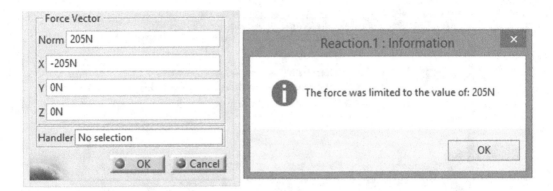

FIGURE 4.214 Information box regarding the limitation of the applied force value.

Figure 4.214 shows the information message (from line 10) regarding the automatic force limitation, but also the three *Force Vector* value fields of the *Distributed Force* dialog box where the force value in the negative direction of the *X* axis changes automatically.

In line 11 the curly brackets used by the three "*if*" conditions (opened in lines 2, 4 and 8) are closed. The syntax of the code sequence must be carefully observed.

If the force is greater than or equal to –150 N (comparison made in line 12), the initially chosen steel is kept, with the yield strength of 2.5×10^8 N/m² (line 13).

Once these steps are completed, the user must update the sensors, for each of them the *Update Sensor* option is available from its own context menu (Figure 4.215). The sensors can be updated one by one, depending on what information is desired by the user at a certain time or all at once (the context menu of the *Sensors.1* feature). In the figure, the user can see the *Update* symbol (a simple vortex) which is present next to the icons of all sensors.

The values of the sensors change, and are displayed in the specification tree. Also, the results and/or images must be activated/deactivated to display only those desired by the user (*Von Mises Stress, Estimated local error, Translational displacement vector,* etc.). The operation takes place from the context menu of each result through the *Activate/Deactivate* option (Figure 4.216).

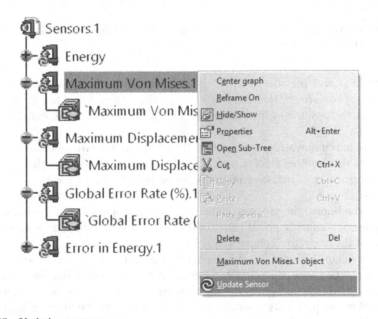

FIGURE 4.215 Updating a sensor.

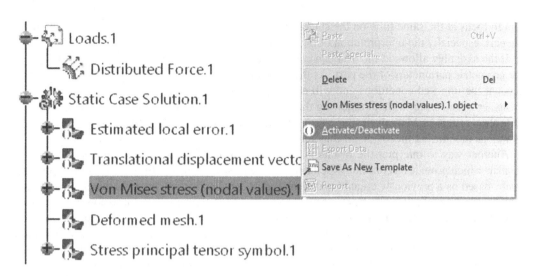

FIGURE 4.216 Activate/deactivate the FEM analysis results.

In conclusion, in the example presented, the user specified a force value of –212 N (Figure 4.212), the program detected a maximum stress greater than the yield strength of 2.5×10^8 N/m^2 of the initially chosen steel and changed this steel with another that has the yield strength of 4.5×10^8 N/m^2.

Performing the FEM computations also led to a maximum value of the displacements, higher than the one accepted by the user (0.37 mm). For this issue, a solution was adopted to limit the force applied to the part, of –205 N, regardless of the value entered by the user. The user is also informed about this limitation in order to take it into account in designing the assembly and establishing the loads accepted by the components.

For the force limited to –205 N and after updating the sensors, the new values of stresses and displacements are displayed with the help of the two palettes and values in Figure 4.217. The user notices that the displacements are below the value of 0.37 mm, and the maximum stress, of 4.22×10^8 N/m^2, is lower than the value of the yield strength of the new material established by the reaction code.

The presented method is a variant to automate the FEM analysis process, relieving the user of the concern of limiting the applied force or of changing the part material in certain situations, allowing

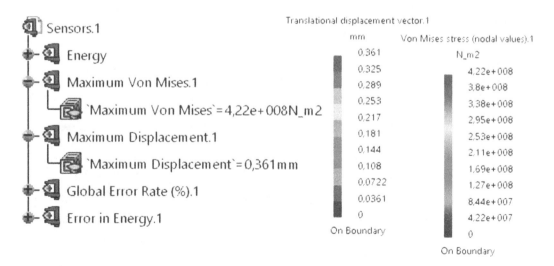

FIGURE 4.217 The values of the computed results for the maximum stress and the maximum displacements.

him to focus at the same time on the stages of design, establishing restraints and loading mode of the part, especially the interpretation of the results obtained.

If the assembly allows it, the user can complete the reaction code with instructions for changing the geometric parameters of the part. As the FEM analysis takes place, the program will take into account the imposed conditions and will automatically compute the maximum stresses, displacements, error rates, etc. without requiring the user's intervention. The method also represents a simplification of this FEM analysis, but it requires the correct design and writing of the reaction code, as well as monitoring the results, if they remain within the limits imposed by the user.

Another way to interpret the results is available with the *Animate* tool on the *Analysis Tools* toolbar, which provides a continuous sequence of successive frames of the model analysis, interactively, based on a previously created and displayed image. Each frame presents the obtained result, with a certain amplitude. The frames follow quickly, giving the user the impression of movement. Of course, the role of this tool is only to facilitate the understanding of the behaviour of the entire system consisting of the model, restraints, loads, etc. the displayed displacements being deliberately exaggerated, just for highlighting, not representing the real deformations of the model.

Clicking the *Animate* icon and selecting an active image result in the specification tree opens the *Animation* dialog box (Figure 4.218), which contains the standard animation controls and a slider to increase/decrease the playing speed.

The animation can be displayed once or multiple times using the *Change Loop Mode* icon button, its refinement (number of key frames) being set by the value in the *Steps Number* field. The minimum number of frames is 5, and the maximum number is 50.

Pressing the *More≫* button expands the options of this dialog box, thus allowing the user to choose the animation mode (the *Animation Mode* button) between asymmetric (default) and symmetric, as

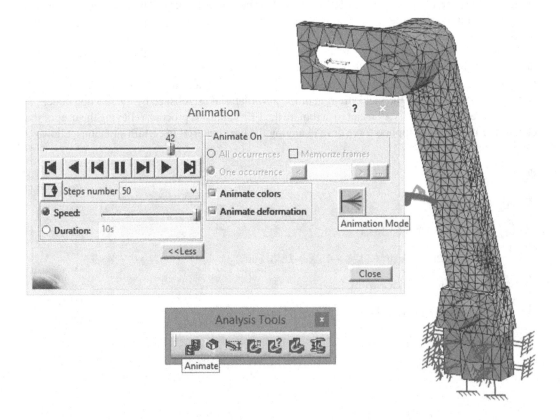

FIGURE 4.218 Animation of an FEM analysis result.

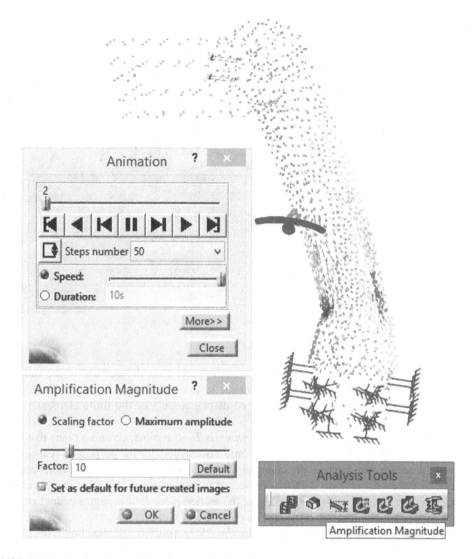

FIGURE 4.219 Setting the amplification magnitude of the animation.

well as the interpolation of stress and displacement values, so that between the frames key chosen to exist a smooth animation.

Figure 4.219 shows a frame taken during the running of an animation, with the purpose of visualizing how the analysed model is deformed (in an exaggerated manner). Of course, the deformation has the effect of changing the positions of certain nodes and changing the dimensions of the finite elements located in the stress zones. In the figure, an image of the *Main Stress tensor symbol* type was used, also displayed in the specification tree.

The amplitude of the animation is controlled using the *Amplification Magnitude* tool of the *Analysis Tools* toolbar.

The tool icon is active only if a fully defined system analysis computation has been created and there is at least one active image result in the *Static Case Solution* feature. In the dialog box in Figure 4.219, the user can check the options *Scaling factor* to set a constant scaling factor or *Maximum Amplitude* to actually enter a value, in millimetres, through which the maximum amplitude of displacements can be visualized.

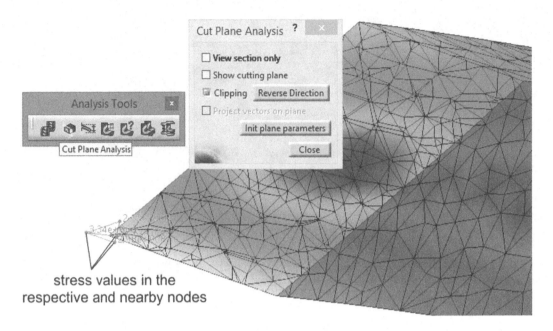

FIGURE 4.220 Sectioning a FEM analysis result.

By default, the user only views the image results placed on the outside of the model, but there is also the possibility of having access inside it, in certain section planes. Thus, the values of the stresses present in certain nodes, the displacements of the finite elements, etc. are highlighted.

The *Cut Plane Analysis* tool, located on the *Analysis Tools* toolbar, creates a plane that the user manipulates with the help of the compass (rotations, movements), for the purpose of real-time visualization of the results. By dynamically changing the position and orientation of the section plane, the interior of the model (Figure 4.220) is presented in two manners, namely, as a portion of it, remaining after the removal of the volume above the section plane (*Show cutting plane*), or only an area of the model actually located in the section plane (*View section only*).

The section plane remains active as long as the *Cut Plane Analysis* selection box is open, closing it using the *Close* button leads to the disappearance of the plane and the return to the initial shape and view of/over the model, while the specification tree remains unchanged.

4.8   ANALYSIS OF A WELDED STRUCTURE

Usually, many welded constructional steel structures consist of beams, channels, angles and plates. These are subsequently joined to each other by bolts, welded joints or rivets and are often subjected to various loads, can be in service for many years and have to deal with environmental and service factors that produce all types of stresses. Any steel component, that is part of such structure, will have an independent structural function. Many of these components are welded together, to form an assembly, which make a contribution to the strength, durability, aspect and life of the structure.

Steel structures are subjected to variable loads and can undergo fatigue failures from the weld. Sometimes, welded joints can represent the weakest part of structures and are susceptible to failure. The quality and strength of welded joints depend on factors such as the design, dimensioning and welding processes applied during manufacturing. It is, thus, necessary to design, analyse and simulate a welded structure, taking into account the loading of the joints, the steel type, welding process and the geometry of the structure.

The application presents the main stages of modelling and FEM analysis of a relatively common and known welded structure, composed of a base plate, a U-profile welded to it and a stiffening element between the base and the profile.

The user considers that the parts are made of sheet metal with reduced thickness. In the practice of finite element analysis, if a certain part is relatively thin and has a simple shape, it is considered to be a 2D solid. This approach is recommended because the model for analysis contains fewer finite elements, is easier to discretize and its thickness is defined as a 2D property.

The three parts considered to be welded to form the analysed structure are easily modelled in the *CATIA v5 Generative Shape Design* workbench, without involving the assembly workbench and its tools. Also, this means that the FEM analysis is simpler and faster, with very similar results.

In the application, the parts have the same thickness, 3 mm, and will be modelled as simple surfaces in the same geometric set and saved in a single *CATPart* file. The user can choose other ways of working, in which each part is modelled separately and inserted as a component in the assembly. This way of working is recommended, however, in cases where the number of parts in such a structure is large and they are assembled both by welding and by standard assembly elements (screws, rivets, etc.). Regardless of the chosen working mode, the user must establish certain assembly conditions necessary for the FEM analysis, some of which are presented and explained in this application.

The 3D modelling of the three components is very simple. According to Figure 4.221, the welded structure is composed of the *Base, UProfile* and *Stiffener* surfaces.

In the *CATIA v5 Generative Shape Design* workbench, in *Sketch.1* of the *XY Plane,* the user draws a square with sides of 100 mm, to be symmetrical to the *H* and *V* axes of the coordinate system. The inside of the square is filled by a surface by applying the *Fill* tool (Figure 4.222). The surface has no thickness and, therefore, no volume, but these properties will be specified in a future stage to be taken into account by the FEM analysis.

In the same *XY Plane,* but in *Sketch.2,* the user draws a circle with a diameter of Ø12 mm, positioned 35 mm from the edges of the *Fill.1* surface. The circle is multiplied on the surface using the *Rectangular Pattern* tool, the distance between the holes in each of the two directions is 70 mm (Figure 4.223).

Thus, the *RectPattern.1* feature is added to the specification tree, and the four circles are arranged on the *Fill.1* surface. The user applies the *Split* tool to extract four surfaces inside the circles. In the *Split Definition* dialog box (Figure 4.224) the *Fill.1* surface was selected in the *Element to cut* field, and the *Cutting elements* are *Sketch.2* and *RectPattern.1.* Initially, the specification tree contains the *Multi Output.1 (Split)* feature, then the user changes its name (from the context menu) to *Base.*

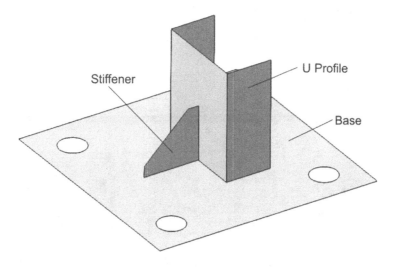

FIGURE 4.221   Isometric representation of the welded structure.

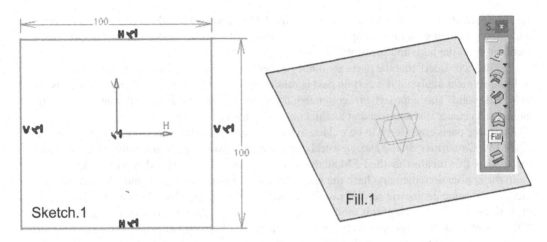

FIGURE 4.222   Representation of *Sketch.1* and *Fill.1* surface.

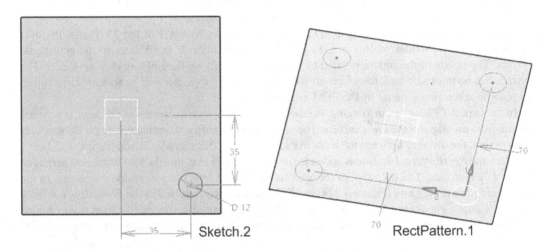

FIGURE 4.223   Representation of *Sketch.2* and *RectPattern.1* features.

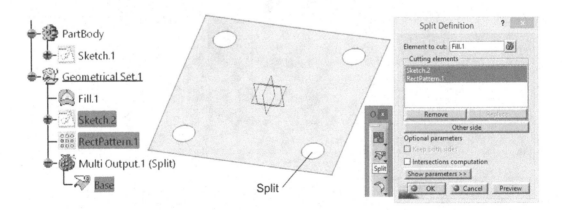

FIGURE 4.224   Extracting circles from the surface *Fill.1*.

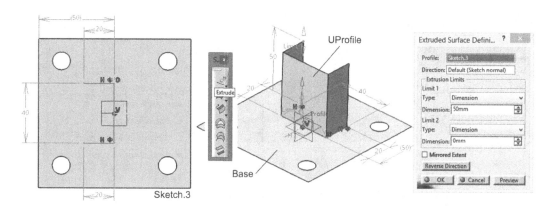

FIGURE 4.225 Drawing the *UProfile* and its extrusion.

The name change will prove to be very important in the next stages of the analysis. For verification, the area of this component of the welded structure is 9547.611 mm^2.

In *Sketch.3* of the *XY Plane,* a simple profile is drawn, composed of three lines (two are horizontal and one is vertical). The horizontal lines are 20 mm long and are symmetrical to the *H* axis of the coordinate system. The vertical line is 40 mm long and coincides with the *V* axis.

In Figure 4.225 the *Base* surface and the three lines, as well as a dimensional constraint of *Reference type:* (50) can be seen. This does not influence in any way the profile or its positioning in the *XY Plane* and on the *Base* surface, but provides important information for the user, regarding this positioning: the vertical line is in the middle of the *Base* surface (which has a side of 100 mm on the surface *Fill.1*).

Such reference constraints are often used in certain sketches and have only an informative role. The respective sketches are already fully and correctly constrained, but it is possible that a certain dimension is not explicitly present. The introduction of that dimension causes an over constrain in the sketch, and the user prefers to keep it as a reference constraint by checking the *Reference* option in the *Constraint Definition* dialog box.

The drawn profile is extruded over a distance of 50 mm in the positive direction of the *Z* axis. The obtained surface is initially called *Extrude.1*, but the user changes its name to *UProfile*.

In the *ZX Plane,* a simple profile *(Sketch.4)* consisting of several lines is drawn, two of them are coincident with the *Base* and *UProfile* surfaces. The dimensions are in accordance with Figure 4.226.

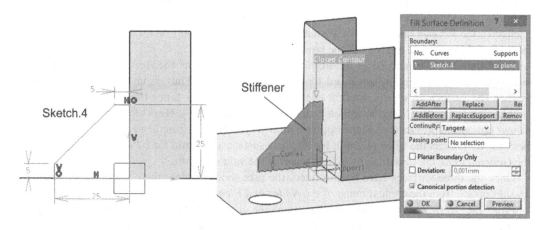

FIGURE 4.226 Drawing of the stiffening component.

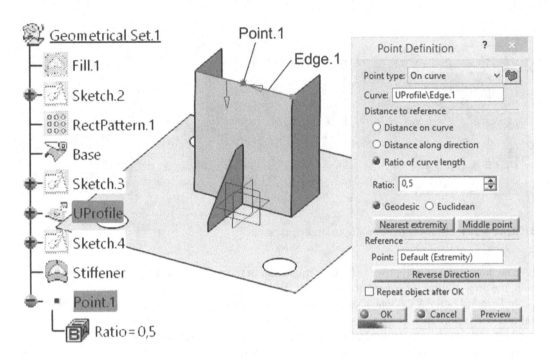

FIGURE 4.227   Inserting a point on the middle of the top edge of the *UProfile* surface.

Being a closed profile, *Sketch.4* is used to create the third surface of the considered structure. The *Fill* tool is applied again, the outline of the sketch is chosen and then the user renames the surface to *Stiffener*.

The three surfaces, due to the way they were created, are in contact along some edges. These will be used in a subsequent stage to establish certain constraints specific to the FEM analysis.

A point is inserted on the long edge at the top of the *UProfile* surface, placed in the middle of this edge. In the *Point Definition* dialog box (Figure 4.227), the user chooses the *On curve* option in the *Point type* drop-down list and selects the edge marked in the figure. Below, the *Ratio of curve length* option is checked and the user enters the value of 0.5 in the *Ratio* field. One of the ends of the edge is chosen as a reference point.

From the *Insert* menu or from the *Tools* toolbar, the user inserts an axis system with the origin in the previously created *Point.1*. In the dialog box in Figure 4.228, the user chooses the *Standard* type, the origin point, then three edges of the *UProfile* surface to establish the X, Y and Z axes. For each axis, the *Reverse* option is also available. This axis system will be used later in the explanations related to the FEM analysis of the structure.

From the *Start → Analysis & Simulation* menu the user opens the *Advanced Meshing Tools* workbench. It contains many advanced tools for mesh creation. Thus, from the *Meshing Method* toolbar, the *Surface Mesher* is used for each of the three surfaces. Since the application uses flat, simple surfaces, in the *Global Parameters* dialog box (Figure 4.229), *Mesh* tab, the user chooses the *Linear* option and presses the second option-button *Set frontal quadrangle method*. In the *Mesh size* field, the user enters the value of 2 mm for the *Base* surface and 1 mm for *UProfile* and *Stiffener*. In the *Geometry* tab, the *Constraint sag* field receives the value 1.5 mm.

Once the three surfaces have been refined, the user should update them using the *Update All Meshes* option from the context menu of the *Nodes and Elements* feature of the context menu.

From the *Mesh Analysis Tools* toolbar, the user clicks the *Free Edges* icon. According to Figure 4.230, some edges belonging to the three surfaces become green (visible on screen). In the *Free Edges* selection box, it is possible to tick the *Autofocus* option. The browse buttons allow the

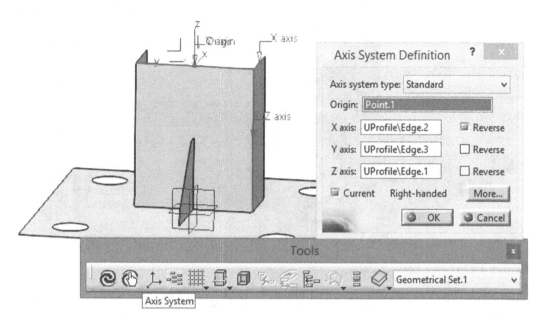

FIGURE 4.228   Inserting a new axis system.

user to scan the free edge subsets. *Current subset* displays the number of subsets of free edges and lets the user visualize a particular subset either by entering a number or by clicking the browse buttons.

Thus, the user is able to display subsets of free edges and thereby quickly visualize incompatible mesh. A subset of free edges is a connected component of free edges. An edge of a 2D mesh that is shared with a 3D mesh is not free anymore.

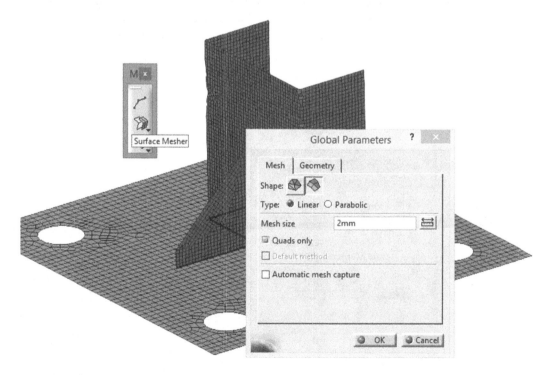

FIGURE 4.229   Discretization of surfaces.

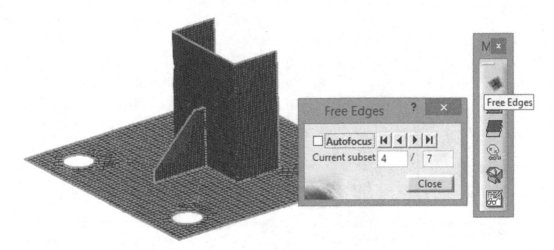

FIGURE 4.230 *Free Edges* checking.

For some of these edges, located at the contact between surfaces, the user will define connection properties that are specific to the application.

To define a material to be applied to this structure, the user returns to the *Generative Structural Analysis* workbench, clicks the *User Material* icon on the *Model Manager* toolbar and selects a standard steel that has the characteristics presented in the previous applications. The *Steel* material is added to the specification tree within the *Materials.1* feature and will be used in the next steps.

The user must set the properties of the three meshes: material and thickness. Thus, from the same *Model Manager* toolbar, the user accesses the *2D Property* icon and the dialog box in Figure 4.231 opens. The same properties are applied to each component of the structure: in the *Supports* field, the user selects each of the three previously created meshes: *Base*, *UProfile* and *Stiffener*. The user checks the *User-defined material* option and chooses the material created and recommended for these components. To establish a thickness of each mesh, the user enters a value of 3 mm in the

FIGURE 4.231 Defining a *2D Property*.

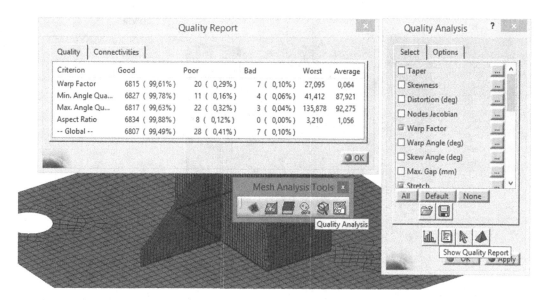

FIGURE 4.232 Some results of the mesh quality check.

Thickness field. In this application, the three components have the same thickness and are made of the same material, but, as a supplementary work topic, the user may try to study the behaviour of the structure when the components have different thicknesses and/or other material properties.

The user can ask the *CATIA v5* program to perform a quality test of the three meshes, components of the structure. From the *Start → Analysis & Simulation* menu, the *Advanced Meshing Tools* workbench opens, then the user presses the *Quality Analysis* icon on the *Mesh Analysis Tools* toolbar. As a result, the selection box with the same name opens and the user has access to numerous options (Figure 4.232).

These options represent a complex set of criteria applied on the meshes and the user is able to change the criteria taken into account for the mesh optimization. The user presses the *Show Quality Report* icon and notices that on the *Good* column the percentages are very high (over 99%), but these values may vary depending on the previously selected options and the manner the surfaces were created.

By clicking the *OK* or *Apply* buttons, the structure is represented in three colours, in a very suggestive way (Figure 4.233). Thus, during the quality check, the mesh elements are assigned green, yellow and red colours, visible on screen. Green, which is dominant, is used when the elements are interpreted by the solver without any problem. Yellow is used when the elements are solved with very few possible problems. Red is used when the elements are hardly properly resolved.

Returning to the report displayed through the three colours, the user can find out information about any element of the structure by pressing the *Analyze an Element* icon (Figure 4.234). The *Analyze Single* dialog box is displayed with the quality type *(Good* – green, *Poor* – yellow, *Bad* – red)* and value for all the selected criteria assigned to this particular element.

Along with the arrow *Analyze an Element,* the *Quality Analysis* selection box also contains the *Worst Elements Browser* icon. The user is initially proposed to identify and display a number of 10 such elements, having, of course, a red colour. Once the problems are reported, their solution is possible by changing the geometry, the type of finite element (*Linear* or *Parabolic*), by refining the mesh, etc. After applying any modification, the stage of checking the quality of the structure's meshes can be resumed.

The three components are assembled by welding, so the user must define the specific connections. They should not be represented within the structure by a certain geometry.

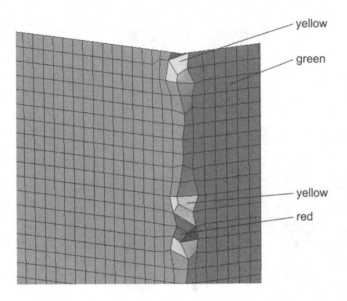

FIGURE 4.233 Mesh elements assigned green, yellow and red colours, visible on screen.

Back in the *CATIA v5 Generative Structural Analysis* workbench, the user clicks the *Line Analysis Connection* icon to open the selection box with the same name (Figure 4.235). In the *Name* field the user can change the name of the connection, then in the *First component* and *Second component* fields he chooses two flat surfaces, according to the figure. The last selection is the line between the two surfaces *(1 Edge)*. In the specification tree, the *Line Analysis Connection.1* feature (the name is taken from the selection box) is added within the *Analysis Connection Manager.1* feature.

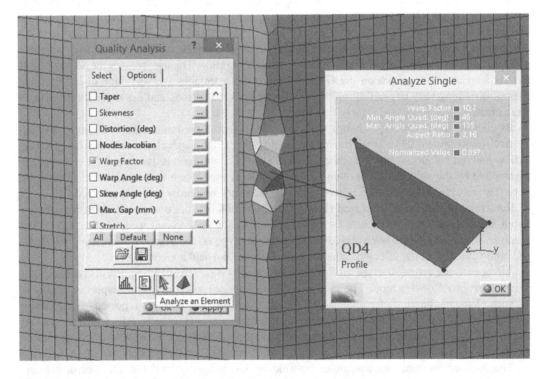

FIGURE 4.234 Analysis of a single element.

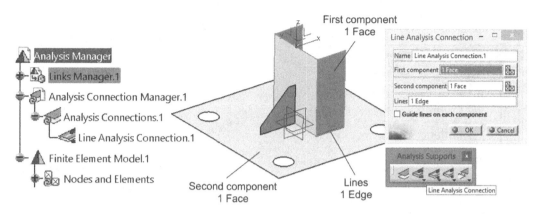

FIGURE 4.235   Defining *Line Analysis Connection*.

In the Figure 4.236, the connection is represented by a thick black line and it is basically the support on which the welding is applied. Thus, on the *Connection Properties* toolbar, the user clicks the *Seam Welding Connection Property* icon (Figure 4.236) and the selection box opens in which the name of the welding property must be specified. By default, this name is quite suggestive, it can be modified by the user, and the *Seam Welding Connection Property.1* feature is added to the specification tree.

In the *Type* drop-down list, the user specifies the type of this connection as, obviously, *Rigid*. The connection symbol is displayed over the support line. Choosing another type of property involves additional selections via the *Component edition* button (not shown in the figure).

A *Seam Welding Connection Mesh.1* part appears, also, under the *Nodes and Elements* feature (Figure 4.237). The user is able to edit the feature by double-clicking on it in the specification tree.

In the dialog box that appears, the user observes the support on which the welding is performed, the number of applied welds (*Welds:* 1), the chosen type of connection (*Rigid*, in a non-editable field) and two parameters: *Maximal gap* and *Mesh step*. The first parameter represents the maximum distance between the elements to be welded. The seam welding connection mesh part is created with a default *Mesh step* value, which is computed as a ratio of the seam length. If this value is much smaller than the size of the connected meshes, *CATIA v5* may find it difficult to solve/compute the problem and this may lead to an *"Out of Memory"* error message.

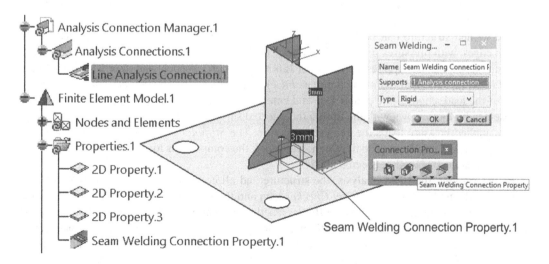

FIGURE 4.236   Defining *Seam Welding Connection Property*.

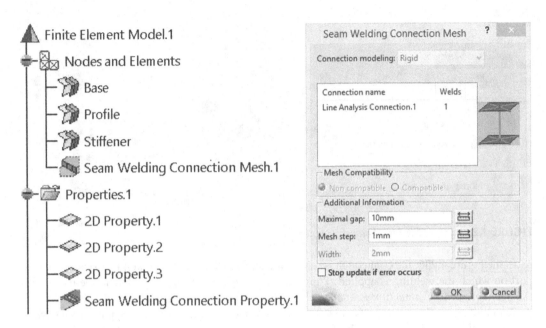

FIGURE 4.237 Defining *Seam Welding Connection Mesh* settings.

Usually, the *Mesh step* value (here it is 1 mm) respects the proportion of the connected meshes, and is around half of the smallest connected mesh. The user should remember that in this application, the *Mesh size* has the value of 2 mm for *Base* (Figure 4.229) and 1 mm for *UProfile* and *Stiffener*.

Similarly, the other features *Line Analysis Connection.2, ..., Line Analysis Connection.5 (LAC 1–5)* are created within the *Analysis Connection Manager.1* feature. Based on them, *Seam Welding Connection Property.2, ..., Seam Welding Connection Property.5 (SWCP 1–5)* are defined and *Seam Welding Connection Mesh.2, ..., Seam Welding Connection Mesh.5* features are automatically inserted, according to Figure 4.238. The figure shows the structure model after completing the definition of all connections with the seam weld symbols for each edge.

The user applies a *Clamp* restriction to the structure on each circular edge of the *Base* component (Figure 4.239). In the *Clamp* selection box, the user specifies the name of the entire set of restraints, then selects the four circular edges by holding down the *Ctrl* key (a multiple selection). As such, the *Supports* field is completed with *4 Edges*.

The welded structure can be loaded in many ways: with forces distributed on surfaces or on certain edges, with pressures positioned on the vertical or horizontal surfaces, etc. Thus, according to Figure 4.240, the user applies a pressure of 120000 N/m² on a vertical and internal surface of the *UProfile* component (marked in the figure as *Face*). In the *Pressure* selection box, the user sets the name of the load, chooses the planar face and enters the pressure value. The applied *Pressure.1* is represented by four arrows with the tips on the selected surface.

The applied pressure loads the analysed structure. The *Stiffener* component will distribute this load to the *Base* component. The five connections link the components together, so that the stresses that appear are distributed between them.

Before running the FEM analysis, the structure and all conditions applied to it must be verified. Thus, the user presses the *Model Checker* icon and the information box with the same name becomes available (Figure 4.241). In the *Bodies* tab, the verification status for the components of the structure is displayed, the user notices *OK* on the *Status* column, the names of the components and the applied material. Nearby, in the *Connections* tab (Figure 4.242), the connections between the components of the structure are verified. The user notices the *OK* status, but also the fact that these connections do not present material.

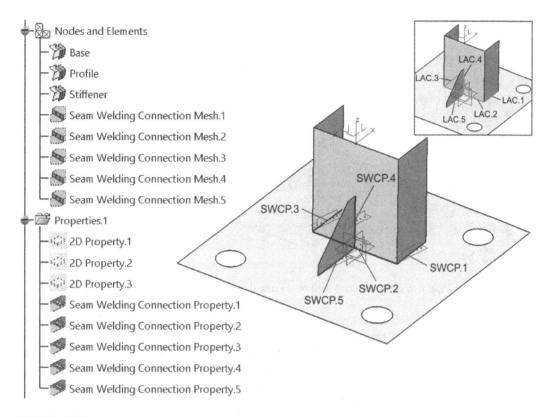

FIGURE 4.238 The structure model with all connections.

The user cannot apply material properties to the *Seam Welding Connection Mesh* features because they have no geometry, as they are defined only as a connection between components (it can also be seen on the *Connected Mesh Parts* column). However, it is possible to specify a material for the support lines based on which the welds were defined. As can be seen in Figure 4.243, the user has assigned the same *Steel* material to each *Line Analysis Connection* feature, without, however, influencing the future results of the FEM analysis too much.

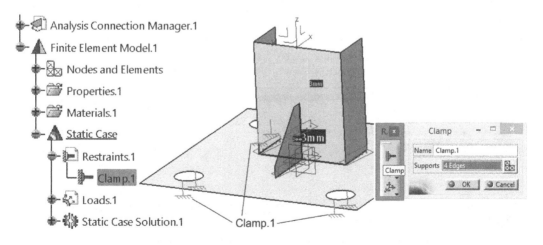

FIGURE 4.239 Applying the *Clamp* restraint to the *Base* component.

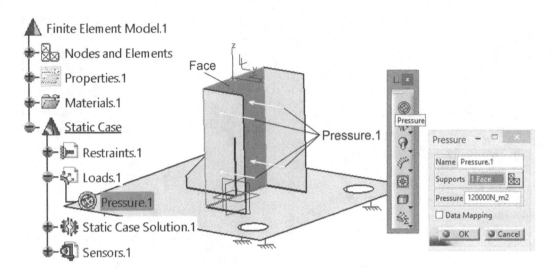

FIGURE 4.240 Applying the *Pressure.1* load on a vertical surface.

Once the structure model has been completed and verified, the user can run the computation stage of the FEM analysis. Due to the fact that the user has defined connections between the structure's components, material properties, the *Clamp.1* restriction and the *Pressure.1* load, etc. the structure behaves as a single entity. The stresses and displacements are distributed in the components. By clicking the *Clamp* icon, the selection box with the same name from Figure 4.244 opens and the computation stage of the FEM analysis begins.

Depending on the desired results, the user chooses one of the four options from the drop-down list of the *Clamp* selection box. The hardware resources required for the FEM computation are displayed after pressing the *OK* button. The program determines the required resources and displays some estimated values, if they are equal to zero the computation is not possible, and the user must identify the previous stage that was not completed successfully.

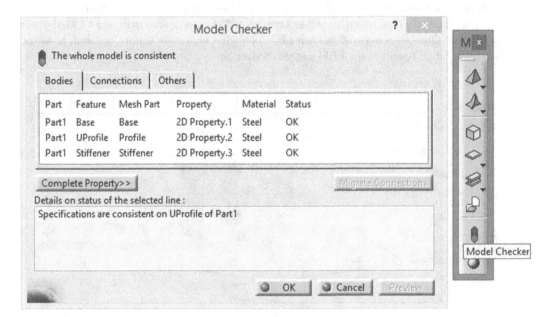

FIGURE 4.241 Checking the components of the structure.

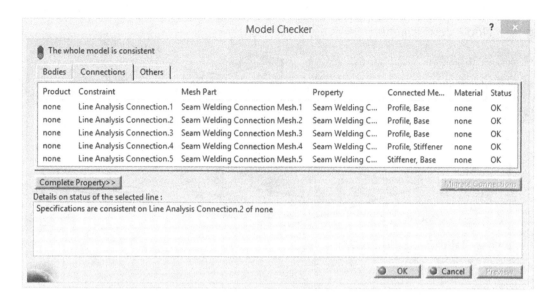

FIGURE 4.242 Checking the connections between components.

Once the computation is finished, the tools from the *Image* toolbar are available to the user to display and interpret the results. The specification tree updates according to the computed and added images.

Figure 4.245 clearly shows how the welded structure deforms. The user chose the *Deformation* image result, then set the *Amplification Magnitude* value to display the deformed mesh. The deformation is, of course, exaggerated and takes into account the connections established between the components of the structure. Thus, the *Stiffener* and *UProfile* displacements/ deformations due to the *Pressure.1* load produce the *Base* deformation. By running the result with the help of the *Animate* tool, the correct behaviour, which is similar to the real one, of the studied structure can be observed.

By default, the structure deforms as a whole, which is correct. The user can view the deformation of a single component or of the entire welded structure. Thus, he double-clicks on the *Deformed mesh.1* feature in the specification tree and the *Image Edition* selection box opens. In the *Selections* tab, according to Figure 4.246, in the top list, *Available Groups,* the user chooses the components that will be represented in the result, then presses the add arrow V in the bottom list, *Activated Groups.* The exaggerated deformation of the *Base* component can be observed.

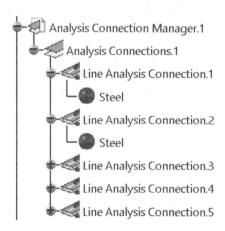

FIGURE 4.243 Applying materials to *Line Analysis Connection* supports.

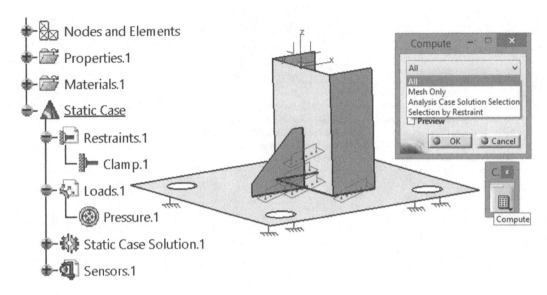

FIGURE 4.244 Launching the computation stage of the FEM analysis.

Similarly, with the help of the *Von Mises Stress* tool on the same *Image* toolbar, the user identifies the stresses that appear in the welded structure, but also in the individual components (double-click on the feature *Von Mises stress (modal values).1* and select the component). Figure 4.247 shows the areas with maximum stresses and the colour palette with stresses for the *Base* component.

The user prefers to display the results for each component of the structure because he can easily observe the areas with minimum and maximum stresses, their values (with the *Image Extrema* tool), but also the stress distribution.

Each component presents a set of such values and it is very important to identify whether or not the maximum stress values exceed the yield strength value of the specified material. Often, in the case of a structure/assembly, the FEM analysis gives a maximum stress value, but the component

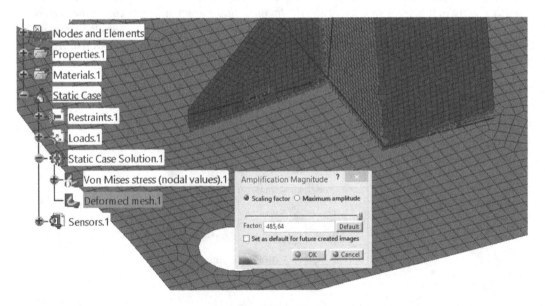

FIGURE 4.245 Changing the amplification magnitude to display the deformed mesh.

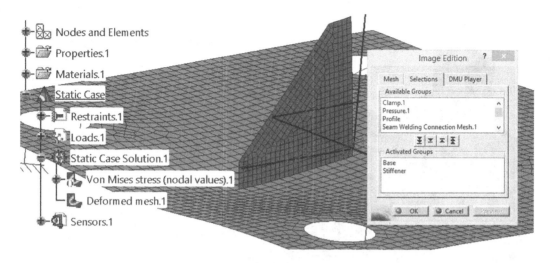

FIGURE 4.246 Selection of components whose deformations will be displayed.

on which it appears is not clearly specified. The graphic representation of the stress distribution is made according to this maximum value. Thus, certain components, which present lower stresses, do not display these values within the structure. The program displays the maximum stresses and their values at the level of the entire structure/assembly.

In Figure 4.247, several areas with maximum (red) and medium stresses (green and yellow) are marked. The maximum stress value that appears in the *Base* component is 1.98×10^7 N/m^2 (19.8 MPa), much lower than the yield strength value of the considered steel (2.5×10^8 N/m^2, 250 MPa).

The *Stiffener* component presents a maximum stress of 2.94×10^7 N/m^2 (29.4 MPa), and the *UProfile* a maximum stress of 2.16×10^8 N/m^2 (216 MPa), that is close to the yield strength value of the specified steel. Different information specific to the final FEM analysis is shown in Figure 4.248.

These components show stresses due to restraints, connections and loading *Pressure.1,* along with the welds. As previously presented, the *Seam Welding Connection Mesh.1, …, Seam Welding Connection Mesh.5* features of the specification tree represent the welds defined by the user between the components of the studied structure.

The user follows the previously presented procedure by double-clicking on the feature *Von Mises stress (modal values).1,* and the *Image Edition* selection box opens. In the *Selections* tab, according to Figure 4.249, the features representing the welds are activated one by one. In the *Activated*

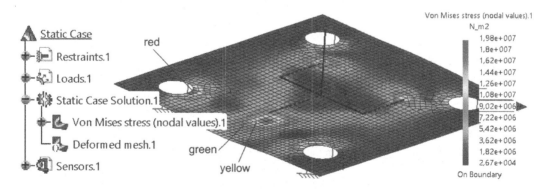

FIGURE 4.247 Displaying the areas with maximum stresses and the colour palette values for the *Base* component.

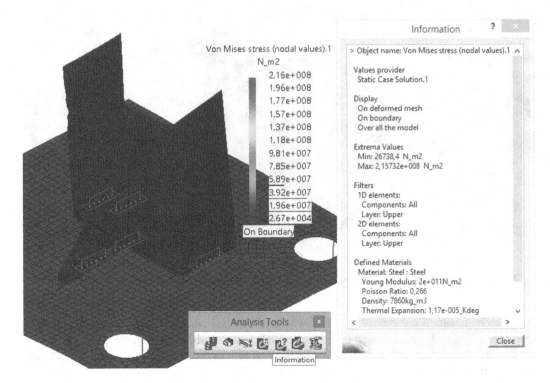

FIGURE 4.248 Display of general information after the FEM analysis.

Groups list, the feature *Seam Welding Connection Mesh.1* can already be observed, and the program displays the stress values.

Thus, the maximum stresses computed in the welds are *Seam Welding Connection Mesh.1*: 1.45×10^8 N/m² (145 MPa), *Seam Welding Connection Mesh.2*: 1.55×10^8 N/m² (155 MPa), *Seam Welding Connection Mesh.3*: 1.55×10^8 N/m² (155 MPa), *Seam Welding Connection Mesh.4*: 2.59×10^7 N/m² (25.9 MPa) and *Seam Welding Connection Mesh.5*: 1.45×10^8 N/m² (145 MPa). To better understand where these welds are positioned, the user should refer to Figure 4.238.

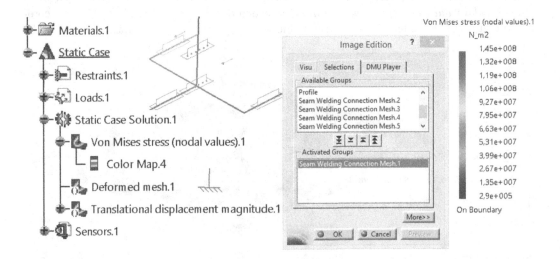

FIGURE 4.249 Displaying the stress values of the *Seam Welding Connection Mesh.1* feature.

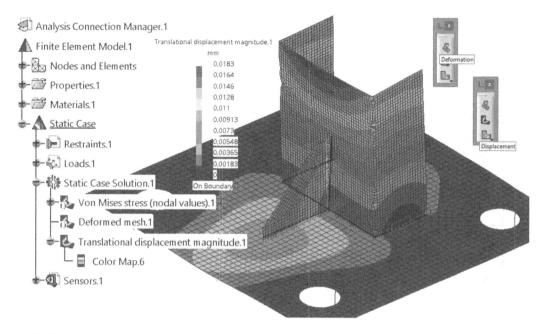

FIGURE 4.250 Display of displacements and their maximum and minimum values.

Important for this type of structure are its deformations. In the specification tree, the user adds the image results *Deformed mesh.1* and *Translational displacement magnitude.1* (Figure 4.250). Of course, the deformations are displayed exaggeratedly, but the minimum and maximum values are also shown. It is observed that the maximum displacement of the welded structure is 0.0183 mm, being positioned at the top of the *UProfile* component. The *Base* component is also deformed, but it also presents extended areas with very small displacements.

Such a displacement value (approximately 0.02 mm) is acceptable for this type of structure, especially because the user's expectations are related to a certain flexibility and possibility to handle certain loads. The presence of the *Stiffener* component is very important for supporting the *UProfile* component and reducing its deformation.

Regarding the precision of the FEM analysis performed, the *Global Error Rate (%)* sensor displays a value of 21.06%, according to Figure 4.251. The value can be accepted, but the user should try to improve the FEM analysis by a better discretization of the components.

The simplest discretization solution, with minimal intervention from the user and in which it exists the possibility to propose/impose an error rate, involves the use of the *New Adaptivity Entity* tool. In the *Global Adaptivity* selection box, the user enters the new value in the *Objective Error (%)* field, then tries to select the components *Base*, *UProfile*, *Seam Welding Connection Mesh.1*,

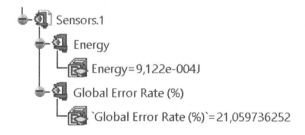

FIGURE 4.251 Display of FEM analysis global error rate.

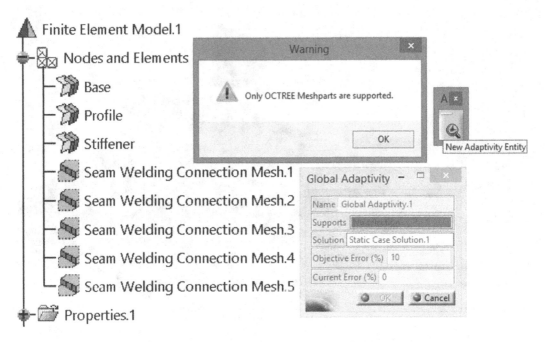

FIGURE 4.252 Attempt to apply the *New Adaptivity Entity* tool on the welded structure.

etc. from within the *Nodes and Elements* feature of the specification tree. The program, however, does not allow this selection for the *New Adaptivity Entity* tool and displays a warning message (Figure 4.252).

So, it is not possible to apply the *New Adaptivity Entity* tool, but, according to Figure 4.229 and the associated explanations, the user may apply a different discretization of the components surfaces. Thus, by double-clicking on each *Base*, *UProfile* and *Stiffener* feature from within *Nodes and Elements*, the *Global Parameters* dialog box opens (Figure 4.253).

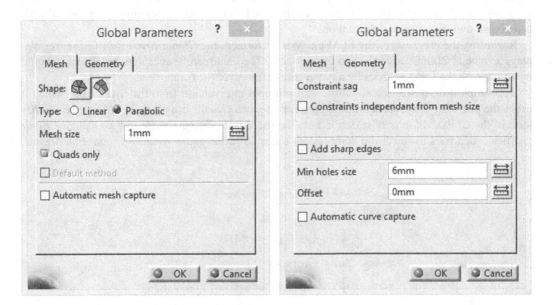

FIGURE 4.253 Advanced discretization of the structure components surfaces.

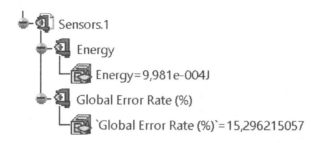

FIGURE 4.254 Display of the FEM analysis error rate after discretization.

In the *Mesh* tab, the user checks the *Parabolic* option, the second option-button *Set front quadrangle method* remains selected, then in the *Mesh size* field it is entered the value of 1 mm for the *Base* surface and 0.5 mm for *UProfile* and *Stiffener*. In the *Geometry* tab, the *Constraint sag* field receives the value 1 mm for each of the three surfaces. Thus, the discretization is improved, and the mesh network is better defined.

The five welds in the specification tree *(Seam Welding Connection Mesh)* are not editable in the same way, but the program automatically changes their discretization because they are linked to the structure components through the *Line Analysis Connection* features (Figures 4.235 and 4.236 together with the explanations related).

The first update of the FEM analysis should refer to the *Global Error Rate (%)* sensor, which, after the new discretization, displays a smaller and more convenient value of 15.3%, according to Figure 4.254. Certainly, for even smaller values, the user can try setting values of 0.5 mm–0.7 mm in the *Mesh size* fields and 0.5 mm in the *Constraint sag* fields.

Together with these changes related to the discretization, the results from the *Static Case Solution.1* will also be updated. Thus, the components of the structure will present other maximum stresses, as follows: *Base*: 2.91×10^7 N/m² (29.1 MPa), *Stiffener*: 1.38×10^8 N/m² (138 MPa), *UProfile*: 1.56×10^8 N/m² (156 MPa), *Seam Welding Connection Mesh.1*: 1.1×10^8 N/m² (110 MPa), *Seam Welding Connection Mesh.2*: 3.19×10^7 N/m² (31.9 MPa), *Seam Welding Connection Mesh.3*: 1.21×10^8 N/m² (121 MPa), *Seam Welding Connection Mesh.4*: 9.77×10^7 N/m² (97.7 MPa) and *Seam Welding Connection Mesh.5*: 1.38×10^8 N/m² (138 MPa).

The areas with maximum and minimum stresses are kept as in the previous case, before the discretization. Changes in these maximum stresses are observed, some have increased, others have decreased, but none of them exceed the yield strength value of the considered steel.

Updating the results-images *Deformed mesh.1* and *Translational displacement magnitude.1* (Figure 4.255) leads to displaying the maximum and minimum displacements. Thus, it can be seen that the maximum displacement of the welded structure is 0.0197 mm, being positioned at the top of the *UProfile* component. The representations in the figure also show the new discretization of the structure components; the user notices the density of the network of nodes and elements, according to the settings in Figure 4.253.

As the main conclusion, the user can assume that the welded structure has a good behaviour at the considered load, the maximum stresses appearing in the components produce only the elastic deformation of the structure, the welds are not in danger of breaking. With the help of the *Generate Report* tool located on the *Analysis Results* toolbar, the user can summarize the results obtained in a simple web page, consisting of an *index.html* file and some images.

In the *Report Generation* selection box (Figure 4.256) the user specifies where the files created for the report will be saved, chooses its name and ticks the *Add created images* option. Thus, after obtaining the report, the user finds out the number of nodes and elements (83186, respectively, 27597) of the structure, their type and quality, the number of degrees of freedom of the elements that were taken over by the conditions imposed to create the structure, the results of forces and moments determined by the load *Pressure.1* on each axis *X*, *Y* and *Z*, etc.

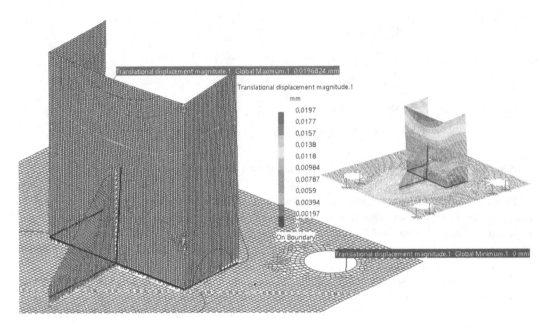

FIGURE 4.255   Display of displacements and their maximum and minimum values after discretization.

Before starting the next phase of the application, the user must save the files involved in the FEM analysis for future reference.

To continue with another approach, the user has the possibility to impose a maximum value of the *UProfile* deformation and the *CATIA v5* program will compute the required total load. This type of FEM analysis is useful when the structure has a certain limitation related to the deformation of one or more of its components.

Thus, from the specification tree, the user removes the *Pressure.1* load by choosing the *Delete* option from its context menu (Figure 4.257, left) and displays the *Axis System.1* coordinate system (by applying the *Hide/Show* option, in Figure 4.257, right).

The user chooses to keep the current geometry and the discretization of the structure's components, the welds, the material applied to the components, the properties related to the sheet thickness, the *Clamp.1* restriction. To obtain results as close as possible to the real situation, the user adds a virtual component to be in contact with *UProfile*. Thus, from the *Virtual Parts* toolbar, the user clicks the *Rigid Virtual Part* icon and opens the selection box with the same name (Figure 4.258).

In the *Supports* field, the internal planar face of the *UProfile* is chosen, and the user notices two lines drawn on that face. If the user specifies a point in the *Handler* field, then the virtual

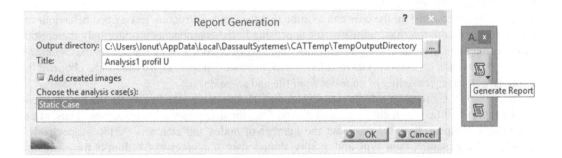

FIGURE 4.256   Generation of the FEM analysis report of the welded structure.

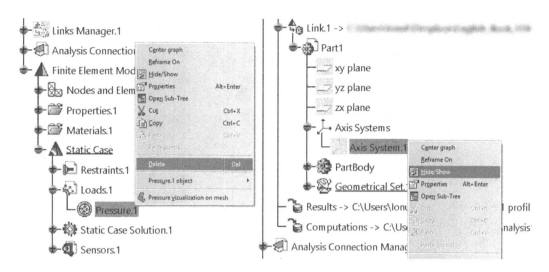

FIGURE 4.257 Deleting the *Pressure.1* load and displaying the *Axis System.1* coordinate system.

part is applied there. The chosen point is the origin *Point.1* of the coordinate system (according to Figures 4.227 and 4.228). If the user does not specifically select a point, the centroid (the point at which the lines meet) will be used as the handler point. This point selected as handler must be a *Part Design* point (a geometric entity).

A *Rigid Virtual Part* is a rigid body connecting a specified point to specified part geometries, behaving as a mass-less rigid object which will stiffly transmit actions (masses, restraints and loads) applied at the handle point, while locally stiffening the deformable body or bodies to which it is attached. It does not take into account the elastic deformability of the parts to which it is attached.

When applied, this rigid virtual part has the following effects: a node is created in coincidence with the specified handle point, each node of the specified geometry supports meshes is connected by a kinematical rig-beam element to the handle node, a set of rig-beam relations is generated between the handle node degree of freedom and the connected nodes degree of freedom.

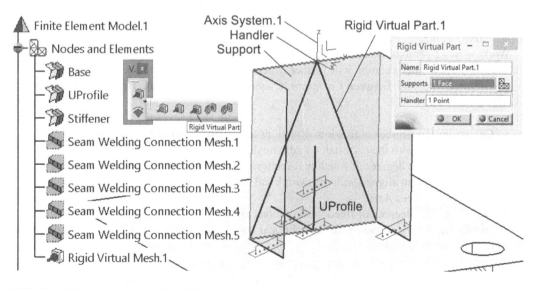

FIGURE 4.258 Inserting the *Rigid Virtual Part.*

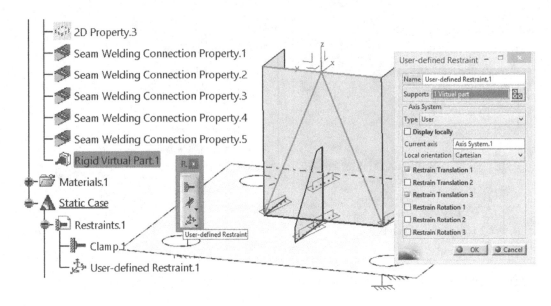

FIGURE 4.259 Defining the *User-defined Restraint*.

Thus, the *Rigid Virtual Part* generates as many rig-beam kinematical elements as there are nodes on specified support meshes, so the FEM analysis may require important resources to obtain the expected results.

As a consequence, two features are added to the specification tree: *Rigid Virtual Mesh.1* in *Nodes and Elements* (Figure 4.258) and *Rigid Virtual Part.1* in *Properties*. Unlike the other elements of *Nodes and Elements,* double-clicking on *Rigid Virtual Mesh.1* has no effect.

The area with maximum deformations of the *UProfile* component is on the edge (Figure 4.255) where the *Axis System.1* coordinate system was defined and which contains the *Handler* point for the virtual part. Thus, a new restraint will be created using the *User-defined Restraint* tool. In the selection box with the same name in Figure 4.259, in the *Supports* field the user selects *Rigid Virtual Part.1,* chooses the *User* type for the coordinate system, then *Axis System.1* (by clicking directly on it within the welded structure or by choosing from the specification tree) in the *Current axis* field. The *Cartesian* orientation is kept, then the user checks the *Restrain Translation 1* and *Restrain Translation 3* options.

Working in a *User* axis system, the degree of freedom directions will be relative to the specified axis system. Their interpretation will further depend on the axis type choice. For orientation, the user is allowed to choose between *Cartesian, Cylindrical* and *Spherical*. The choice is important as follows:

- *Cartesian*: The degrees of freedom directions are relative to a fixed rectangular coordinate system aligned with the cartesian coordinate directions of the *User-defined Axis*.
- *Cylindrical*: The degrees of freedom directions are relative to a local variable rectangular coordinate system aligned with the cylindrical coordinate directions of each point relative to the *User-defined Axis*.
- *Spherical*: The degrees of freedom directions are relative to a local variable rectangular coordinate system aligned with the spherical coordinate directions of each point relative to the *User-defined Axis*.

For the available degrees of freedom, the user may impose or release them; six degrees of freedom per node: *Translation 1* = translation in the *X* direction, *Translation 2* = translation in the *Y*

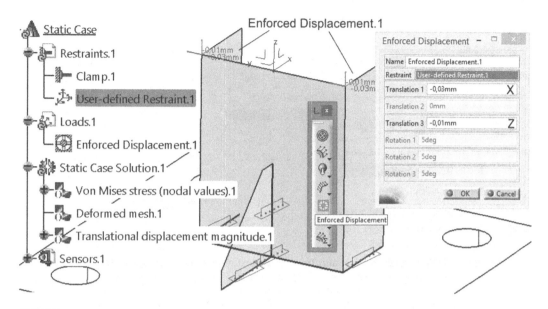

FIGURE 4.260 Entering the maximum displacement values on the *X* and *Z* axes.

direction, *Translation 3* = translation in the *Z* direction, *Rotation 1* = rotation around the *X* direction, *Rotation 2* = rotation around the *Y* direction and *Rotation 3* = rotation around the *Z* direction.

By ticking the *Restrain Translation 1* and *Restrain Translation 3* options, in the next step the user will impose possible displacements on the *X* and *Z* axes. Practically, the points on the selected upper edge will move along the respective axes, without, however, having a rotational movement, but only translation.

From the *Loads* toolbar, the user clicks the *Enforced Displacement* icon and opens the selection box with the same name in Figure 4.260.

In the *Restraint* field, *User-defined Restraint.1* is selected, then the values −0.03 mm are entered in the *Translation 1* field for *X* axis displacement and −0.01 mm in the *Translation 3* field for *Z* axis displacement. Practically, these are the maximum values accepted by user for the analysed structure on each *X* and *Z* axis. The minus sign means that these displacements of the upper edge of the *UProfile* (and of the *Handler* point, by default) will take place in the opposite direction to the *Axis System.1* axes. It is observed that the user does not want, for example, to have displacements on the *Y* axis because the *Restrain Translation 2* option is not checked in Figure 4.259, nor has a value been set in the *Translation 2* field (Figure 4.260).

To resume the computation stage, the user right-clicks on each result in the specification tree (Figure 4.261) and chooses the *Activate/Deactivate* option of the context menu.

The maximum values of the stresses and displacements that appear in the structure are displayed nearby. Thus, the maximum stresses for each component are *Base*: 7.2×10^7 N/m^2 (72 MPa), *Stiffener*: 3.38×10^8 N/m^2 (338 MPa), *UProfile*: 1.65×10^8 N/m^2 (165 MPa), *Seam Welding Connection Mesh.1*: 1.51×10^8 N/m^2 (151 MPa), *Seam Welding Connection Mesh.2*: 6.23×10^7 N/m^2 (62.3 MPa), *Seam Welding Connection Mesh.3*: 1.65×10^8 N/m^2 (165 MPa), *Seam Welding Connection Mesh.4*: 7.52×10^7 N/m^2 (75.2 MPa) and *Seam Welding Connection Mesh.5*: 3.18×10^8 N/m^2 (318 MPa). Also, in *Rigid Virtual Mesh.1* a maximum stress of 3.06×10^7 N/m^2 (30.6 MPa) appears.

From the context menu of the *Sensors.1* feature (Figure 4.262) in the specification tree, the *Energy* and *Global Error Rate (%)* sensors are updated, then the *Reaction Sensor* is added. It provides information related to the loads (forces and moments) required to fulfil the condition of maximum displacement on the *X* and *Z* axes according to the values imposed in Figure 4.260.

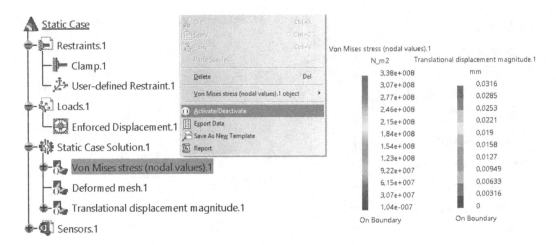

FIGURE 4.261 Activating the image results from the context menu in the specification tree.

It is observed that the value of *Global Error Rate (%)* is a very convenient one, of 8.48%, so the analysis is accurate. In the *Reaction Sensor* selection box (Figure 4.263) the user selects *User-defined Restraint.1* in the *Entity* field, the type *User,* the *Axis System.1* coordinate system and the *Cartesian* orientation. Once the selections are made, the *Update Results* button becomes available, which calculates the force and moment values for each *X, Y* and *Z* axis. The negative values represent the direction of application of these loads relative to the coordinate system. The values of these loads are displayed in the selection box in the two *Force* and *Moment* tabs, but also in the specification tree within the *Reaction Sensor.1* feature.

Figure 4.264 graphically presents the distribution of displacements on different areas of the welded structure. It can be easily observed that the most displaced area is at the top of the *UProfile*. The displacement values for *Point.1* can be observed in detail, as well as the maximum value of 0.0316 mm placed at one of the edge ends. Also, the *Base* component is stressed in the area of contact with the *Stiffener,* the displacement value being approximately 0.0205 mm.

According to the condition imposed by the user, the structure shows a displacement of 0.03 mm, but the stress generated by the loads (3.38×10^8 N/m², 338 MPa) exceeds the yield strength of the considered steel. Therefore, it can be considered that the structure deforms plastically under the imposed conditions. To deal with this problem, the user has the possibility to thicken the *Stiffener*

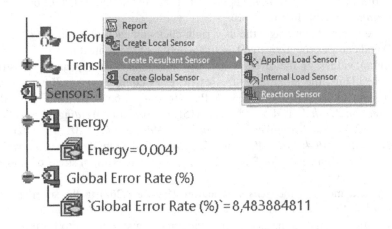

FIGURE 4.262 Activation of the *Reaction Sensor.*

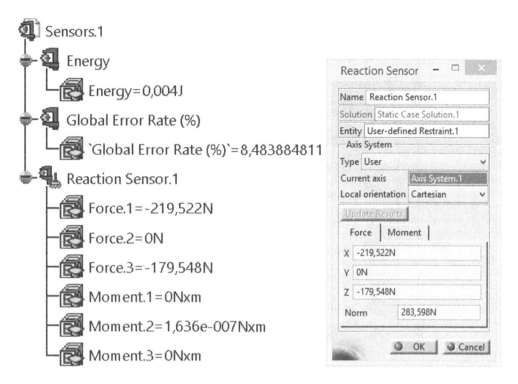

FIGURE 4.263 Values of the loads applied on the structure.

or to add a second such component, to change the material from which the structure is made, to change the assembly method, etc.

The results of the FEM analysis, however, also depend on how the user establishes the initial conditions. For example, keeping the welded structure, the discretization of the components and the condition of the imposed maximum displacements, the user recomputes the FEM analysis, without using, however, a virtual part, not being interested in the stresses and deformations that may appear

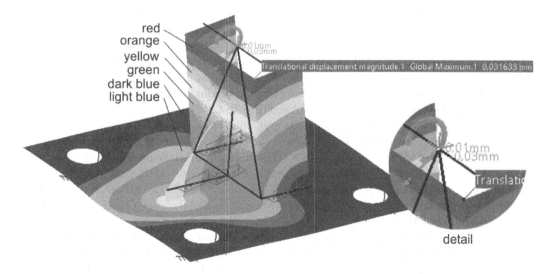

FIGURE 4.264 Displacement results when a virtual part is used.

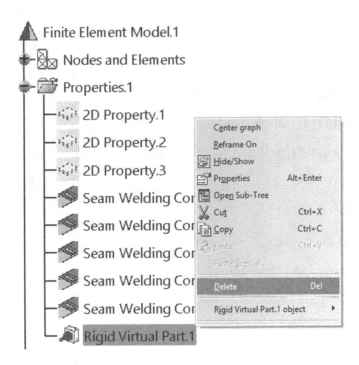

FIGURE 4.265   Removing the *Rigid Virtual Part.1* from the specification tree.

in it. This decision greatly simplifies the FEM analysis, the computation time is shortened, but the results lose precision.

Thus, in the specification tree the user removes the *Rigid Virtual Part.1* feature by the *Delete* option of its context menu (Figure 4.265). Simultaneously, from *Nodes and Elements* the program also deletes *Rigid Virtual Mesh.1*.

In the specification tree, the features of *Static Case* require updating because *User-defined Restraint.1* no longer contains/is no longer defined by the virtual part, the loads depend on this restraint, and the results of the FEM analysis cannot be computed. Thus, the user must edit *User-defined Restraint.1* by double-clicking on it, and the selection box from Figure 4.266 opens. In the *Supports* field, instead of the virtual part, the user selects the upper edge of the *UProfile* and the program creates and places two restraint symbols at the ends of the edge. The rest of the settings are kept, according to the figure, including the initial restraint, *Clamp.1*.

The *Enforced Displacement* values are also kept, as in Figure 4.260. To recompute the results, the user right-clicks on each one in the specification tree (similar in Figure 4.261) and chooses the *Activate/Deactivate* option of the context menu. Thus, the maximum values of the stresses and displacements appearing in the structure are displayed (Figure 4.267).

In the new conditions, it is observed that the maximum stresses for each component are very different than if the user applied a virtual part: *Base*: 4.77×10^7 N/m² (47.7 MPa), *Stiffener*: 2.09×10^8 N/m² (209 MPa), *UProfile*: 2.16×10^8 N/m² (216 MPa), *Seam Welding Connection Mesh.1*: 1.36×10^8 N/m² (136 MPa), *Seam Welding Connection Mesh.2*: 4.57×10^7 N/m² (45.7 MPa), *Seam Welding Connection Mesh.3*: 1.47×10^8 N/m² (147 MPa), *Seam Welding Connection Mesh.4*: 1.91×10^8 N/m² (191 MPa) and *Seam Welding Connection Mesh.5*: 2.09×10^8 N/m² (209 MPa).

For the same proposed objective (according to the condition imposed by the user, the structure shows a displacement of 0.03 mm), the maximum stress generated by the loads drops to 2.16×10^8 N/m², 216 MPa, a value lower than the yield strength of the considered steel. In these conditions, without the use of a virtual part, it can be considered that the structure deforms elastically.

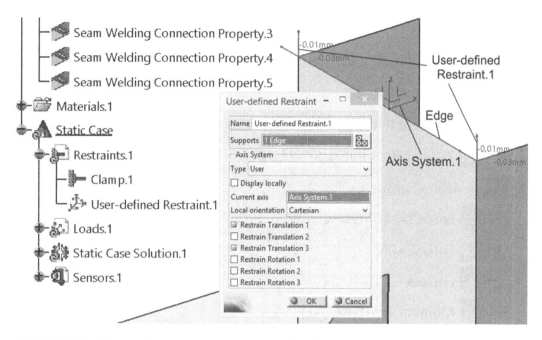

FIGURE 4.266 New setup of the *User-defined Restraint.1*.

The maximum stress value appears in the *UProfile* component as in the case of Figure 4.248. The stresses appearing in the other components and in the welds are, however, different.

In the specification tree, in the feature *Sensors.1,* the user finds the value *Global Error Rate (%)* = 18.92% (Figure 4.268), higher than in the previous case, when a virtual part was used. Also, the values of the loads required for a displacement of 0.03 mm are lower in this case.

The comparative presentation of the two situations (with and without a virtual part) shows that the results of the FEM analysis depend on how the user specifies the loading conditions, the

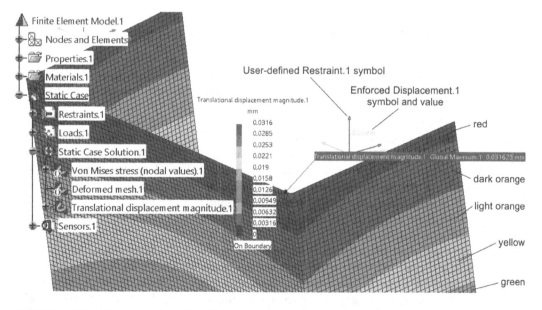

FIGURE 4.267 Displacement results with no virtual part.

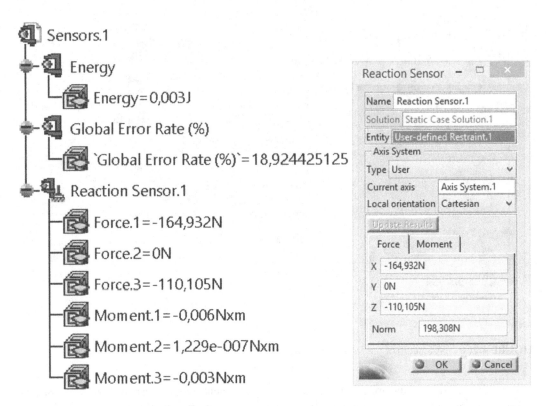

FIGURE 4.268 Values of the loads applied on the structure with no virtual part.

restraints, the mesh discretization and the use of a virtual part. The choice of optimal conditions, but also the interpretation of the results, are influenced by the user's experience, the geometry of the parts, their role in a certain structure, the type and manner of its use and the precision required for the FEM analysis.

4.9   ANALYSIS OF AN ASSEMBLY FROM A CONTROL DEVICE

In the application, the user will perform the finite element analysis of a subassembly of three parts. The subassembly is quite simple; it belongs to a positioning and control device with mechanical actuation. The constructive shapes of the parts are very different, according to the functional roles of each one.

Considering the representation indicated in Figure 4.269, the mobile support (2) is assembled with the plate (1) by a profile guide rail. The role of the rail is to guide the mobile elements of the device during its operation and to ensure their precise movement.

The mobile support has only one degree of freedom (translation along the guide), due to the shape and position of the guiding surfaces (two pairs of conjugate surfaces).

The axle (3) is tightly mounted in a hole (diameter Ø18 mm) of the mobile support (2) in the position shown in the figure. On the axle at its free end acts a force whose value and direction are known during the operation of the device.

The 2D drawings of the parts are presented in Figures 4.270–4.272. Many other components of the positioning and control device are missing from this assembly. The reason was to simplify the explanations related to finite element analysis, but the user may continue later the application: add new components, change the value of the force and its direction of application, apply other loads, change certain dimensions of the three components, etc.

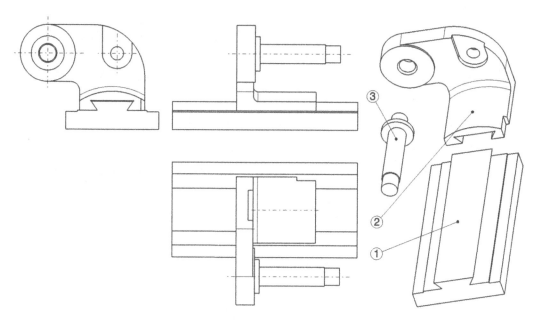

FIGURE 4.269 Projections of the analysed assembly.

A pull rod (not shown and not considered in the application) mounted in the other hole (diameter Ø14 mm) of the mobile support is used to operate the support. The axis of that hole is parallel to the guide. In the analysis considered in the application, the influence of the action of the rod was not considered, studying only the axle loading and the way in which it creates stresses in the components of the assembly.

The parts are modelled individually in the *CATIA v5 Part Design* workbench and saved with the names shown in the previous figures. Also, these components are available for download. To create the assembly, the user accesses the *CATIA v5 Assembly Design* workbench and inserts the three components using the *Existing Component* option from the *Insert* menu.

In the specification tree in Figure 4.273, these three components (plate with guide, mobile support and axle) can be observed in the order they were added to the *Product1* assembly. Its name can be changed from the context menu, the *Properties* option, the *Product* tab and the *Part Number* field. The component parts can be inserted in any order, but the plate with guide component must be assigned a first *Fix* constraint to take over its six degrees of freedom. The plate is considered to

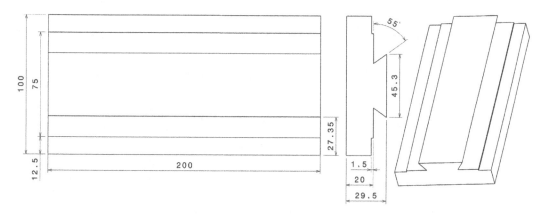

FIGURE 4.270 Two-dimensional drawing of the plate with guide.

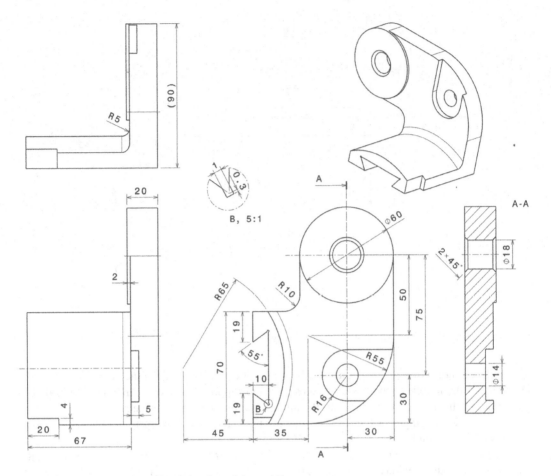

FIGURE 4.271 Two-dimensional drawing of the mobile support.

be the main component of this assembly and, once fixed, the mobile support is assembled on this component, then the axle fits in.

From the same *Insert* menu, the user adds the assembly constraints (*Coincidence, Contact, Offset, Angle, Fix,* etc.) between the components of the assembly.

Moving the components in the assembly can be done with the specific options in the menu *Edit → Move → Manipulate* or with the help of the compass (Figure 4.274). The figure shows how the

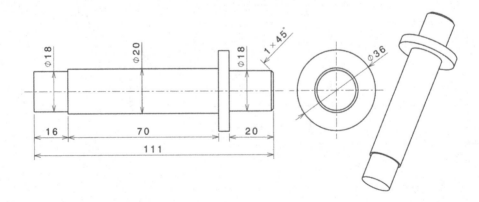

FIGURE 4.272 Two-dimensional drawing of the axle.

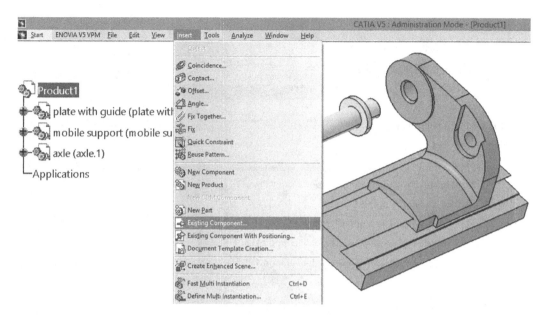

FIGURE 4.273 Inserting the components into the assembly.

Surface contact constraint is added between two flat surfaces (one surface belongs to plate with guide, the other to mobile support). The constraint symbol is displayed on each of the two surfaces. The numbering of constraints set between components starts with 1 for *Fix,* continues with 2 for this *Surface contact* constraint, etc. The names of the constrained components are written in parentheses. The first entry is the first component selected.

The assembly rules in *CATIA v5* say that the user should select the first constraint element on the fixed component or on an already constrained one, and the second element on the component to be assembled. These elements are of type: surface, point, edge, axis, plane, etc.

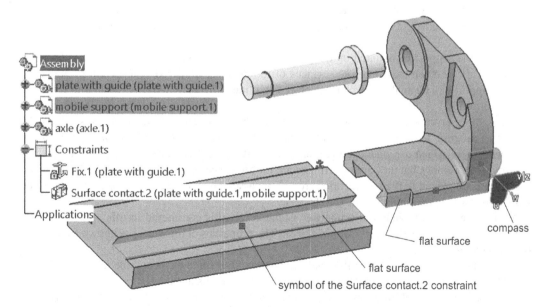

FIGURE 4.274 Adding the first two assembly constraints: *Fix.1* and *Surface contact.2*.

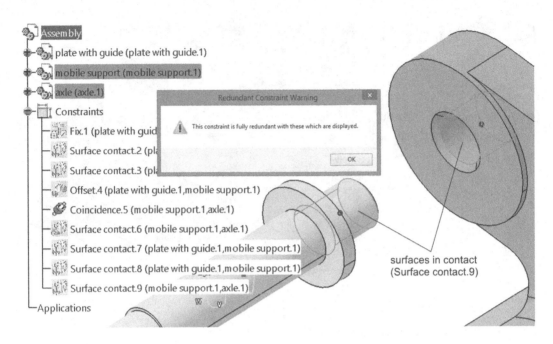

FIGURE 4.275 The analysed assembly and the applied constraints.

Next, the user adds other constraints (such as coincidence, surface contact and linear distance). Constraints are established based on the functional role of each part and its position in the assembly. The assembly constraints are important and necessary, but the user will add numerous constraints that are specific to finite element analysis in the following steps.

Thus, in Figure 4.275 the user has the components in assembled representation and the applied constraints. According to this figure, a coincidence constraint (its symbol is a circle) was imposed between the axle axis and the axis of the hole in the mobile support *(Coincidence.5)*. A surface contact constraint *(Surface contact.9)* was established between the cylindrical surface of the axle and the cylindrical surface of the hole. Also, between the corresponding flat surfaces of the two parts (axle and mobile support) the user creates a surface contact constraint *(Surface contact.6)*.

The mobile support is mounted on the plate guide by four surface contact constraints *(Surface contact.2, .3, .7, .8)* on the conjugate surfaces of the guide. Considered in a certain position, the mobile support will be analysed when it is placed at a linear distance *(Offset.4)* of 45 mm from one of the plate ends.

Thus, the user creates surface contact constraints between pairs of surfaces. This causes the *CATIA v5* program to display an information and warning message: *This constraint is fully redundant with these which are displayed* (Figure 4.275). It means that additional constraints are not required for assembly, but they are very important for finite element analysis. Finally, all the applied constraints are placed in the specification tree of the assembly, they are saved together with it and will be the basis of the physical constraints.

Within the assembly, material properties (standard *Steel* is considered in this application) are added to each part using the *Apply Material* icon. The properties of this steel are known from previous applications.

To set the parameters and start the analysis process with finite elements, the user accesses the *CATIA v5 Generative Structural Analysis* workbench. The first stage consists of choosing the location to save the files resulting from the analysis process. The user applies the *External Storage* tool on the *Solver Tools* toolbar.

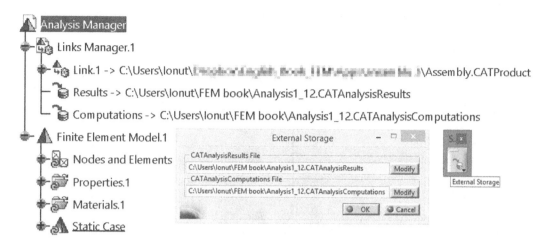

FIGURE 4.276 Establishing the storage location of the analysis process files.

Figure 4.276 shows the toolbar with the *External Storage* icon, the destination of the files to be stored, but also the structure of *Links Manager.1* features. The user has to note that the analysis and computations results will be saved/stored in a different folder than the assembly files. This option is recommended to work efficiently, the user stores the 3D geometry of the assembly in one location and the analysis files in another location. Often, after completing the FEM analysis and processing/interpreting the results, the user frees up the storage space occupied by certain of these files due to their very large size.

So, the user will perform the FEM analysis of this assembly. Each component is an independent solid body which will be defined as mesh and connected to the other components. Several specific restraints are required.

Thus, the network of nodes and elements of each component of the assembly is discretized. In the specification tree, *Nodes and Elements* contains the *OCTREE Tetrahedron Mesh* features for the three components (Figure 4.277).

By double-clicking on each of these meshes, the user displays a dialog box to choose the size and type of the finite element, the tolerance between the real model and the discretized model, etc.

Figure 4.278 shows two dialog boxes, each titled *OCTREE Tetrahedron Mesh*. The settings in the left box apply to the mobile support and plate with guide, and those in the right box apply to the axle. It is observed that the axle is better defined (discretized) because the load is applied on its surface and it will deform the most.

For all parts, the element type was chosen as *Linear*, in order to simplify the analysis process and shorten the computation time. As the user obtains the first results, with a certain precision, he can opt to change the element type to *Parabolic* and/or for a higher refinement of the network.

Based on the assembly constraints, in the next stage the physical constraints are established, that are necessary to simulate the transmission of stresses, generated by the axle loading, in the

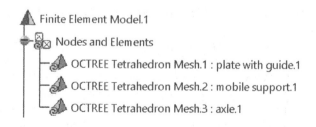

FIGURE 4.277 The finite elements of the components.

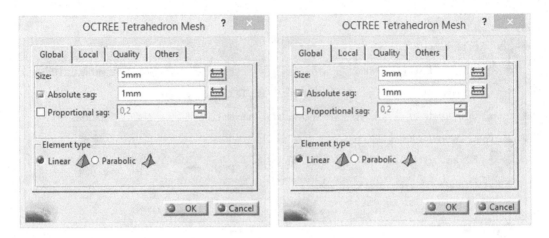

FIGURE 4.278   Discretization of nodes and elements network.

entire assembly of the three components. The optimal choice of physical constraints is a rather difficult stage because the user must understand how the assembly works and the relations between its components.

For the example considered in the application, the physical constraints are applied in the order in which the assembly constraints were imposed. *Fastened Connection Property, Pressure Fitting Connection Property* and *Contact Connection Property* constraints are proposed. The user can try other options, depending on the results obtained.

For the contact surfaces between the plate with guide and the mobile support, three possible variants of physical constraints/connections can be applied, with interesting results:

- *Slider Connection Property*: A slider connection is the link between two parts that are constrained to move together in the local normal direction at their common boundary. These parts will behave as if they were allowed to slide relative to each other in the local tangential plane.
- *Contact Connection Property*: A contact connection is the link between two parts that are prevented from inter-penetrating at their common boundary, and will behave as if they were allowed to move arbitrarily relative to each other as long as they do not come into contact within a user-specified normal clearance. When they come into contact, they can still separate or slide relative to each other in the tangential plane, but they cannot reduce their relative normal clearance.
- *Fastened Connection Property*: A fastened connection is the link between two parts that are fastened together at their common boundary, and will behave as if they were a single body. In the FEM applications this is equivalent to the situation where the corresponding nodes of two compatible meshes are merged together.

All these three connection relations take into account the elastic deformability of the interfaces between the respective two parts.

If the user applies the *Slider Connection Property,* the mobile support will move under the action of the loading force along the plate guides. In the case of *Contact Connection Property,* the user can define the friction ratio between the support and the plate, including the *No sliding* option. The case study presented in the application uses the *Fastened Connection Property* (Figure 4.279) by which the mobile support is fixed on the plate guides, so there is no movement between these two components. Thus, it is considered that after positioning the components in the assembly, the mobile support is blocked by the user on the plate, then the loading force is applied.

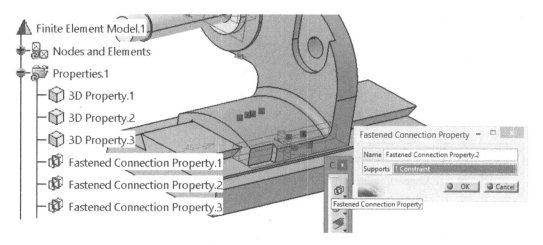

FIGURE 4.279   Establishing a *Fastened Connection Property* between the plate guides and the mobile support.

A *Fastened Connection Property* is also established between the axle and the corresponding hole in the mobile support because there must be a coincidence between the axes of the axle and of the hole. Also, between the flat surfaces of the axle and of the mobile support, the *Contact Connection Property* is chosen (Figure 4.280). The introduction of a friction ratio of 0.2 is observed, but the user may try, in certain situations, to check the *No sliding* option.

For the contact between the cylindrical surfaces of the axle and the hole (Figure 4.275), *Pressure Fitting Connection Property* is chosen because the axle is assembled by pressing in the hole of the mobile support. There is no interference between the two components, so the value in the *Overlap* field is 0 mm (Figure 4.281). If the assembly was done by threading, using the *Bolt Tightening Connection Property* would have been recommended.

The *Pressure Fitting Connection* uses assembly surface contact constraint as a support. A *Pressure Fitting Connection* is the link between two parts which are assembled in a *Pressure Fitting*

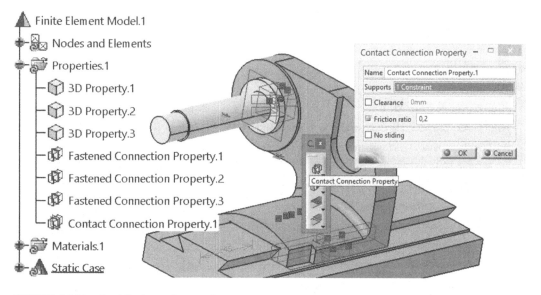

FIGURE 4.280   Establishing a *Contact Connection Property* between the corresponding flat surfaces of the axle and the of mobile support.

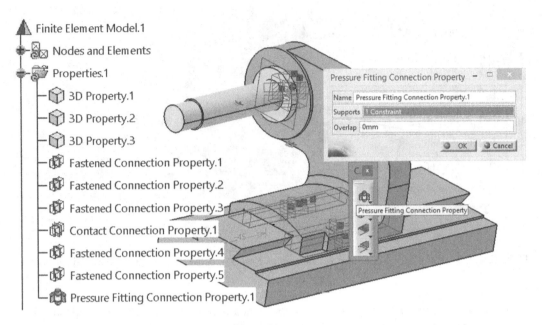

FIGURE 4.281   Establishing a *Pressure Fitting Connection Property* between the corresponding flat surfaces of the axle and of the mobile support.

configuration, more precisely when there are interferences or overlaps between both parts. Along the surface normal, the connection behaves as a contact connection with negative clearance value (positive overlap). The difference lies in the tangential directions where both parts are linked together. The *Pressure Fitting Connection* relations take into account the elastic deformability of the interfaces.

For a good understanding of how to use these constraints, Figure 4.282 presents a list of pairs of assembly constraints–physical constraints, required for the FEM analysis. There is a lack of correspondence for the *Fix.1* constraint and the linear distance (constraint *Offset.4*), which only has a role in the assembly process. In choosing the optimal variants for the correspondence of

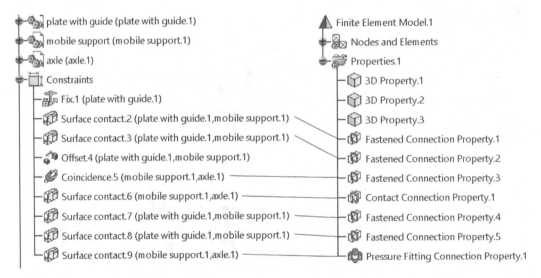

FIGURE 4.282   Correspondence between the assembly constraints and the constraints required for the FEM analysis.

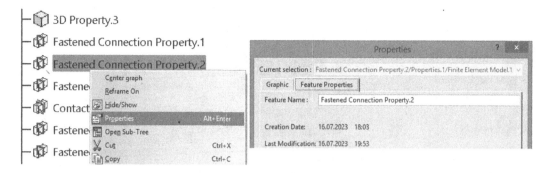

FIGURE 4.283 Changing the name of a physical constraint using the context menu.

the constraints, the user can try several variants and observe the differences between the results obtained, especially regarding the way of transmission of stresses between components and the appearance/presence of critical areas.

All the physical constraints appear in the specification tree under the feature *Properties.1*. Once applied, by double-clicking on any of them, the user can change the name and some options (for example, in the case of *Contact Connection Property* and *Pressure Fitting Connection Property*). Thus, for assemblies with several components (including many assembly constraints) it proves to be useful to rename the physical constraints for easier identification. However, neither when creating that physical constraint nor when double-clicking it later in the specification tree, the paired assembly constraint is displayed. Changing the name of the physical constraint can be done in the dialog box during the creation stage (example: Figure 4.281, the *Name* field) or from the context menu (Figure 4.283).

After applying the physical constraints, the model of the assembly is completed with the respective symbols, as shown in Figure 4.284. Other such symbols can also be observed in Figure 4.281.

Next comes the stage of adding a *Clamp* restraint, positioned on the two front and base faces of the plate with guide. The use of the restraint on the respective surfaces ensures the immobilization of the mobile support, which is necessary for completing the finite element analysis. In an assembly,

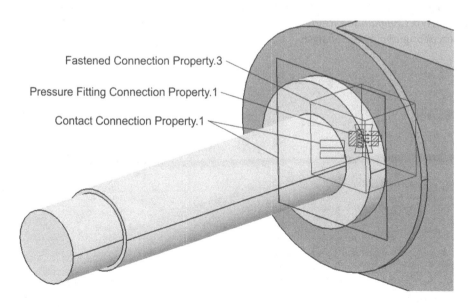

FIGURE 4.284 Examples of constraints symbols applied between the axle and the mobile support.

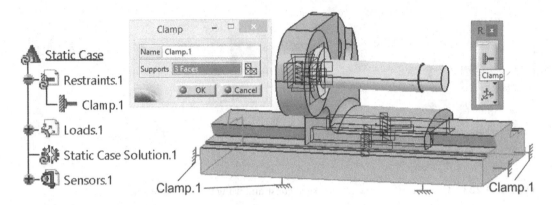

FIGURE 4.285 Applying the *Clamp* restraint.

at least one of the components must have an established restraint. For the assembly studied in this application, the *Clamp* restraint is the most suitable to be applied to the mobile support.

Also, at this stage, the user applies a load on the free end of the axle with a *Distributed Force* having a value of 500 N. It is observed that the force is applied along the *X* axis after the user has selected the cylindrical surface from the end of the axle component. The force is represented on screen by eight yellow arrows pointing in the same direction. Note that depending on how the components were modelled and assembled, the axis may be different.

Figure 4.285 shows the symbols of the *Clamp* restraint applied to the three flat faces of the mobile support, as well as the specification tree containing the *Clamp.1* feature.

Figure 4.286 shows how the distributed force was applied to the cylindrical surface at the axle end and the *Distributed Force* dialog box. Depending on the values set by the user in the *X, Y* and *Z* fields (from the *Force Vector* area), the arrows receive a certain direction, and the value in the *Norm* field is updated.

Once the stages of imposing the initial conditions are completed (discretisations, restraint, loading with a distributed force, etc.), the user should save all the analysis files (components, assembly, analysis). As there are several files and of different types, using the *Save Management* option (in the menu *File*) is mandatory. The finite element analysis can then be run using the *Compute* icon. The information box *Computation Resources Estimation* (Figure 4.287) presents some details regarding the resources required by the analysis.

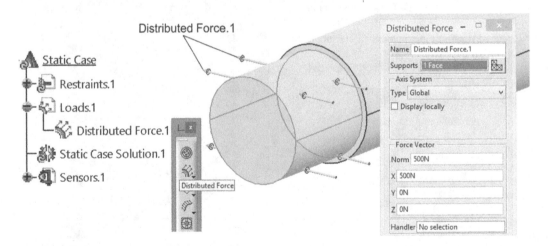

FIGURE 4.286 Establishing the direction and value of the distributed force.

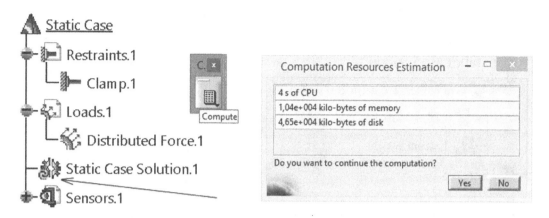

FIGURE 4.287 Starting the finite element analysis process.

The user should note that if the values in the fields of the information box are null, then he has omitted something or made a mistake in preparing the analysis settings. So, the analysis will not be performed. The computation process in the case of assemblies demands important hardware resources, the most requested being the RAM memory and the processor. Depending on their availability in the computer system, the duration of the computation varies from a few seconds to several tens of minutes, during which time the user cannot work in *CATIA v5*.

Figure 4.287 shows that the specification tree is completed with the feature *Static Case Solution.1*, which can/will contain various solutions, depending on the use of the following tools: *Deformation, Von Mises Stress, Displacement, Principal Stress* and *Precision*. Also, as long as the computation is not yet ready, the *Static Case Solution.1* feature displays the *Update* symbol, the meaning that the solution is being identified.

For this application, the *Von Mises Stress* result is computed first in order to determine the stresses induced in the parts of the assembly by the loading distributed force applied on the end of the axle. In parallel with the highlighting of the *Von Mises* results, the *Image Extrema* tool is also applied in order to locate the minimum and maximum values of the stresses both globally and locally. At first glance, the most stressed component is the axle. The generated stresses are transmitted in all components due to the correctly applied physical constraints.

In the specification tree in Figure 4.288, the feature *Global Maximum.1* appears on screen. It displays on the assembly model a (red) label containing the type of the extreme value, along with

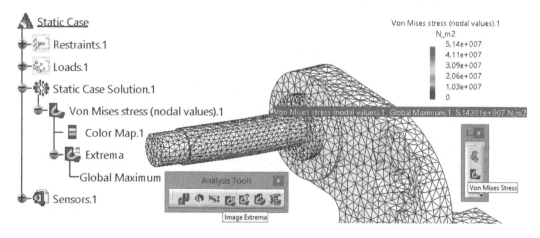

FIGURE 4.288 Representation of *Von Mises* results and location of maximum stress.

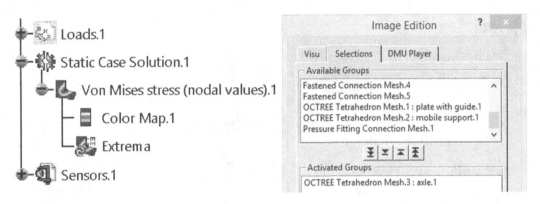

FIGURE 4.289 Selection of the component for which the analysis results will be displayed.

its value. This label is connected to a certain node of the network, in the area where the respective extreme value has been reached.

As expected, the maximum stress appears on the axle, in the area of its assembly with the mobile support, due to the sudden reduction of the diameter.

To display the analysis results for each component of the assembly, the user double-clicks on the feature *Von Mises stress (nodal values).1* in the specification tree and the *Image Edition* selection box opens (Figure 4.289). From the *Available Groups* list in the *Selections* tab, the user chooses the component and adds it to the list below, *Activated Groups*. In the case presented in Figure 4.289, the axle component is activated and the other components of the assembly are no longer displayed. Thus, Figure 4.290 shows the selected component, the maximum stress value and its location.

The maximum and minimum stress values are displayed in the palette of colours and values that accompanies the *Von Mises* representation, but also in the *Information* box (left) in Figure 4.291. The axle is stressed with a maximum value of 5.14201×10^7 N/m², but it is lower than the yield strength of the material (2.5×10^8 N/m²) from which it was considered to be manufactured.

To determine the error rate the *Precision* tool is used and is located on the *Image* toolbar, together with the *Information* tool. Thus, the *Estimated local error.1* feature appears in the specification tree. The *Information* box on the right is displayed after selecting this feature with the *Information* icon enabled.

For the finite element analysis process considered in the application, *CATIA v5* computes the results with an estimated global error rate of 35.9808%. The value is also visible in the feature *Sensors.1 → Global Error Rate (%)* in the specification tree.

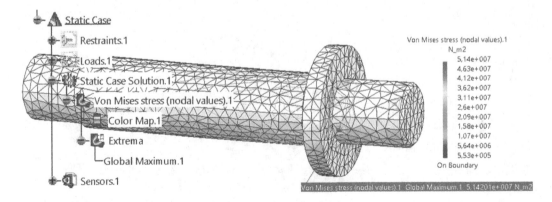

FIGURE 4.290 Displaying the *Von Mises* stress and locating the maximum value on the axle component.

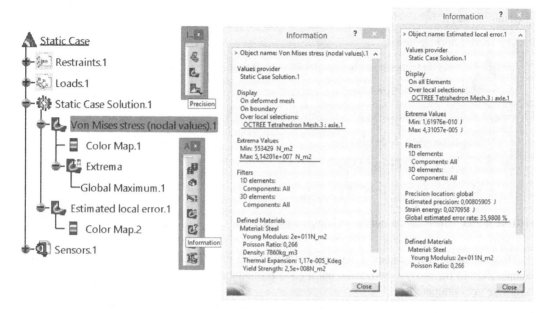

FIGURE 4.291 Displaying additional information.

Based on the finite element analysis practice, it is considered that the percentage value is too high and does not provide an appropriate degree of precision. Thus, it is necessary to refine the assembly and resume the analysis process in order to reduce the value of the global error rate by applying the *New Adaptivity Entity* tool.

In the *Global Adaptivity* dialog box, in the *Supports* field (Figure 4.292) the user selects the *OCTREE Tetrahedron Mesh* components of the assembly (3 *Mesh* parts), then, in the *Objective Error (%)* field, it is entered the desired error rate: 20%.

Once the value of the error rate is established/imposed by the user, the analysis process is resumed using the *Compute with Adaptivity* tool. In the *Adaptivity Process Parameters* dialog box (Figure 4.293) the number of computation iterations to which the assembly will be subjected (*Iterations Number* field) is set. The user then enters the value of 1 mm in the *Minimum Size* field.

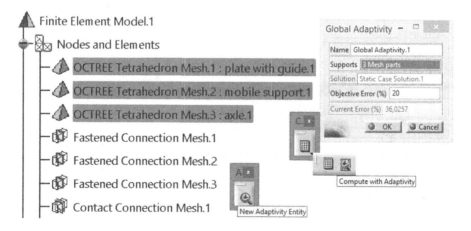

FIGURE 4.292 Establishing/imposing the desired error rate.

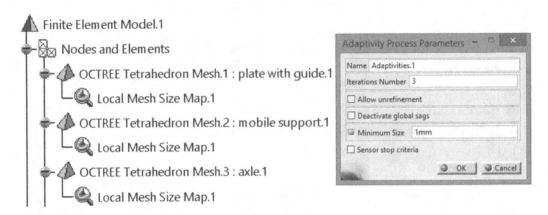

FIGURE 4.293 Choosing the number of computation iterations.

Thus, the higher the number of iterations and the lower the *Minimum size* value is, the longer the computation of the analysis will take, but a good refinement of the components meshes and a low error rate are obtained.

The end of this second analysis process results in a new refinement of the network of nodes and finite elements, and the specification tree is completed with the *Local Mesh Size Map* features for each component of the assembly.

The objective is achieved (Figure 4.294), the global error rate value drops to 17.0713%, but a significant increase in the maximum stress value is noted, namely: 9.2045×10^7 N/m², compared with 5.14201×10^7 N/m², obtained at the end of the first analysis, with an error rate of 35.9808%. Figure 4.294 also shows the different refinement of the components, especially in the stressed areas of the axle (compared with the axle represented in Figure 4.290).

The maximum stress value is also located on the axle, but according to Figure 4.289, the user can determine the extreme stresses for each component. Figure 4.295 shows these stresses separately for the three components.

Also, with the help of the *Image Extrema* tool, the areas where the minimum and maximum stresses values of each component in the assembly are located are identified, both globally and locally.

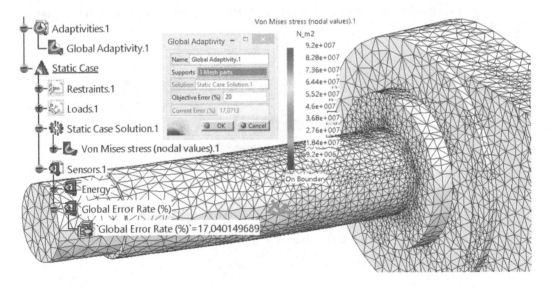

FIGURE 4.294 Results of the analysis process.

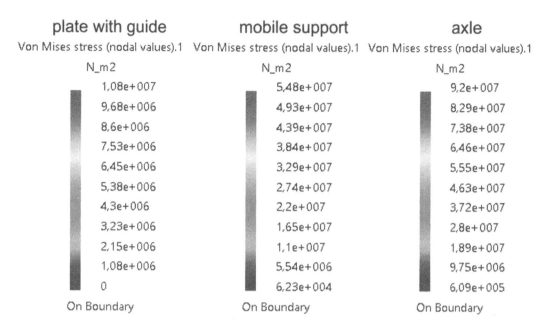

FIGURE 4.295   The extreme stresses values of each component.

The separate display of stresses on components proves to be useful especially when they are manufactured from different materials. Thus, a better analysis of the considered assembly is obtained. There are situations in which the yield strength of the material applied to a certain part is lower than the maximum stress computed for that part.

So, the part will suffer a plastic deformation, leaving the elastic zone. In this elastic zone a part with a specified material will return to its original shape for a given amount of stress. A part with a specified material has plastic behaviour when the computed stress is larger than the elastic limit. In the plastic zone, the part does not come back to its original size or shape when the loads stop and it acquires a permanent deformation. Plastic behaviour ends at the breaking point. So, there is the possibility of cracks appearing in the structure of the analysed model and, of course, then in the structure of the real model. In such cases, the user has the possibility to change the shape of the part or the material applied to it. In both cases, however, the user imposes the resumption of the analysis process with finite elements, also considering a reduced rate of the global error.

There are cases in which the shape or material of the respective part cannot be modified, because they are imposed by the design/production team(s), and their optimizations are transferred to the conjugated parts.

In the previous figures (Figures 4.288, 4.290 and 4.294) the axle is visualized in a deformed representation. For a better understanding of how the components of the assembly are influenced by the application of constraints and loading force, the program proposes an exaggerated deformation of the parts. The actual displacement values of the nodes of the finite element mesh are found using the *Displacement* tool on the *Image* toolbar.

As expected, the value of the maximum displacement (0.145 mm) is located on the axle in the area of its free end (Figure 4.296) on which the distributed force acts. Figure 4.297 presents the maximum displacement values for each component. Mobile support shows displacements especially in the end area (it has a diameter of Ø60 mm) where it is assembled with the axle. The plate with guide presents displacements in the area of its assembly with the mobile support. These displacements have very small values (0.04 mm for mobile support and 0.0004 mm for plate with guide) and do not influence the sliding or the assembly of the two components.

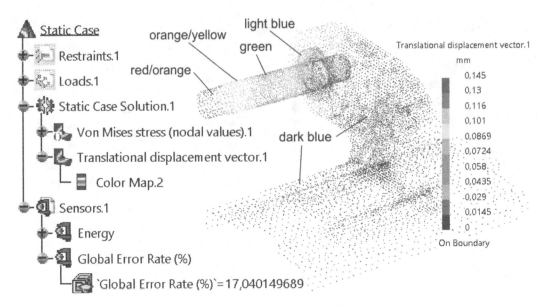

FIGURE 4.296   Representation of the displacement vectors of the assembly and the value of the maximum displacement.

Figure 4.298 shows the displacements of the network nodes from the free end of the axle and some of their values. For each vector, three values are available, according to the three axes X, Y and Z, in the considered reference system. These values are displayed by clicking on any of the arrows of the represented vectors field. The specification tree is completed with the feature *Translational displacement vector.1* (Figure 4.296). It is observed that the feature *Von Mises stress (nodal values).1* was disabled by default with the insertion and activation of the feature *Translational displacement vector.1*. The user notes that only one result of the analysis can be active at a certain time.

With the help of the *Sensors* feature, also located in the specification tree, the user can find out various concise information about the results of the analysis process. Thus, by right-clicking on this

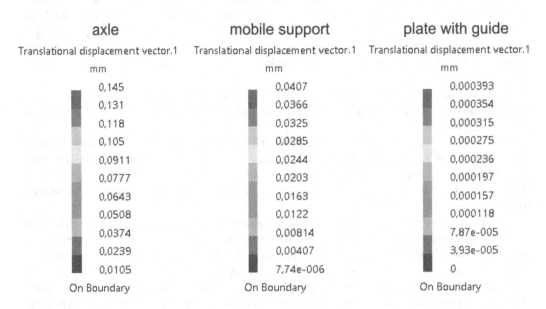

FIGURE 4.297   Maximum displacement values for each component.

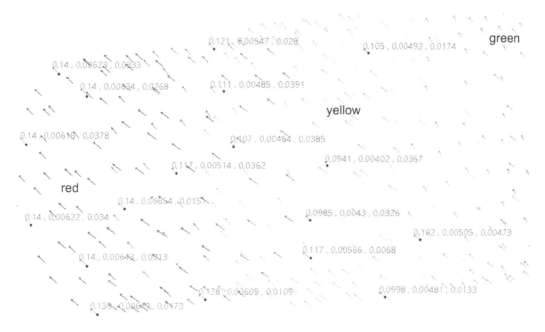

FIGURE 4.298   Some values of the mesh nodes displacements for the axle component.

feature, the user chooses the option *Create Global Sensor* from its context menu that appears. In the *Create Sensor* dialog box, from the available list (Figure 4.299) the user can select certain sensors to be inserted in the specification tree.

In the case of the assembly considered, the user may select all the sensors from the available list, making a multiple selection using the *Shift* key, then he presses the *OK* button for validation. The higher the complexity of the analysis and the larger the number of selected sensors, the longer their creation time increases. Any modification done by the user to the assembly or to the conditions in which the finite element analysis is computed leads to the need to update the sensors.

In the specification tree, within the feature *Sensors.1* (Figure 4.300), the selected sensors become available and the list shows the name and value of each one. The values are computed based on the analysis results and are not editable. The significance of the sensors is obvious: the deformation energy (in J), the global error rate (in %), the maximum displacement of the network nodes (in mm),

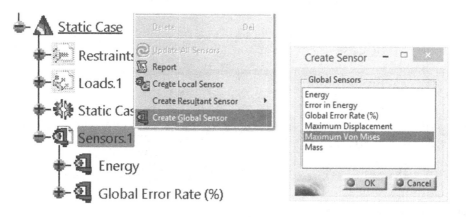

FIGURE 4.299   Creating *Global Sensors.*

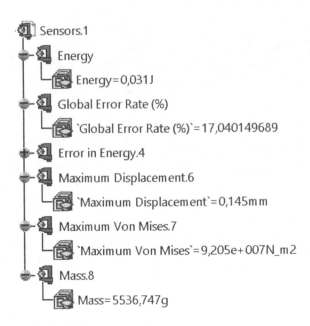

FIGURE 4.300 Display of sensors and their values.

the maximum stress (in N/m²) and the mass of the assembly (according to the materials applied to each component).

The user may continue the application with the creation of sensors for the reactions resulting from establishing the physical constraints between the components of the assembly.

Thus, the user chooses the *Create Resultant Sensor → Reaction Sensor* option in the context menu of the *Sensors* feature (Figure 4.301) and the *Reaction Sensor* dialog box opens (Figure 4.302). The user creates the sensors by choosing one by one all the available physical constraints (*Fastened Connection Property*, *Pressure Fitting Connection Property* and *Contact Connection Property*) from the *Properties.1* feature. In the *Entity* field, the user chooses the constraint, then the axis system (*Global or User*). In the *Force* and *Moment* tabs, the *X*, *Y* and *Z* fields are filled with values computed by pressing the *Update Results* button. The values are then displayed in the specification tree by the *Reaction Sensor* feature (1 for *X* axis, 2 for *Y* axis and 3 for *Z* axis).

Each reaction represents the resulting force and moment at restraint and connection specifications. Reaction sensors are only available for static analysis solutions.

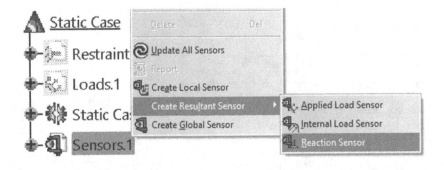

FIGURE 4.301 Creating the *Reaction Sensors*.

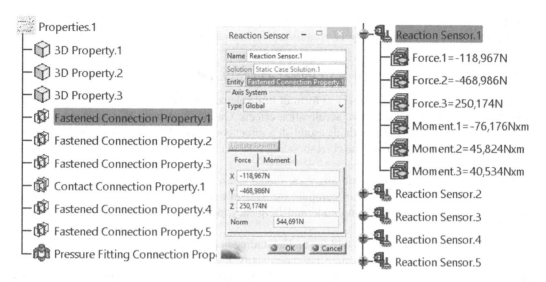

FIGURE 4.302 Creating and displaying sensors for physical constraints.

Pressing the *OK* button leads to the completion of the specification tree with the respective sensors, continuing the previously created ones. The values in Figure 4.302 can also be viewed by double-clicking on any of these sensors. This displays the *Reaction Sensor* information box.

In Figure 4.303 a sensor (of force and moment) of the physical constraint *Fastened Connection Property.1* is symbolized by the green colour on screen.

If the user selected a connection property in the *Reaction Sensor* dialog box, the direction of the reaction vector is given by the effects of the first component of the connection properties on the second component.

In this manner, it is possible to determine how a component of the assembly influences another component with which it is in contact. The forces and moments have an impact on the components,

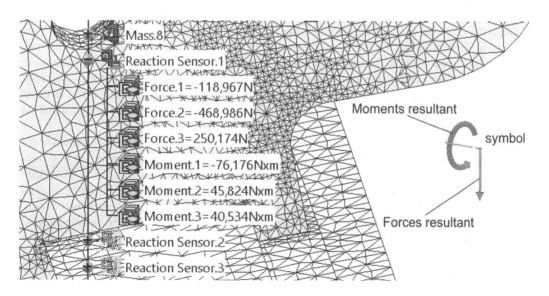

FIGURE 4.303 Symbolization for force and moment of a physical constraint sensor.

producing stresses and deformations. Also, the identified values confirm the correctness of the way in which the physical constraints were established between the components.

4.10   MODAL ANALYSIS OF A CONNECTING BEAM

Quasi-static analysis, the most prevalent form of finite element analysis, involves applying loads at an extremely slow rate to minimize acceleration effects. On the other hand, dynamic analysis is a more intricate approach. Both methods establish a direct correlation between a specific input (like a force exerted on a system) and the corresponding system response (such as the system's displacement caused by the applied load).

Unlike quasi-static and dynamic analyses, modal analysis offers a comprehensive view of a system's response limits. Users can compute the system's response for a specific input, such as an applied load with a defined amplitude and frequency, determining results like the maximum displacement.

Modal analysis is an important tool for understanding and computing the vibration characteristics of mechanical structures, no matter their complexity. It converts the vibration signals of excitation and responses, measured on a complex structure that is difficult to perceive, into a set of modal parameters that can be easily anticipated. Although the mathematical and physical aspects involved in determining the structural behaviour may be complex, the fundamental concepts and their practical application can be easily understood.

The modal domain is one important perspective for understanding structural vibrations. When structures are excited at their natural frequencies, they vibrate or deform in distinct shapes known as mode shapes. During regular operating conditions, a structure undergoes a complex vibration comprising all mode shapes. This enables users to understand and determine all the types of possible vibrations. The process of determining a structure's natural frequencies, modal damping, and mode shapes through specialized tools and simulations is referred to as *Modal Analysis*.

All structures, ranging from simple objects and mechanical components to vehicles, aircrafts, bridges and buildings, possess an inherent frequency known as the resonant frequency. This is the frequency at which the object naturally vibrates and facilitates the transfer of energy, with minimal loss, from vibrational to kinetic. As the frequency approaches the resonant frequency, the response amplitude gradually increases, eventually reaching infinity. Therefore, the results of modal analysis reveal these critical frequencies where the amplitude tends towards infinity.

For instance, if the user has a look under the hood of a car he can see that the running engine is vibrating. Depending on the design and dynamics of the engine components, they may resonate at the natural frequencies of the engine vibrations. This resonance can result in undesirable noise and vibration. When combined with the heat generated by the engine operation and some bad road conditions it may lead to premature fatigue and failure of the components and discomfort in using the car. By knowing the natural frequencies of a component, the engineers can design it and its assembly to avoid specific ranges of vibrations to prevent resonance.

When the user is aware that the structure vibrational behaviour is close or about to coincide with one of the natural frequencies identified in the modal analysis, they should consider a new design or change of the components to shift the natural frequency away from the excitation frequency. New or improved structural elements can be added to increase the stiffness of the structure, also its mass can be increased or decreased. Consequently, the excitation frequency will no longer closely match the natural frequency of the structure. Other techniques of vibration suppression may include increasing the damping of the structure by changing the material or adding viscoelastic material to the surface of the structure. Introducing vibration absorbers, tuned to the frequency of the excitation force, can also be effective, inducing substantial vibrations in the absorber and diminishing vibrations in the structure.

In this application, the user studies the vibration behaviour of the connecting beam part presented through its 2D drawing in the first application (Figure 4.1). The 3D modelling of the part is

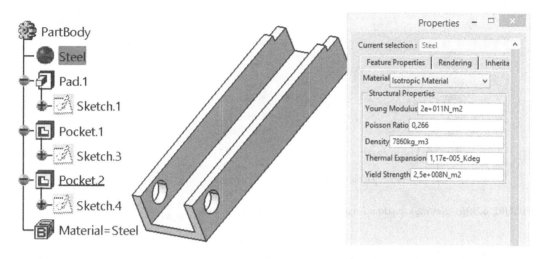

FIGURE 4.304 The part 3D model and some settings of material properties.

simple, and the present analysis can even use a copy made after applying the material properties (Figure 4.304).

As an option, the user can change the properties of the material in the *Properties* dialog box, activated by right-clicking on the material name *(Steel)* in the specification tree. Changing the properties required for the analysis with finite elements is done in the *Analysis* tab, but the user is also able to modify the name of the chosen steel and different properties of its representation *(Rendering* tab).

The user accesses the *Generative Structural Analysis* workbench from the *Start → Analysis & Simulation* menu. If the *New Analysis Case* selection box opens, the *Frequency Analysis* option should be selected. This type of analysis involves the determination of the natural modes of vibration of the part considered under the conditions imposed by the operation of its assembly.

If the specification tree contains the *Static Case* feature, it can be removed (choose the *Delete* option from the context menu). Then, from the *Insert* menu, *Frequency Case* is added to the specification tree, according to Figure 4.305.

In the *Frequency Case* selection box (Figure 4.306) the options for inserting new *Restraints* and *Masses* are selected. In other situations, these settings can be retrieved from existing analysis cases

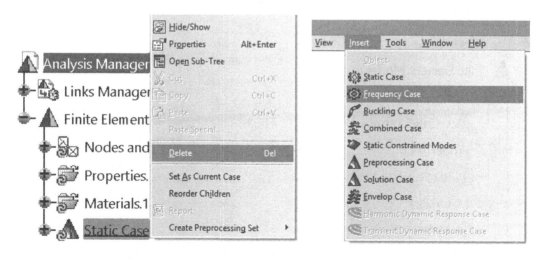

FIGURE 4.305 Removing the static analysis and inserting the *Frequency Case.*

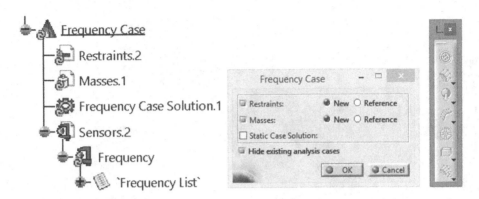

FIGURE 4.306 Settings to add a *Frequency Case* and *Loads* toolbar.

by choosing the *Reference* options. With the addition of *Frequency Case,* the other analyses are hidden by selecting the *Hide existing analysis cases* option. These cases are independent, although they have certain settings in common: the geometry of the structure (part or assembly), the applied material, the discretization of the finite element network, applied restraints, etc.

Also, in Figure 4.306 it is observed that all the icons on the *Loads* toolbar are disabled by default because loads are not required in any modal analysis. The *Sensors.2* feature in the specification tree contains *Frequency* which presents the list of frequencies. For now, all these elements require updating.

Although the *CATIA v5* program implicitly defines the network of nodes and elements, the user will edit this mesh and establish the size of the finite element (*Size* = 5 mm), the maximum tolerance between the real model and the discretized model used in the analysis (*Absolute sag* = 1.5 mm), the finite element type (*Element type = Linear*), etc. Thus, from the *Nodes and Elements* feature of the specification tree, the user double-clicks on *OCTREE Tetrahedron Mesh.1* and opens the selection box with the same name (Figure 4.307). The type of the finite element was chosen as *Linear* because the geometric shape of the part is simple, with flat faces. Thus, most of the settings used are the same as in the first application so that the user can make comparisons between the behaviour of the part in the two situations.

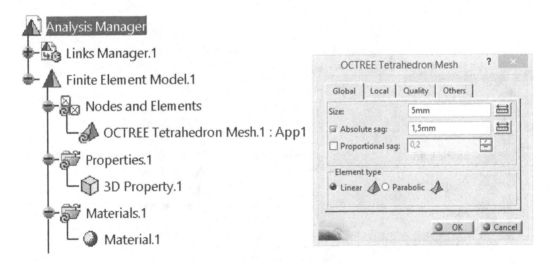

FIGURE 4.307 Discretization mesh characteristics for the part model.

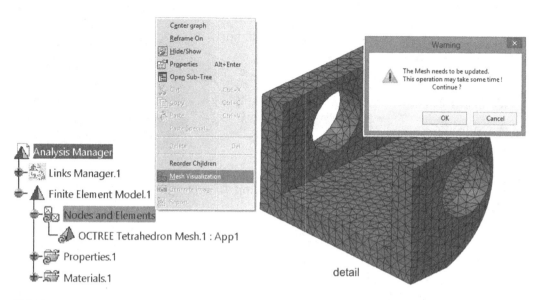

FIGURE 4.308 Mesh visualization, detail.

To display the mesh of the part model, the user accesses the *Mesh Visualization* option from the context menu of *Nodes and Elements* (Figure 4.308). *CATIA v5* displays an information/warning box informing the user that the mesh update requires a period of time, and he clicks the *OK* button. The figure also shows a detail with the part model in the new representation.

A high density of the mesh is observed. According to Figure 4.309, it contains 14318 nodes and 57602 finite elements. The information box is displayed after selecting the *Mesh.1* feature in the specification tree with the *Information* icon on the *Analysis Tools* toolbar enabled. Such checks are useful and necessary to estimate the discretization level of the mesh.

Along with the mesh displayed and the *Mesh.1* feature appearing in the specification tree, it is noticed that the *OCTREE Tetrahedron Mesh.1* feature has been updated (the *Update* symbol next to it has disappeared), and the user can apply the *Clamp* restraint. The selection of the surfaces to be fixed/clamped is possible if the user deactivates (permanently or temporarily) the mesh visualization from the context menu of the *Mesh.1* feature, through the *Activate/Deactivate* option. In Figure 4.310 the restrained surfaces are highlighted: the two cylindrical surfaces of the holes and the flat surface at the end of the part.

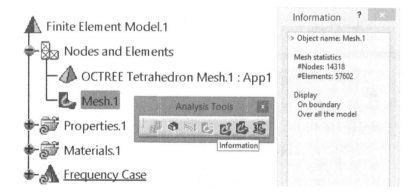

FIGURE 4.309 Displaying information about the obtained mesh.

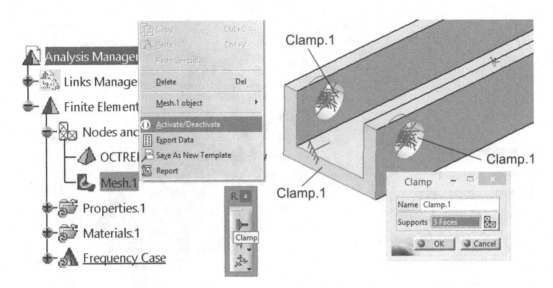

FIGURE 4.310   Applying the *Clamp.1* restraint.

The *Clamp.1* feature is added to the specification tree. Next, the user should double-click on *Frequency Case Solution.1* that is also in the specification tree to open the parameters dialog box in Figure 4.311. By default, in the *Number of Modes* field, the program is set to compute 10 modes using the *Iterative subspace* method.

This method (with both its options: *Iterative subspace* and *Lanczos*) is only available if the *ELFINI Structural Analysis (EST)* product is installed, otherwise, the default method is *Iterative subspace*. If the user selects the *Lanczos* option, a new *Shift* option appears, set on *Auto*. It computes the modes beyond a given value: *Auto, 1Hz, 2Hz* and so forth. *Auto* means that the computation is performed on a structure that is partially free and it is the default option.

The *Mass Parameter* option lets the user and the program take into account the structural mass of the part. It has an option, *Exclude*, that, once checked, allows the user to exclude the structural mass from the total mass summation when computing the solution of a frequency case with additional mass. If this structural mass parameter in a frequency case without additional masses is excluded (frequency case without masses set), an error occurs while computing

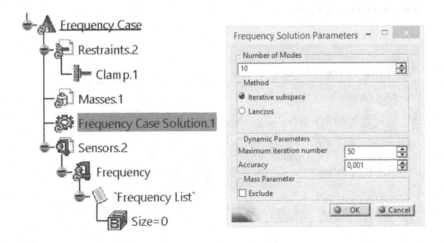

FIGURE 4.311   *Frequency Solution Parameters* dialog box.

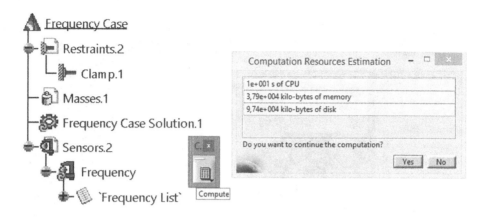

FIGURE 4.312 Starting the computation process.

the solution. Also, if the frequency case does not contain any masses set, this option should not be selected.

In this moment, the *Frequency List* feature in the specification tree has the value 0 within the *Size* parameter, but its value will change to 10 once the computation step starts.

The user calculates the frequencies by pressing the *Compute* icon and the *Computation Resources Estimation* information box from Figure 4.312 opens. Updating the *Frequency* sensor by opening its context menu has the same effect. The hardware resources required for the analysis are shown in the figure. Note that the user has not added any loads from the *Loads* toolbar (all icons are inactive by default). The results are, in fact, the natural frequencies of the part.

From the *Image* toolbar, the *Deformation* tool is used to display the first results: how the part deforms for each frequency. The user double-clicks on the *Deformed mesh.1* feature in the specification tree to open the *Image* information box, the *Occurrences* tab. Each mode in the left list has a frequency (in Hertz) in the right list, according to Figure 4.313. The first value, 441.161 Hz, represents the frequency from which the part starts to vibrate.

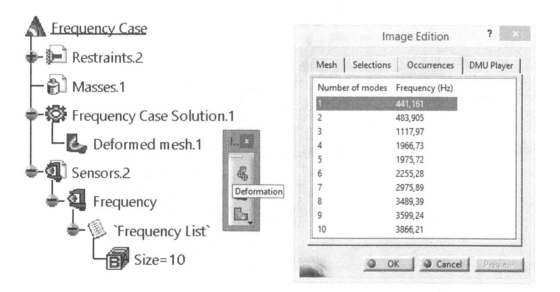

FIGURE 4.313 List of natural frequencies with values.

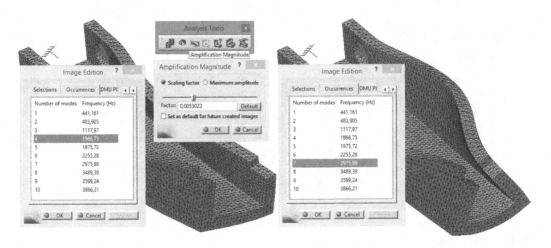

FIGURE 4.314 Deformation of the part for two frequencies.

Figure 4.314 shows two types of deformation of the part when it is subjected to excitations with frequencies of 1966.73 Hzand 2975.89 Hz. The deformations are exaggerated represented, but the amplitude can be set by pressing the *Amplification Magnitude* icon (*Analysis Tools* toolbar) and choosing a value for the *Scaling factor* parameter. From the same *Image* toolbar, the user applies the *Von Mises Stress* tool to study the stress distribution on the part's solid.

Figure 4.315 shows the distribution of the stresses appearing in the part's solid 3D model for the frequency of 1966.73 Hz. Depending on the frequency value selected in the list in the *Occurrences* tab, the part will deform differently, and the values in the palette change. Also, the areas of maximum stress change their position from case to case.

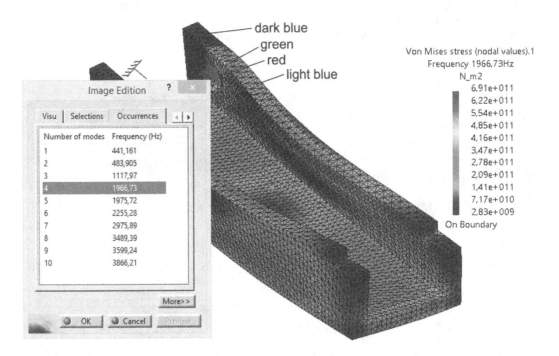

FIGURE 4.315 Distribution of stresses in the part for the frequency of 1966.73 Hz.

For the situation presented in Figure 4.315, it is observed that the value of the maximum stress is 6.91×10^{11} N/m², a value much higher than the yield strength of the material (2.5×10^8 N/m²). However, this value is reached only for the frequency of 1966.73 Hz, so, the part should be subjected to an excitation with this frequency.

For any frequency value in the list, with the help of the *Image Extrema* and *Information* tools, the user can determine the area with maximum stress and its value.

In the specification tree, according to the previous figures, it can be seen that the *Global Error Rate (%)* feature is missing from *Sensors.2*, and the icons *Precision*, *New Adaptivity Entity*, etc. are inactive.

By using the *Animate* tool on the *Analysis Tools* toolbar, the user observes how the part deforms and the stress distribution. In the *Animation* selection box (Figure 4.316) the user clicks the *More»* button to expand the options on the right. Then, he ticks the *One occurrence* option, selects the fourth frequency and presses the *Animation Mode* button to obtain a symmetrical animation (in the palette, the presence of negative values, that are symmetrical with the ones above, can be observed).

With the help of the [<] and [>] buttons, it is possible to quickly choose other frequencies, and by pressing the [...] button, the *Frequencies* list is opened for the selection of a certain frequency. A simpler animation, which briefly displays each frequency mode, is obtained by selecting the *All occurrences* option. If in the case of *One occurrence* the user had the possibility to select the number of frames (drop-down list *Steps number*, 50 frames for the most fluid animation), then in the case of *All occurrences* one frame is available for each frequency.

The explanations are similar for the other frequencies in the list. In practice, the part can perform very well in a static case, supporting various loads, but it will show cracks/breaks for certain frequencies. In operation, these must be avoided, but also certain stiffening elements can be applied to the solid part, or it can be assembled with other elastic elements that will absorb/dampen the vibrations, etc.

The information mentioned earlier, as well as many others, are presented in detail by means of a report file that can be created by the user with the help of the *Generate Report* tool. In the

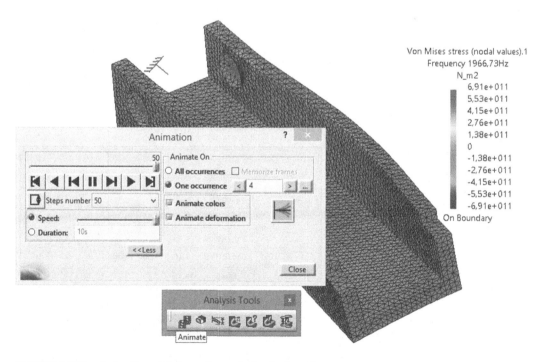

FIGURE 4.316 Animation of deformation and distribution of stresses.

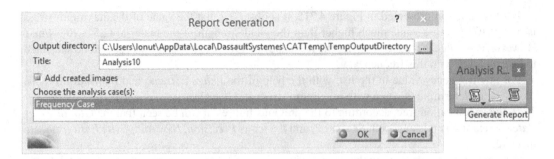

FIGURE 4.317 Creating a modal analysis report.

Report Generation selection box (Figure 4.317) the user chooses the destination of the files, selects *Frequency Case* and ticks the *Add created images* option.

Figure 4.318 shows a fragment of this report, obtained in the form of an *html* page.

A known solution of vibration reduction is to add an additional mass to the implicit one of the part. This part is fixed at one of its ends by the *Clump.1* restraint, according to Figure 4.310. At the other end a mass distributed on the two free flat surfaces is applied.

In Figure 4.319 the user disabled both results (*Deformed mesh.1* and *Von Mises stress (nodal values).1*) in the specification tree from their context menu. A *Distributed Mass* load is added from the *Masses* toolbar. In the dialog box with the same name, in the *Supports* field the user selects the two flat faces, and in the *Mass* field sets its value, in grams. Entering the respective value can be done in grams (50000 g) or in kilograms (50 kg). In the second case, the program converts the value into grams.

Number of computed modes   : 10
Boundary condition for modes computation :   clamped
Number of iterations already performed   : 0
Total Number of iterations performed   : 4
Relative eigenvalues tolerance required   : 1 . 000e-003
Relative eigenvalues tolerance obtained   : 3 . 591e-004

Mode number	Frequency Hz	Stability
1	4.4116e+002	1.0897e-013
2	4.8390e+002	3.2097e-013
3	1.1180e+003	1.3830e-009
4	1.9667e+003	2.7445e-007
5	1.9757e+003	7.7137e-008
6	2.2553e+003	2.6843e-006
7	2.9759e+003	3.6249e-006
8	3.4894e+003	6.4903e-005
9	3.5992e+003	7.2385e-005
10	3.8662e+003	3.5909e-004

Modal participation :

Mode	Frequency Hz	Tx (%)	Ty (%)	Tz (%)	Rx (%)	Ry (%)	Rz (%)
1	4.4116e+002	0.00	0.03	57.04	10.02	0.00	0.00
2	4.8390e+002	41.82	0.00	0.00	0.00	18.46	6.40
3	1.1180e+003	17.80	0.00	0.00	0.00	38.87	3.69
4	1.9667e+003	10.39	0.00	0.01	0.00	3.32	2.41
5	1.9757e+003	0.67	0.00	0.20	0.06	0.17	0.15
6	2.2553e+003	0.00	0.54	17.46	2.90	0.00	0.00
7	2.9759e+003	0.00	0.33	0.47	0.07	0.00	0.00
8	3.4894e+003	0.00	74.65	0.05	0.00	0.00	0.00
9	3.5992e+003	8.04	0.00	0.00	0.00	9.56	2.36
10	3.8662e+003	1.16	0.00	0.00	0.00	0.51	0.24
	Total	79.86	75.55	75.23	13.06	70.88	15.25

FIGURE 4.318 Fragment of the generated report.

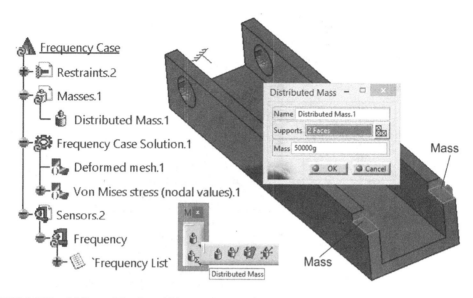

FIGURE 4.319   Adding a *Distributed Mass* at the end of the part.

As in other cases that the user has encountered while working with *CATIA v5,* the measurement units are very important, they must be carefully specified, and always be visible next to the entered values. From the menu *Tools → Options → General → Parameters and Measure → Units* tab (Figure 4.320) these measurement units, but also their display mode, precision, etc. are set. The mass used in the application is displayed in grams regardless of how the user entered it because the default setting is grams.

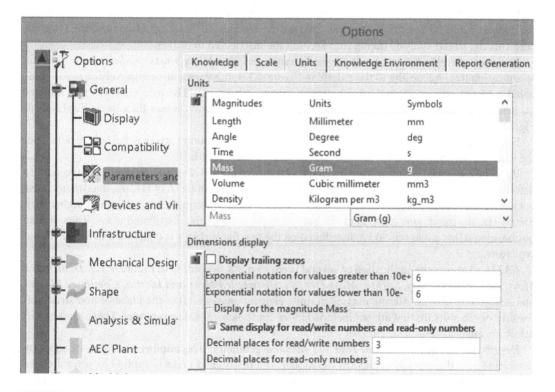

FIGURE 4.320   Some settings regarding the choice of measurement units.

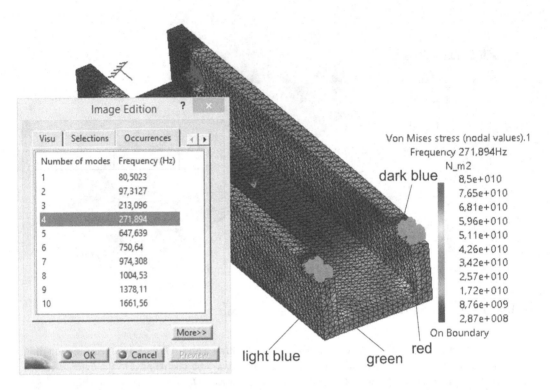

FIGURE 4.321 Distribution of stresses in the part for the frequency of 271.894 Hz.

In the specification tree in Figure 4.319, it can be seen that many features require updating. From the context menu of *Sensors.2,* the user chooses the *Update All Sensors* option, and the program performs the modal analysis taking into account the distributed mass.

The user activates the results *Deformed mesh.1* and *Von Mises stress (nodal values).1* in the specification tree. According to the palette in Figure 4.321, the values are completely changed compared with the previously studied situation, without added mass. The user observes that the frequencies are between 80.5 Hz and 1661.56 Hz, significantly lower values than those displayed in Figure 4.315.

For example, for the fourth frequency (271.894 Hz) the maximum stress is 8.5×10^{10} N/m^2, but the value is, however, much higher than the material yield strength.

By creating animations for each frequency, the user notices that the tendency of the part to deform has been reduced. For frequencies with values lower than 647.639 Hz, the maximum stress is located in the application area of the *Clamp.1* restraint. For frequencies with values higher than 750.64 Hz, the maximum stress is in the area of application of the distributed mass. Thus, it may be considered that a mass of 50 kg distributed on the two flat surfaces is equivalent to a new *Clamp* type restraint.

CATIA v5 allows the application of several types of mass, according to Figure 4.319: *Distributed Mass, Line Mass Density, Surface Mass Density, Distributed Mass and Inertia, Combined Masses* and *Assembled Masses.* Adding masses in finite element analyses takes the place of non-structural features or other details that are not included in the finite element analysis model. These masses will affect the analysis results.

Distributed masses are placed over the selected geometry. The number and type of supports determine how the mass is applied to the model. If a distributed mass is applied to several edges it means that the whole value of the mass is divided into mass per unit length and applied evenly along the edges. Also, a distributed mass applied to several surfaces is divided into mass per unit area. Distributed masses can also be used as supports for gravity loads. Distributed masses can be applied to points, vertices, curves, edges and surfaces. All supports for a single distributed mass must be of the same type.

Linear masses are distributed as mass per unit length over the selected geometry; they can be applied to curves, edges or line groups.

Surface masses are distributed as mass per unit area over the selected geometry, and can be applied to surfaces, faces or surface groups.

Distributed masses and inertias are applied to all selected geometry. The mass is applied as mass per unit volume. The inertia is applied as specified for each translational and rotational component. Distributed masses can also be used as supports for gravity loads.

Distributed masses and inertias can be applied to virtual parts or to points or vertices that are associated with a node in the mesh, edges and faces can also be selected.

At the end of the application, the user can try to apply other types of masses or add elements to stiffen the structure of the part.

A video solution of this application can be studied at https://youtu.be/zJysrd3FuIQ.

4.11   OPTIMIZING THE ANALYSIS PARAMETERS OF A HINGE

This application proposes a finite element analysis of a part in the structure of a hinge assembly, followed by the optimization of its geometry to meet dimensional and mass requirements while respecting two conditions: one related to a defined Safety Factor and another about a maximum displacement.

The goal is to introduce the user to the interesting and useful techniques for advanced design of parts using the *CATIA v5 Product Engineering Optimizer* workbench. It provides engineers with easy-to-use but powerful tools that can be used to optimize almost any type of data.

Thus, the *Product Engineering Optimizer* workbench takes the engineer beyond the basic solid and surface modelling. Combining the modelling skills with FEM simulation and parametric optimization proves the *CATIA v5* capabilities and its recognition as a leader in most industries. An optimally designed product, no matter the complexity, enhances the end-user experience while also improving its functions.

With the help of this workbench, the user can explore different design alternatives for a certain product and identify reasonable compromises to improve its quality, in parallel with decreasing the costs allocated to find the best solution. Through this workbench, *CATIA v5* reduces the number of design iterations, shortens the time needed to identify the optimal/imposed value of a parameter and validates a certain solution found intuitively. Intrinsic and user-defined parameters, sensors, constants and results obtained by computation are involved; optimization is, in fact, an optimal combination of them all, while respecting certain constraints.

Such an optimization approach begins with the creation of free parameters, called design parameters or design variables. These parameters are used to model the analysed part. There are dimensional links between them, but none depends on the parameters that will be identified through optimization. The relations established between these parameters lead to the creation of design constraints. Through the automatic/optimized editing of some of the free parameters and following some complex computation steps, the optimization objectives will be achieved; these are clearly presented to the user.

The part analysed in the application is presented by its 2D drawing in Figure 4.322. Its modelling is relatively simple and is done in the *Part Design* workbench following the video tutorial at https://youtu.be/Hhl_R7UF24o.

Also, the file containing the 3D model of the part can be downloaded and used in the analysis because it contains exactly the parameters that will be presented during the application.

Figure 4.323 shows the main sketch *(Sketch.1)* of the part. The user observes the complete definition of all lines and arcs contained by the sketch. The profile was created by a symmetrical offset of

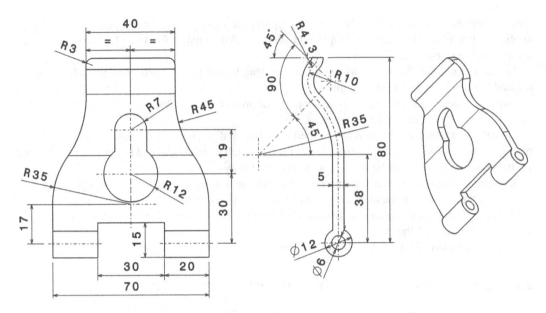

FIGURE 4.322 Two-dimensional drawing of the hinge part.

the two horizontal lines and the distance between them is 5 mm. The value is stored in the *Offset.13* parameter, but the name may be different depending on the user's approach. This parameter, as well as the others of the part, are used for solid modelling. This thickness of the part directly influences the mass, a parameter that will be studied in the application in order to optimize the part's geometry.

There are, however, some constraints imposed to the part 3D model, namely, the load applied to certain surfaces must not generate stresses above the yield strength of the material, the error rate of

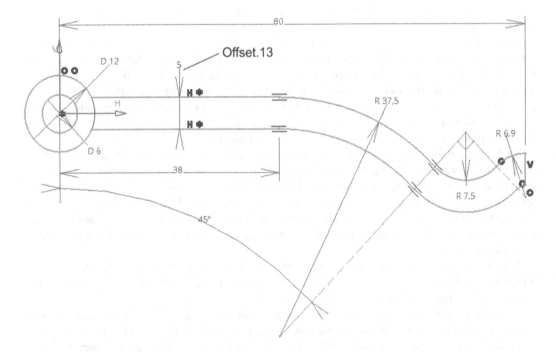

FIGURE 4.323 Example of free parameter, offset between two parallel lines.

the FEM analysis must be low, and there should be others (constraints, conditions) that that the user can add (especially to limit the displacements).

Thus, the user must identify and go through three main steps in this application: the correct modelling of the studied part, the finite element analysis and the optimization of the part to meet certain constraints. They are set by the user depending on the field of applicability of the part/assembly.

For example, in the application it is necessary to compute and respect a Safety Factor, defined as the ratio between the yield strength of the material and the maximum stress calculated in the solid part. If the yield strength is a constant in the analysis, the maximum stress is computed by *CATIA v5* for each constructive variant, in a complex iterative process.

Also, the application requires the minimization of the part's weight. The *Offset.13* parameter directly influences the weight and indirectly the stiffness of the part, and, therefore the maximum stress. The problem is of medium complexity and cannot be solved by successive attempts by the user, thus requiring a methodology and dedicated work tools, such as those integrated in the *Product Engineering Optimizer* workbench.

Once the solid modelling is complete, the user must verify the correctness of the 3D model. Thus, the sketches must be correctly constrained, and the order of creation of the features in the specification tree must be logical (Figure 4.324). Finally, the part is obtained with a volume of 20461.766 mm^3, an area of 10960.791 mm^2 and a mass of 160.829 g. All these values will change when the part thickness (parameter *Offset.13*) is modified. The user can calculate and display them in the specification tree if he wants to consider them, for example, as constraints for the stage optimization.

Correct and clear modelling of the part solid is very important because it involves intrinsic and user-created parameters that can be used in the finite element analysis and part optimization steps.

The preparation of the part for analysis follows the steps presented in the previous applications. Thus, a *Steel* material is applied to the part (density 7860 kg/m^3, yield strength 2.5×10^8 N/m^2).

For the mesh definition of the part, a more complex discretization is used (*Size* = 1 mm, *Absolute sag* = 0.8 mm and *Element type* = *Parabolic*). The goal is to obtain a reduced error rate following the analysis, but the time required to perform it will increase significantly. Also, the computations required in the optimization stage will involve significant system resources.

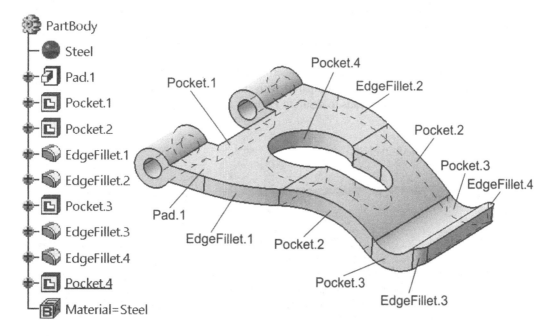

FIGURE 4.324 Features of the part.

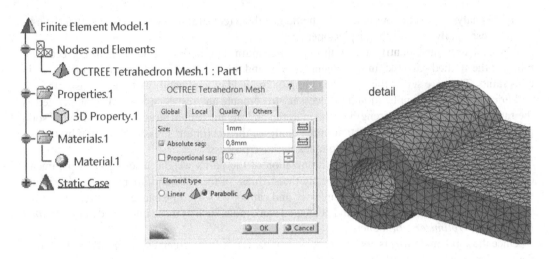

FIGURE 4.325 Mesh definition of the part.

Figure 4.325 shows these settings, but also a detail to observe the discretization of the part (*Mesh Visualization* option from the contextual menu of *Nodes and Elements* or using the *Deformation* tool).

The user applies a *Clamp* restraint on the cylindrical surfaces of the Ø6-mm holes of the part, according to Figure 4.326. The main effect is that the part cannot rotate around the common axis of the two holes.

At the other end of the part, on the two surfaces indicated in Figure 4.327, a distributed force of 120 N is applied in the negative direction of the Z axis.

Following the FEM analysis (using the *Compute* tool), the first results are displayed (Figure 4.328): the maximum stress is 9.2×10^7 N/m^2, located near the cylindrical surfaces (*Clamp.1* restraint is applied in their holes). The maximum displacements, of 0.192 mm, are located in the area of the surfaces on which the distributed force was applied. Of course, according to the experience gained, the user can predict the behaviour of the part under the action of the force and under the established conditions (*Clamp.1* restraint, mesh discretization, etc.).

Figures 4.329 and 4.330 show the location of the maximum stress and the maximum displacement, respectively. Most likely, the maximum stress occurs in that corner because it is at the intersection of three straight edges. By adding a connecting radius on these edges, the position and value of the maximum stress may change. Compared with the yield strength of the part material, the maximum stress is lower, which leads to a Safety Factor > 2.

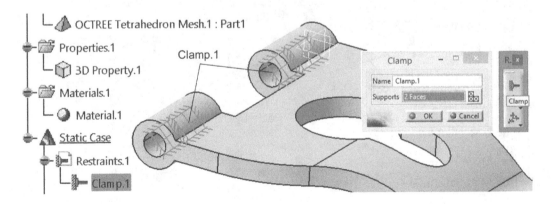

FIGURE 4.326 Applying the *Clamp* restraint.

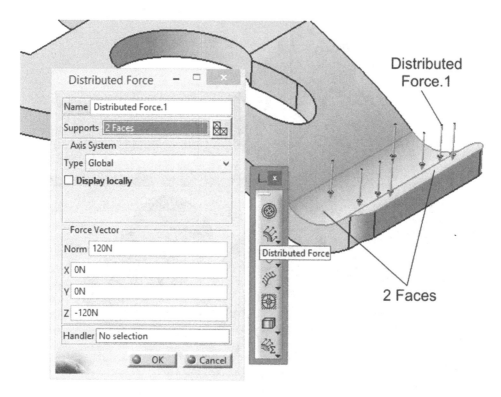

FIGURE 4.327 Applying the *Distributed Force* load on two surfaces.

Figure 4.330 also shows a list of sensors needed in the application. They can be added one by one by the user from the context menu of feature *Sensors.1,* option *Create Global Sensor. Maximum Von Mises* and *Maximum Displacement* sensors are required later in the application.

The user notices a very low value of the sensor *Global Error Rate (%)* of 3.788%, which means a high accuracy of the analysis results and this is good. So, a new discretization of the part is no longer

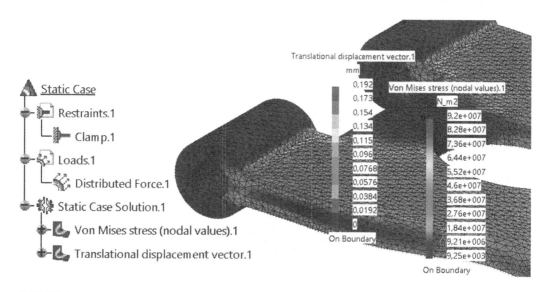

FIGURE 4.328 *Von Mises stress* and *Translational displacement vector* results.

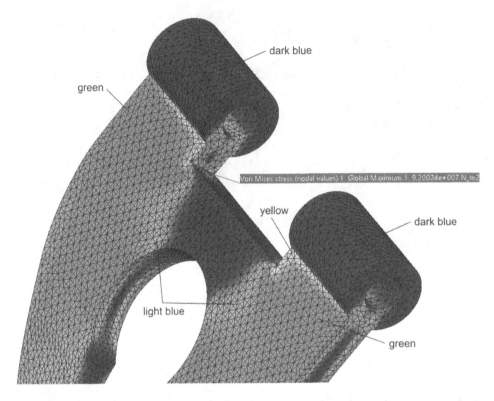

FIGURE 4.329 Locating the maximum stress.

necessary nor the application of optimization tools, such as *New Adaptivity Entity* and *Compute with Adaptivity*.

These first results of the analysis are satisfactory, as the part shows good rigidity for loading with a distributed force of 120 N.

In the next stage of the application, the user will optimize the geometry of the part so that certain constraints are accomplished.

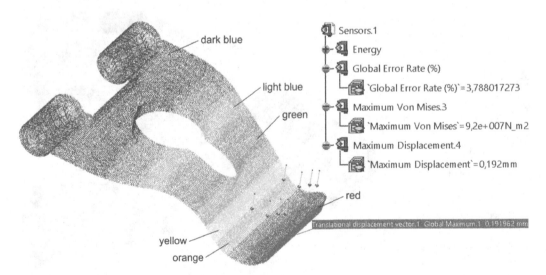

FIGURE 4.330 Locating the maximum displacement and the list of sensors.

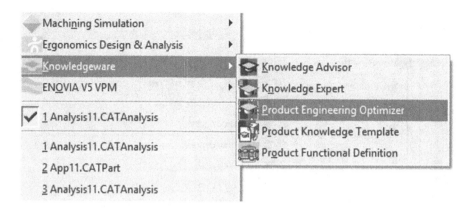

FIGURE 4.331 Opening the *Product Engineering Optimizer* workbench.

Thus, the *Product Engineering Optimizer* workbench opens from the *Start → Knowledgeware* menu, as shown in Figure 4.331.

The application requires the creation of user parameters and of some relations/formulas. By default, *CATIA v5* does not display these features in the specification tree.

Thus, from the menu *Tools → Options*, the user has access to a complex set of options for customizing the program, displaying the working environment, the presence of features in the specification tree, etc. In order to be able to display his own parameters in this specification tree, the user must check the *With value* and *With formula* options from the *General → Parameters and Measure* category, in the *Knowledge* tab (Figure 4.332).

From the same menu *Tools → Options → Infrastructure → Part Infrastructure → Display* tab (Figure 4.333), the user also selects the options *Constraints*, *Parameters* and *Relations*. These features, important in any stage of parameterization/optimization of the part, become visible in the specification tree.

Also, another important setting for how the part behaves during the successive changes it will receive in the optimization stage is in the menu *Tools → Options → Infrastructure → Part Infrastructure → General* tab (Figure 4.334). The user checks *Automatic* in the *Update* area.

Using the *Formula f(x)* icon on the *Knowledge* toolbar, the user creates two parameters: *SafetyFactor* and *Mass*. Thus, in the *Formulas: Analysis Manager* dialog box (Figure 4.335) the

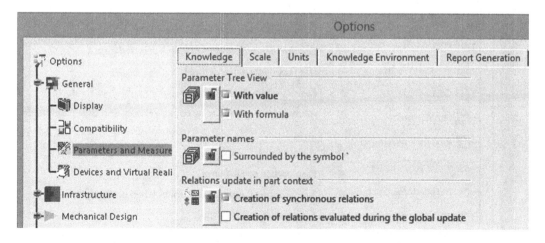

FIGURE 4.332 Settings in the *Knowledge* tab related to the display of parameters in the specification tree.

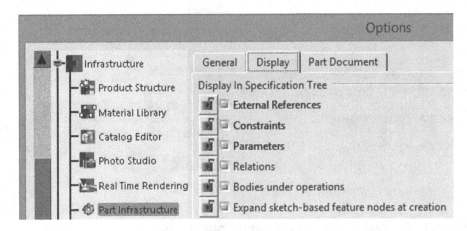

FIGURE 4.333 Settings in the *Display* tab related to which features will be displayed in the specification tree.

user can establish filters for displaying existing and newly inserted parameters through the options in the *Filter Type* list, then choose the type of *SafetyFactor* parameter, using the *Real* option in the drop-down list near the button *New Parameter of type*. The parameter will have a single value.

The user presses the button to create the new parameter and enters the name in the *Edit name or value of the parameter* field. Initially, the program offered a simple name (example: *Real.2*), but it is changed to *SafetyFactor* to be easier to use in the application. By default, this new parameter has the value 0.

By pressing the *Add Formula* button, the user opens the *Formula Editor* editing box (Figure 4.336). The upper field, which contains the *SafetyFactor* parameter, is not editable. To its right is the equal sign (=), and the user adds a relation between two parameters in the field below. It is observed that the parameters have complex names, and the simplest/fastest method of selection and addition in the formula is by choosing them from the specification tree.

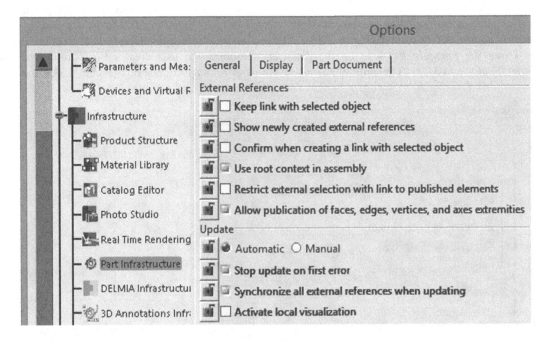

FIGURE 4.334 Settings in the *General* tab related to automatically updating the part after a change.

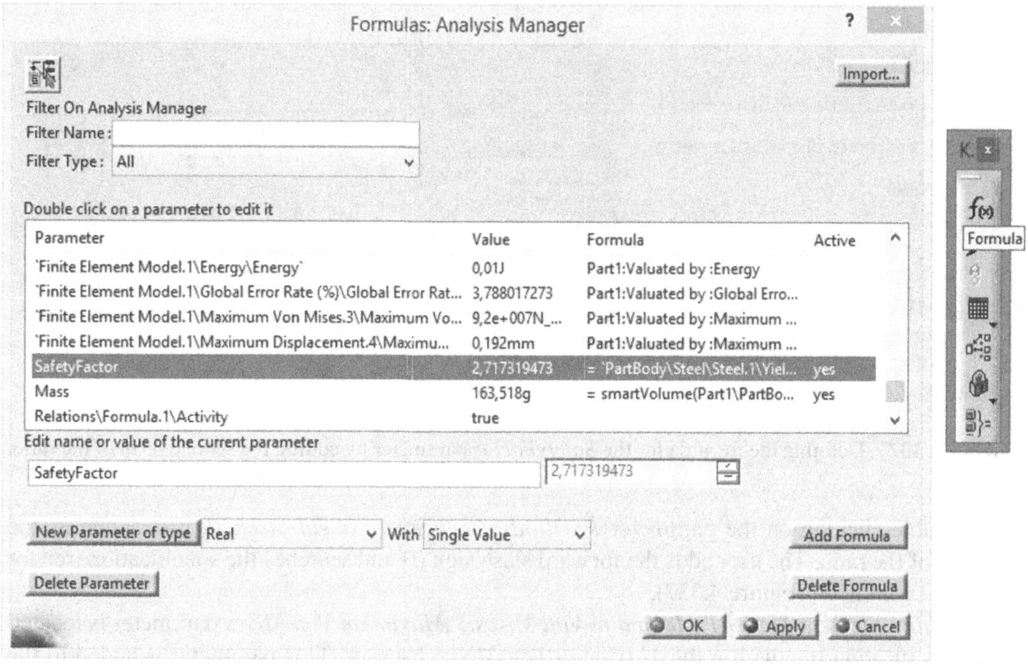

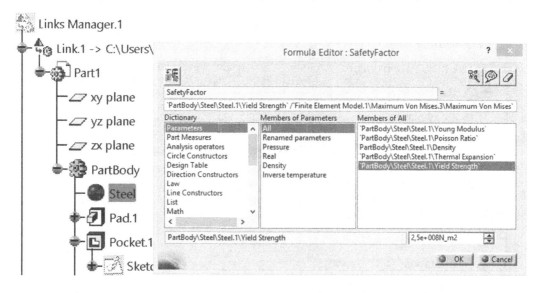

FIGURE 4.335 Creating the *SafetyFactor* user parameter.

The first element of the ratio is the yield strength of the part material. From the specification tree the user looks in the *Links Manager.1,* and clicks the plus sign (+) to expand the part tree structure. *Links Manager.1* is hidden, but this is not important as the selection is made in the specification tree. *Steel* is selected from *Link.1* → *Part1* → *PartBody*, and in the *Members of All* column in the editing box (Figure 4.336) all parameters (regardless of type) of the *Steel* material become available.

FIGURE 4.336 Defining the formula for the *SafetyFactor* parameter by adding the first term of the ratio.

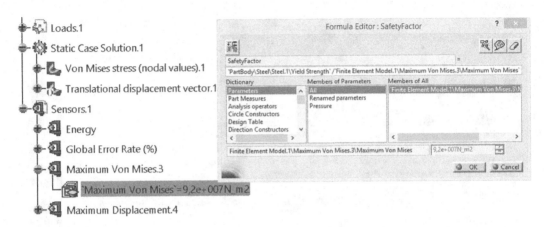

FIGURE 4.337 Defining the formula for the *SafetyFactor* parameter by adding the second term of the ratio.

By double-clicking on the parameter *PartBody\Steel\Steel.1\Yield Strength* it is added as the first term of the ratio. The user adds the forward slash sign (/) and searches the specification tree for the second parameter (Figure 4.337).

The *Finite Element Model.1\Maximum Von Mises.3\Maximum Von Mises* parameter is located in the specification tree within the *Maximum Von Mises.3* sensor. This parameter is added to the formula. In Figures 4.336 and 4.337 it can be seen that for each parameter the program also displays its value. The first parameter *Yield Strength* is editable representing a material characteristic, and the second parameter *Maximum Von Mises* is determined by computation, so it is not editable by the user.

Depending on many other parameters (dimensions of the part, position of the *Clamp.1* restraint, value of the distributed force, etc.) this parameter is computed by the program, its value changing with each such iteration. However, the user has defined this *SafetyFactor* parameter to ensure that the stress value remains within a certain ratio to the yield strength of the part material.

Similarly, the user inserts the *Mass* parameter. This is, of course, of type *Mass* with a single value. *Mass* is defined as the product of the part's volume and the density of its material. Thus, in the *Formula Editor* editing box (Figure 4.338) to easily find the volume, the user chooses the *Part*

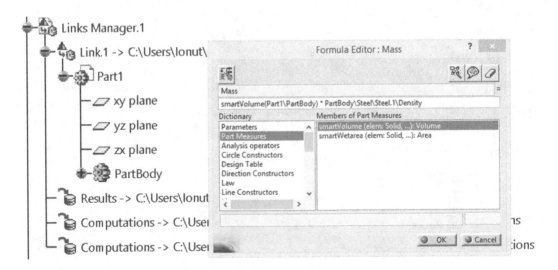

FIGURE 4.338 Defining the formula for the *Mass* parameter.

FIGURE 4.339 Display of parameters and formulas in the specification tree.

Measures option in the *Dictionary* column and the two options in the *Members of Part Measures* column become available.

By double-clicking on *smartVolume (elem: Solid, ...): Volume* it is added to the formula as *smartVolume ()*, and the user must specify between brackets the solid part whose volume needs to be extracted. From the specification tree, also by double-clicking on *PartBody,* the solid completes the first term of the formula: *smartVolume(Part1\PartBody).* The user adds the asterisk symbol (*) as a multiplication sign, then searches for the density of the material. According to Figure 4.336, this is found by selecting *Steel* in the specification tree.

If the two parameters *SafetyFactor* and *Mass* are correctly defined, their values are added to the specification tree together with the relations that calculate them (Figure 4.339). For the initial geometric configuration, the two parameters have the values: *SafetyFactor* = 2.717 and *Mass* = 160.829 g. The user notices the complexity of the names of the two parameters, which is the main reason why they must select them using their features from the specification tree.

If the user wants to edit a formula, he must double-click on it twice. The first double-click switches to the *Knowledge Advisor* workbench, and the second double-click opens the editing box of the respective formula. From this workbench the user must return to the *Product Engineering Optimizer* workbench using the *Start* menu.

Another variant of editing a formula is by double-clicking on the defined parameter, and the editing box from Figure 4.340 opens. The user notices that the *Edit Parameter* box has two fields: the one on the left contains the name of the parameter (which can be edited), and the one on the right has the value calculated by the relation (cannot be edited). Clicking the *f(x)* button/icon opens another formula editing box, like the one in Figure 4.337.

These first two values of the *SafetyFactor* and *Mass* parameters are considered as reference. By applying the part optimization methodology/step to meet certain constraints, these parameters, and many others, will receive various computed values.

Thus, in the *Product Engineering Optimizer* workbench, the user clicks the *Optimization* icon and the dialog box with the same name opens (Figure 4.341). In the *Optimized parameter* field, with the help of the *Select* button, the user chooses the *Mass* parameter and the current value of the part mass (160.829 g) is displayed. The *Minimization* option is chosen from the *Optimization*

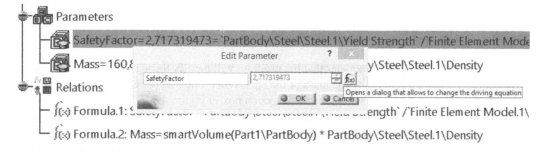

FIGURE 4.340 Editing the formula that calculates the *SafetyFactor* parameter.

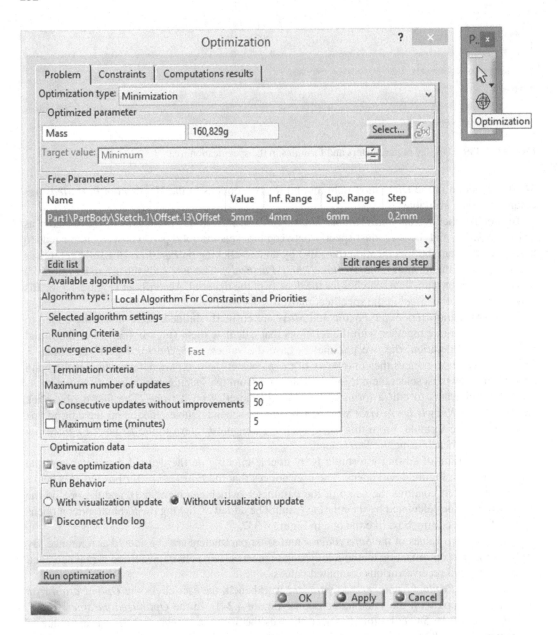

FIGURE 4.341   Choosing the parameter that will be optimized and the parameter that the program will modify through successive iterations.

type list so that the user imposes on the program the need to reduce the mass of the part, but under certain conditions.

Thus, the user declares the parameter that the program will modify through successive iterations so that the condition of minimizing the mass of the part is met. The user presses the *Edit list* button under the *Free Parameters* area and opens the selection box in Figure 4.342.

The *Offset.13* parameter belongs to the *Sketch.1* of the *Pad.1* extrusion. With the *Select the free parameters* box open, the user chooses the sketch from the specification tree so that only its parameters are displayed in the *Parameters* column. Choosing the sketch is, in fact, a first selection filter. If the user wants to apply a second filter, he can select the *Length* option from the *Filter Type* list.

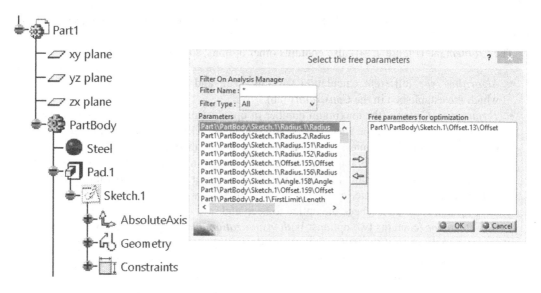

FIGURE 4.342 Selecting the *Offset.13* parameter in *Sketch.1*.

Once identified, the *Offset.13* parameter (with its full name) is added by double-click (or by pressing the arrow pointing right) to the *Free parameters for optimization* column (Figure 4.342).

The parameter is added to the *Free Parameters* list. By default, the program displays its initial value from the 2D drawing (5 mm, Figure 4.322). Several columns and options are available for this parameter, as follows: the value of 5 mm is on the *Value* column, and the *Auto* option is on the *Step* column. Thus, the program would have the possibility of identifying and trying any value. However, the user specifies the limits (lower and upper) between which the parameter can take values during optimization.

By pressing the *Edit ranges and step* button, the editing box in Figure 4.343 opens, the user checks the options and enters the values of 4 mm and 6 mm, respectively, in the *Inf. Range* and *Sup. Range* fields. In the *Step* field, the step value for changing the *Offset.13* parameter is set within this range (*Inf. Range, …, Sup. Range*). The value of the step comes to the aid of the *CATIA v5* program,

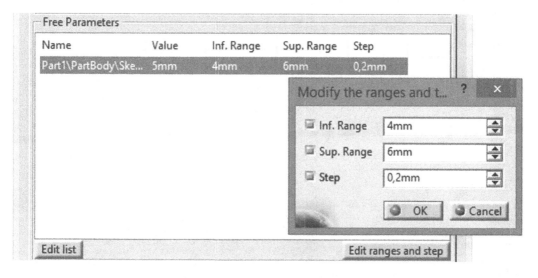

FIGURE 4.343 Specifying the lower and upper limits between which the parameter can take values.

reducing and simplifying the number of variants that it must calculate, and implicitly, the analysis will be of shorter duration.

The *Problem* tab (Figure 4.341) also contains other options, such as:

- *Algorithm type* (different calculation variants, with and without imposed constraints, which are established in the *Constraints* tab).
- *Termination Criteria* (the maximum number of updates/computation iterations is chosen, but also the maximum time allocated to them if the user ticks the *Maximum time (minutes)* option). The time, in minutes, does not speed up the optimization process, but limits the number of results/solutions found in a certain period. For a complete analysis it is recommended to uncheck the option. In this application, the user forces the program to find the optimal solution faster, entering the value 20 in the *Maximum number of updates* field.
- *Save optimization data* (check the option to keep and save the analysis results).
- *Run Behavior* (contains two options: *With visualization update* and *Without visualization update*). In the case of the first option, the analysis is slower because the program displays the results of the computation/optimization iteration as it progresses. In the case of the second option, the results are displayed at the end and it is faster.

In the *Constraints* tab (Figure 4.344) the user chooses the constraints that the program must fulfil while processing the iterations necessary to minimize the mass of the part. Not every dimensional configuration suitable for a low mass value also meets the set constraints.

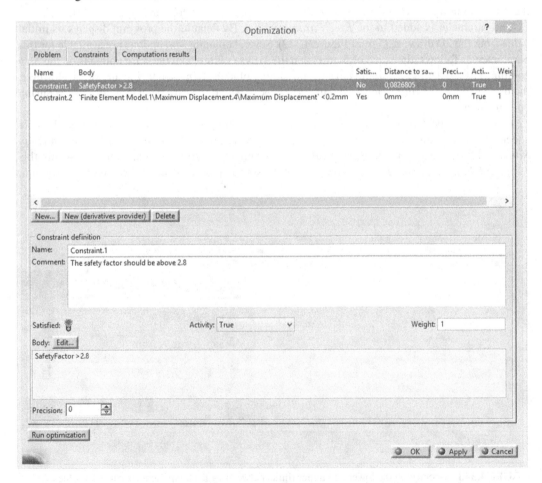

FIGURE 4.344 Insertion of the constraints to be taken into account for minimizing the part mass.

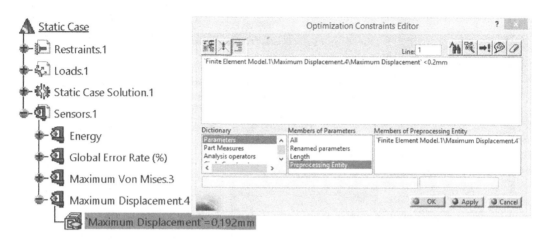

FIGURE 4.345 Writing a constraint.

In this application, the optimization of the part requires two constraints related to the FEM analysis: the *SafetyFactor* parameter must reach values greater than 2.8, and the maximum displacement of the part must be less than 0.2 mm. This last parameter is extracted from the analysis by the sensor *Maximum Displacement.4* (Figure 4.330).

Adding a constraint is simple: the user clicks the *New...* button and the *Optimization Constraints Editor* box opens (Figure 4.345).

Writing the constraint involves choosing the parameter and comparing it with a value. The maximum value of the deformation is stored in the parameter *Finite Element Model.1\Maximum Displacement.4\Maximum Displacement.* It is added in the editor by simply selecting the *Maximum Displacement* feature from the *Maximum Displacement.4* sensor. In the *Members of Parameters* column the user can select the *Preprocessing Entity* option. The respective parameter is displayed next to it, and the user writes the relation comparing it with 0.2 mm. By default, this constraint is fulfilled (Figure 4.344), and *Yes* confirmation appears on the *Satisfied* column.

The *SafetyFactor* parameter is stored in the specification tree in the *Parameters* feature and it is added to the list of constraints compared with the value 2.8. By default, its value is 2.717 (Figure 4.340), so the constraint is not met, and the *Satisfied* column contains *No.*

The *Distance to satisfaction* column contains missing values for the constraints to be met. If a constraint is fulfilled (*Constraint.2*) that value is 0. The *Activity* column can contain *True* or *False* values for an active or inactive constraint, respectively. Changing the state of a constraint can be done from the *Activity* drop-down list, located below in the *Constraint definition* area. Also from here the user has the possibility to specify the importance of a constraint (*Weight* field), to change the name of the constraint and to add a comment, according to Figure 4.344.

The constraints must be set very carefully, the optimization must be logically possible, otherwise the computation step will require a lot of time, and the results will not be correctly identified.

Also, the more complex the part is, the more parameters and constraints are involved, depending on the algorithm chosen and the update variant, the more precise the optimization analysis is, but also the longer it takes.

Launching the optimization is done by pressing the *Run optimization* button. In the *Existing output file* selection box (Figure 4.346) it is observed that the user already has a set of results obtained during a previous optimization and he is asked to choose a location where an *Excel* file with the results of the current analysis will be saved. The file will contain several variants of results, some that will fulfil the constraints and some that will not.

During the optimization computation, the program displays an information box (Figure 4.347) with its progress and a *Computation Status* box. From the figure it can be seen that iteration 11 of 20 is in progress, both constraints are active and fulfilled and the identified value of the mass is

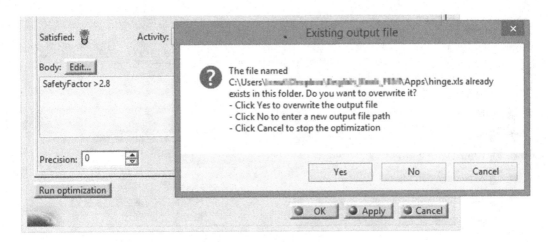

FIGURE 4.346   Saving the optimization results in an *Excel* file.

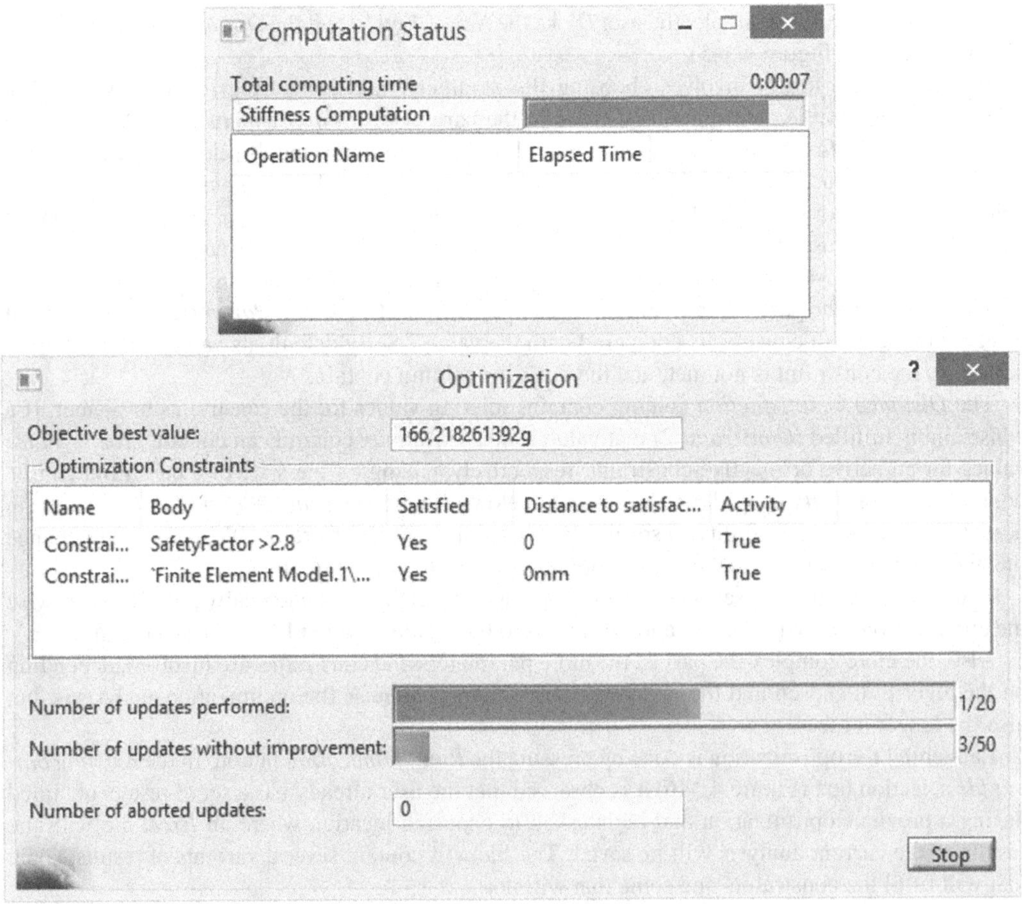

FIGURE 4.347   Running the optimization stage.

A	B	C	D	E	F	G	H	i
Nb Ev	Best (g)	Mass (g)	Analysis Manager\Relations\	Analysis Manager\Rela	Part1\PartBody\Sketch.1\Offset.13\Offset (mm)			
0	0	160,829484	0,082680527	0	5			
1	0	160,829484	0,082680527	0	5			
2	0	160,829484	0,082680527	0	5			
3	0	161,3682408	0,034955681	0	5,02			
4	0	160,829484	0,112985049	0	5			
5	166,2182614	166,2182614	0	0	5,2			
6	166,2182614	166,2182614	0	0	5,2			
7	166,2182614	166,7572892	0	0	5,22			
8	166,2182614	166,2182614	0	0	5,2			
9	166,2182614	166,2182614	0	0	5,2			
10	166,2182614	166,7572892	0	0	5,22			
11	166,2182614	166,2182614	0	0	5,2			
12	165,8000222	165,8000222	0	0	5,184481016			
13	165,1096119	165,1096119	0	0	5,158861622			
14	164,9903148	164,9903148	0	0	5,154434639			
15	164,6876339	164,6876339	0	0	5,143202258			
16	164,1898656	164,1898656	0	0	5,124729563			
17	164,0971694	164,0971694	0	0	5,121289417			
18	163,6467726	163,6467726	0	0	5,104573841			
19	163,2995511	163,2995511	0	0	5,091686929			

FIGURE 4.348   Fragment of the file with the results obtained after the optimization.

166.218 g. As a new iteration enters the computation, the value of the mass changes. For example, iteration 14 has a mass of 164.99 g, iteration 23 a mass of 162.818 g, etc.

The file with the extension *.xls* (Figure 4.348) contains columns with the parameters chosen by the user involved in the optimization, and on rows several variants/solutions identified by the program. An optimal variant is chosen and the value of the *Offset.13* parameter (Figure 4.349) is displayed. After viewing the data stored in the file and if the user does not agree with the variant proposed by the program, he can extract and apply the values of a certain row.

In Figure 4.348 the user notices that the first five results have the value 0 g on the *Best* column and the *SafetyFactor* is less than 2.8 (values other than 0 on the fourth column). Those results that do not simultaneously meet both constraints can be automatically excluded by the program. Thus, in the *Computation results* tab of the *Optimization* dialog box (Figure 4.350) the user selected the *Historic sort* option, and from the *Results to display* list he chose *All constraints satisfied only*. It

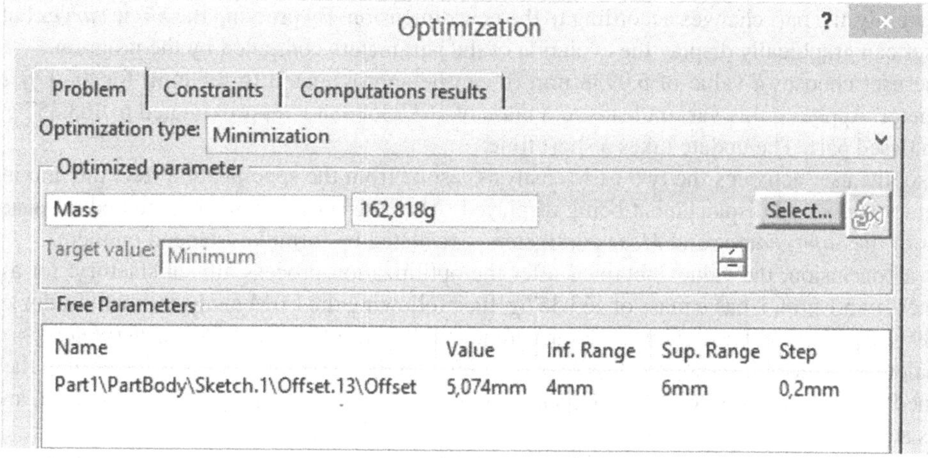

FIGURE 4.349   The variant with optimal results chosen by the program.

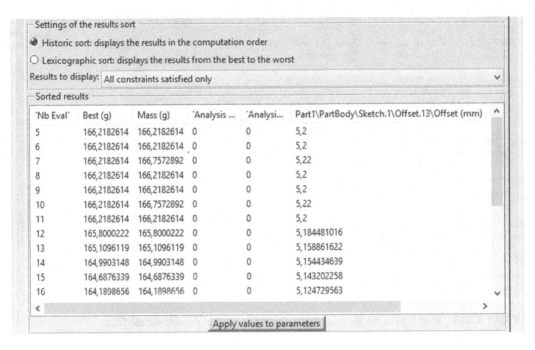

FIGURE 4.350 Optimization results that meet the constraints set by the user.

can be seen from the figure that the iterations with the *Best* value equal to zero are no longer found in the list.

The display order of the available variants is, thus, chronological (in the order in which they were computed by the program), but it can be from the best solution to the worst (by checking the *Lexicographic sort* option).

The solution identified by the program is not always the most suitable because it assumes values with many decimals for most parameters. It would result, therefore, in a part with surfaces and dimensions that are almost impossible to manufacture. So, the user will/may choose a convenient variant and impose it (its parameters) to the part.

In the selection box in Figure 4.350, a variant is chosen, the user checks the dimensions, and then he presses the *Apply values to parameters* button. The optimized parameter of the part, *Offset.13,* will receive the value found on the last column (according to the figure), the other values being calculated.

Certainly, the part changes according to the new dimension. By pressing the *Show curves* button, the user can graphically display the evolution of the parameters computed by the program.

The user chooses a value of 5.0938 mm (it can be approximated to 5.1 mm) for the *Offset.13* parameter. Applying this variant leads to a mass of 163.3568888 g (approximated to 163.357 g) for the analysed part. The update takes a short time.

Also, the user activates the two FEM analysis results from the specification tree, the maximum values for stress and displacement being displayed. Also, the sensors must be updated because the values of the *SafetyFactor* and *Mass* parameters, calculated by formulas, depend on them.

As a conclusion, the values obtained after the optimization process are satisfactory: for a part thickness of 5.1 mm, it has a mass of 163.357 g, then following the FEM analysis, a maximum stress of 8.89×10^7 N/m^2 and a maximum deformation of 0.182 mm (Figure 4.351). *SafetyFactor* = 2.81 was calculated, the deformation is less than 0.2 mm, and the maximum stress is lower than the one obtained initially, of 9.2×10^7 N/m^2. This is normal, the thickness of the part has increased, and so has its rigidity.

To continue the application, the user can add other constraints, search for a new increased value of the distributed force so that the deformation is less than 0.3 mm or less than 0.15 mm, etc.

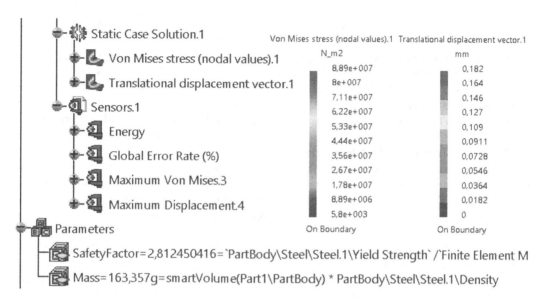

Von Mises stress (nodal values).1   Translational displacement vector.1

N_m2	mm
8,89e+007	0,182
8e+007	0,164
7,11e+007	0,146
6,22e+007	0,127
5,33e+007	0,109
4,44e+007	0,0911
3,56e+007	0,0728
2,67e+007	0,0546
1,78e+007	0,0364
8,89e+006	0,0182
5,8e+003	0
On Boundary	On Boundary

FIGURE 4.351   FEM analysis results for the new variant of the optimized part.

The methodology presented in the application is complex, it requires a lot of attention from the user, and some knowledge of parameterization and programming, but the results are interesting and useful.

4.12 THERMAL ANALYSIS OF AN EXHAUST MANIFOLD

In automotive engineering, an exhaust manifold collects the exhaust gases from multiple cylinders into one pipe. Basically, it refers to the grouping together of multiple outputs. In contrast, an inlet/intake manifold supplies air to the engine, it is mixed with fuel and a spark produces the ignition pushing down the cylinders.

Generally, exhaust manifolds are of simple to very complex shapes, made of cast iron or stainless steel. The most common types of exhaust manifold are made of mild steel or stainless steel, they may be coated with a ceramic-type finish (sometimes both inside and outside), or painted with a heat-resistant finish. The exhaust manifolds collect exhaust gas from multiple engine cylinders and deliver it to the catalytic converter (for petrol engines), the DPF (for diesel engines), the silencer/muffler and the tailpipe.

Because the exhaust manifold is the first component through which the high temperature exhaust passes, it works under the harsh condition of an alternating state between high temperature, normal temperature and very low temperature in winter conditions when the engine starts. In a very short period of time, the exhaust manifold heats up and expands/deforms, and the entire exhaust system, including its fasteners/supports, must allow a normal operation. Problems, such as thermal fatigue cracking, fracture and leakage may occur due to an extreme stress alternation caused by the heating and cooling process.

Regardless of the manufacturer, the exhaust system is complex, it contains filters, sensors, a fastening system on the vehicle, support elements, etc. The failure of this exhaust system causes noise, vibrations, pollution, increased fuel consumption and poor engine performance.

Depending on the working temperature, the exhaust system can be divided into low temperature, medium temperature and high temperature components.

Components, such as the middle muffler, connecting pipe and main muffler, mainly require materials resistant to medium corrosion due to water condensation, whereas the exhaust manifold, extension pipe and catalytic converter are subject to high temperatures and the phenomena of oxidation and fatigue occur, therefore, high corrosion steel materials are required. In the automotive industry, stainless steel can be a good material to meet all these conditions.

Some examples of materials used for these components include the following: 321, SUH 409L, SUS 436LM, SUS 439L, SUS 436L, SUS 436J1L, SUS 429LM, SUS 441L and SUS 430J1L. These materials provide sufficient coefficient of thermal expansion, service temperature, strength, heat and corrosion resistance.

Alloy 321 is a titanium-stabilized chromium nickel austenitic stainless steel with corrosion resistance similar to 304/304L, and 321 is the AISI designation for this material, S32100 is the UNS number and the British Standard designation is 321S12.

Alloy 321 has an excellent resistance to intergranular corrosion after exposure to temperatures in the chromium carbide precipitation range of 800–1500°F (427–816°C). The alloy also has a higher creep and stress rupture properties than alloys 304 and 304L. It also possesses good low temperature toughness. Alloy 321H (UNS S32109) is the higher carbon (0.04–0.10) version of alloy 321. It was developed for enhanced creep resistance and for higher strength at temperatures above 1000°F (537°C). Alloy 321 cannot be hardened by heat treatment, only by cold working. It can be easily welded and processed by standard shop fabrication practices.

Main applications of this 321 steel alloy may be aerospace and automotive (piston engine exhaust manifolds), chemical processing, expansion joints, food processing (equipment and storage), petroleum refining (polythionic acid service), waste treatment (thermal oxidizers), etc.

The operation mode of an exhaust manifold is particularly complex, because it includes temperature, pressure, and vibration requirements.

The application does not solve/propose this complex study, but a simplified one in a static analysis, for the user to understand how temperature and pressure affect such a part of medium complexity. Thus, the approximate conditions are mapped into a finite element model that is used to calculate the stress and deformation of the part structure under these thermal and pressure loads.

The part analysed in the application is presented in Figure 4.352. The user notices that it has a two-hole clamping flange at the top end. The flange ensures the positioning and assembly of the part with the engine block. In the case of modern engines, the part is usually located between the engine cylinder head and the catalytic converter.

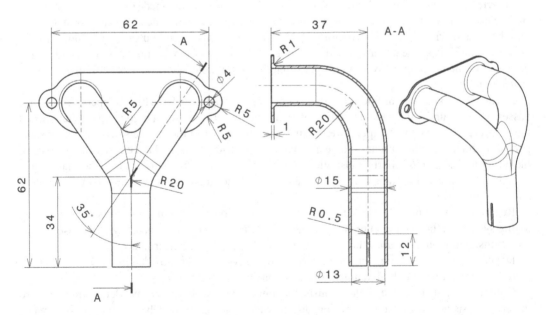

FIGURE 4.352 Two-dimensional drawing of the exhaust manifold part.

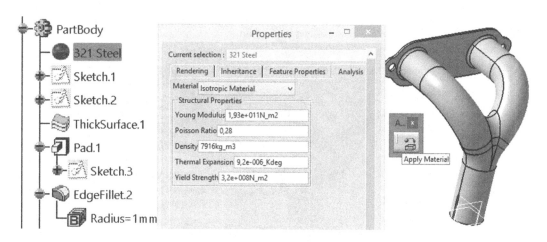

FIGURE 4.353 Properties of the material applied to the analysed part.

Modelling the part is not very complicated, and a helpful video tutorial to follow is at https://youtu.be/TBS5TmRT4hk. There is also the possibility of downloading its 3D model. The user notices the relatively small dimensions of the studied part; it belongs to a radio-controlled car model, propelled by a thermal engine. The part has a wall thickness of 1 mm, which ensures a low weight, but also gives it a quite low rigidity.

Once the surface and solid modelling is finished, the user determines the material applied to the part, which is 321 steel having the properties in Figure 4.353: *Young Modulus* = 1.93×10^{11} N/m^2, *Poisson Ratio* = 0.28, *Density* = 7916 kg/m^3, *Coefficient of linear Thermal Expansion* = 9.2×10^{-6}/°K and *Yield Strength* = 3.2×10^8 N/m^2. These properties are edited by the user in the *Analysis* tab, and in the *Feature Properties* tab the name of the steel is changed. The figure also shows a fragment of the specification tree with features of the solid part.

In the *Rendering* tab the user can determine how the part's material is displayed during the analysis with finite elements, but also when a rendering of this part occurs. It is recommended to save the part at this point, before starting the FEM analysis.

In the *Generative Structural Analysis* workbench the user defines the mesh discretization. Thus, by double-clicking on the *OCTREE Tetrahedron Mesh.1* feature, the dialog box from Figure 4.354 opens. The user chooses small values for the parameters in the fields *Size* = 2 mm

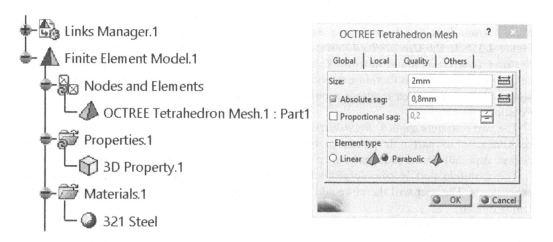

FIGURE 4.354 Mesh definition.

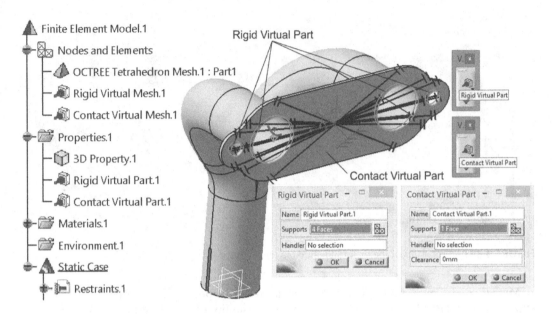

FIGURE 4.355 Creating the virtual parts.

and *Absolute Sag* = 0.8 mm. Also, for a more precise analysis, the finite element type is chosen to be *Parabolic*. The specification tree contains the *3D Property.1* feature and the material applied to the part.

Because the part changes dimensions by heating while it is fixed on the engine block, the user must carefully set the restraints. Thus, from the *Virtual Parts* toolbar, a *Rigid Virtual Part* is added for the cylindrical surfaces of the four holes represented on the clamping flange (Figure 4.355) and a *Contact Virtual Part* for the flat surface of the flange that is in contact with the engine block.

Rigid Virtual Part was chosen because the part is not allowed a large deformation in the area of the holes. Two of these correspond to the exhaust holes of the engine, the other two (smaller) have the role of fixing the flange on the engine block with the help of two screws. Coaxiality with the corresponding engine holes must be maintained for both pairs of holes. *Contact Virtual Part* creates a contact between the flat surfaces of the part and the engine.

The user applies two restraints to the virtual parts, as follows: *Clamp.1* is applied to the *Rigid Virtual Part* and *User-defined Restraint.1* to the *Contact Virtual Part,* according to Figure 4.356. In the *User-defined Restraint* selection box the user has not selected the options *Restrain Translation 2* and *Restrain Translation 3,* they correspond to the *Y* and *Z* axes. The deformation of the part is therefore restricted to the *X* axis (which is perpendicular to the flange), also they are not allowed rotations around the three axes *X, Y* or *Z* by checking the *Restrain Rotation* options.

The two restraints applied, *Rigid Virtual Part* and *User-defined Restraint.1,* do not affect the deformation of the part in the pipes area, but only of the clamping flange. The whole part can thus deform, especially in the end area of the pipe, as it will result from the FEM analysis.

The manifold part is considered to be stressed by the temperature and pressure of the exhaust gas. The user will therefore add two loads, *Temperature Field* and *Pressure* from the *Loads* toolbar.

Regarding the temperature, by default, *CATIA v5* considers the reference temperature of the environment to be 20°C. This means that the part to which this temperature is applied will not show deformations or stresses. The reference can be modified by the user by double-clicking on

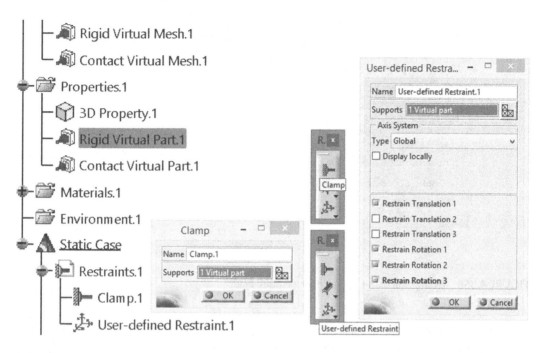

FIGURE 4.356 Applying the restraints to the virtual parts.

the *Environment.1* feature from the specification tree. The option is useful, especially in complex analyses where the studied part has a certain initial temperature.

In the *Temperature Field* dialog box (Figure 4.357) the user selects the *OCTREE Tetrahedron Mesh.1* feature from the specification tree in the *Supports* field, then enters the value of the applied temperature: 80°C. It can be seen that the symbol T appears on part of the solid, the whole solid is affected by the same temperature.

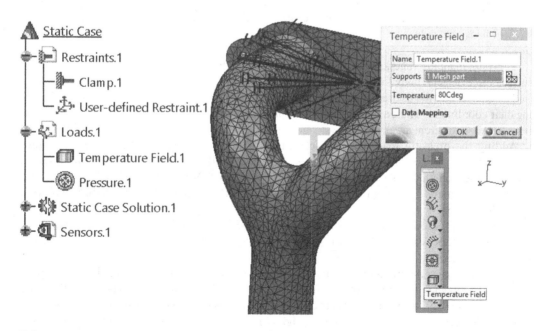

FIGURE 4.357 Applying the *Temperature Field* of 80°C.

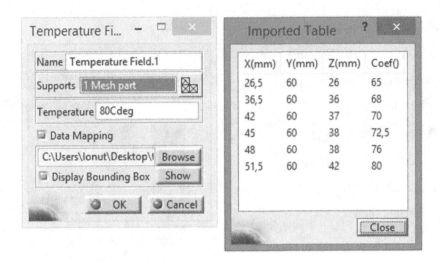

FIGURE 4.358   Example of data mapping a temperature field.

Although not used in this application, a more complex variant of setting the temperature is possible by checking and using the *Data Mapping* option. This is a special functionality that allows the user to load and import a scalar field from a text (*.txt*) or an *Excel* (*.xls*) file. This file must respect a pre-defined format as presented in Figure 4.358:

- It has four columns.
- Each cell contain a single numerical value.
- The first three columns allow the user to specify *X*, *Y* and *Z* point coordinates in the global axis.
- Unit symbol (e.g. mm) written between round brackets must be specified.
- The last/fourth column contains the amplification coefficient. No unit symbol must be specified for the last column because the amplification coefficient is not assigned to a dimensional value.

This data mapping functionality with four columns can be used for pressure, line force density, surface force density, body load, temperature and shell property.

This means that the user can add different temperatures in certain points of the mesh, by knowing their coordinates. Thus, a real and more complex temperature field could be created, with variations between different areas of a part.

Adding the pressure load involves selecting all the interior surfaces of the part. Figure 4.359 shows the *Pressure* dialog box, in the *Supports* field the user has selected the internal surfaces, then sets a pressure of 3 bar = 2.96 atm = 3×10^5 N/m^2 = 0.3 MPa. Using the *Data Mapping* option and knowing the points on the pipe axes, the user can establish different pressures along them.

Figure 4.360 shows a detail with pressure added to the interior of the part. The user notices the symbols of this load, the support surface, but also the two loads added in the specification tree.

Thus, the exhaust manifold part is subjected to a complex set of conditions and the user decides to check the model with the help of the *Model Checker* tool. In the information box with the same name in Figure 4.361, the status *OK* is observed, which means that the FEM analysis can be started. However, before clicking the *Compute* icon, the user saves the analysis steps completed up to this point using the *Save Management* option in the *File* menu.

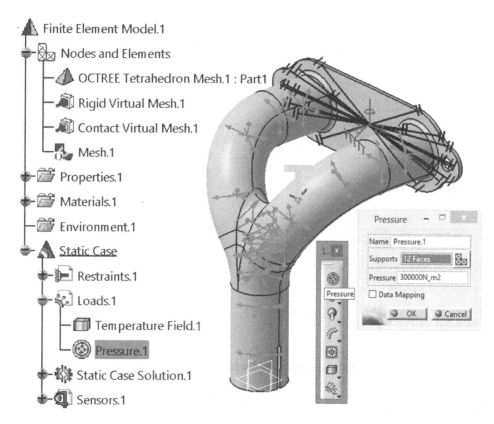

FIGURE 4.359 Applying the *Pressure* load.

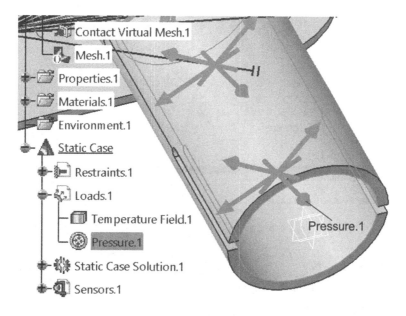

FIGURE 4.360 Detail of the *Pressure* load applied inside the pipe.

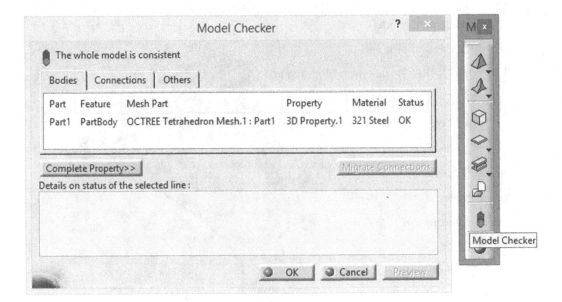

FIGURE 4.361 Checking the model.

The analysis results are added to the specification tree under *Static Case Solution.1* (Figure 4.362). The user notices that the maximum deformation of the part occurs in the end zone of the pipe, on one of the edges of the 12 × 1 mm cut out. The value of this deformation is 0.058 mm, very small considering the role of this part.

Figure 4.362 shows the exaggeratedly deformed part and the distribution of colours (on screen) according to the values of the displacements of the network nodes.

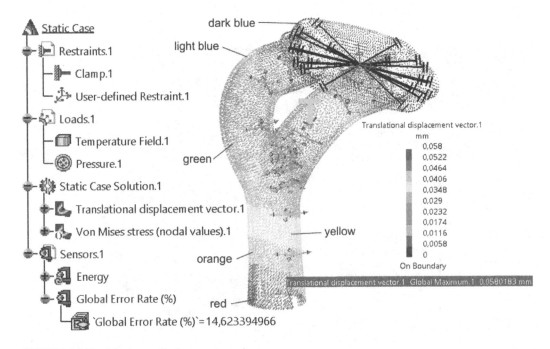

FIGURE 4.362 Maximum displacement result.

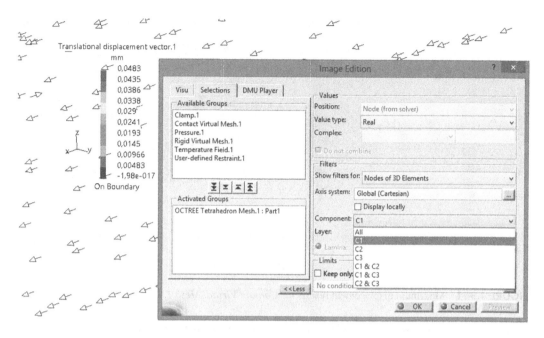

FIGURE 4.363 Maximum displacement results on the *X* axis.

The user observes a convenient value of the *Global Error Rate (%)* = 14.62%. For a mesh refinement that after restarting the FEM analysis provides a lower error rate, the user can use smaller values in the *Size* and *Absolute sag* fields in the dialog box in Figure 4.354. The *New Adaptivity Entity* tool can also be used to demand the FEM analysis to refine the model for a certain error rate.

In the specification tree, by double-clicking on the feature *Translational displacement vector.1,* the user opens the *Image Edition* selection box in Figure 4.363. In the *Visu* tab, by unchecking the *Deform according to Displacements* option, the part is no longer represented deformed. The *Animate* tool no longer works either. By pressing the *More* ≫ button, the box expands to the right and is completed with the options in Figure 4.363.

The user chooses the coordinate system in which the deformations of the part will be represented, then from the field/drop-down list *Component* he selects the axes along which to make this representation. Thus, *C1, C2* and *C3* replace the *X, Y* and *Z* axes (if the default axis system is used).

In Figure 4.363 the choice of component *C1* (so the displacements of the network nodes along the *X* axis will be computed and displayed), the coordinate system and the direction of the arrows can be observed. The value of the maximum displacement on the *X* axis is 0.048 mm. The ability to choose an axis/direction to compute mesh node displacements is particularly useful when the part has certain restrictions (fasteners or assembly elements) on that direction.

In the *Selections* tab, the user makes another selection in the *Activated Groups* field: *Contact Virtual Mesh.1* instead of *OCTREE Tetrahedron Mesh.1.* For the setting in Figures 4.362 and 4.363, the displacements of the exhaust manifold part are displayed, and for the setting in Figure 4.364 the displacements of the virtual part (which can be the engine block, but exactly in the contact area with the studied part) are also displayed.

The displayed values would be correct if the user could apply temperature to this virtual part *(Contact Virtual Mesh.1),* but this is not possible. The *Supports* field in the *Temperature Field* dialog box (Figure 4.357) allows the selection of only *Mesh* entities defined by a network of nodes and elements, not virtual ones. The restriction is logical and obvious because the virtual parts have no geometry and no applied material.

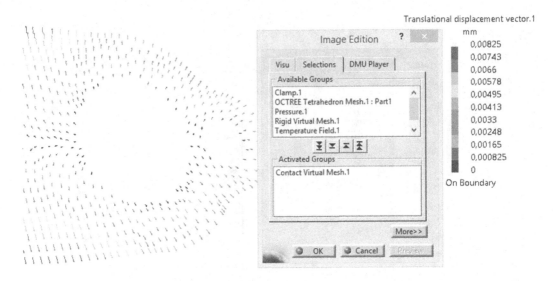

FIGURE 4.364 Maximum displacement results for *Contact Virtual Mesh.1.*

In Figure 4.364 it can be seen that on this virtual part the low and very low values (marked by dark blue and light blue arrows on screen) are placed around the exhaust and assembly holes, which is correct.

Regarding the stresses that temperature and pressure cause in the part, they are located, as expected, in the area of the gas exhaust holes and assembly holes, according to Figure 4.365. The maximum stress value is 2.78×10^8 N/m^2, lower than the *Yield Strength* of the chosen material

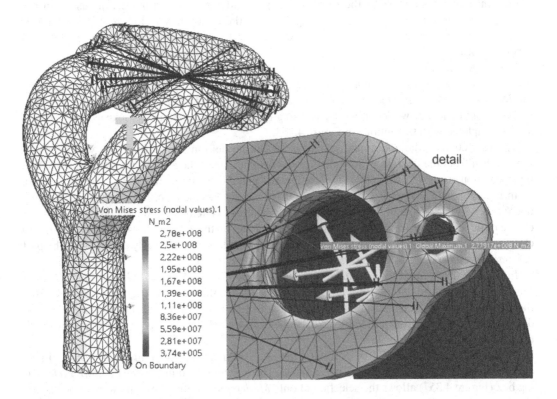

FIGURE 4.365 Maximum *Von Mises stress* result.

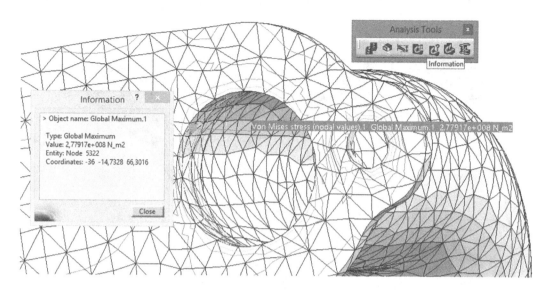

FIGURE 4.366 Locating the node with maximum stress.

$(3.2 \times 10^8$ N/m^2 for 321 steel). Even for the parameter *Global Error Rate (%)* = 14.62%, the value is satisfactory.

On the outside of the pipe, in the contact area with the flange, the manufacturer provided a fillet of R1 mm, according to the 2D drawing in Figure 4.352. That fillet can mean a welding seam, but it also has the role of stiffening the contact area between the two pipes and the flange.

If the user wants to precisely locate the point/node where the maximum stress is reached, the program provides the number of this node but, more importantly, its coordinates in the part's coordinate system. Pressing the *Information* icon on the *Analysis Tools* toolbar and selecting the point marked in Figure 4.366 opens the *Information* box. Finding the precise location proves to be useful, for example, in the case of automatic placement of a temperature, pressure or deformation sensor.

Regardless of the resulting image displayed (Figure 4.362 or Figure 4.365), the deformations of the part, under the influence of the temperature and pressure loads, are exaggerated. By pressing the *Animate* icon, the user observes the part mesh in animated mode (Figure 4.367). Its geometry changes depending on the animation settings (amplitude, number of frames, speed) and gives the user a clear picture of the part behaviour under the considered conditions.

In case the user needs to resume the FEM analysis of the part with some changes (e.g. other restrictions and/or other loads), but without losing the previously obtained results, *CATIA v5* allows inserting a new static analysis. From the *Insert* menu, the user accesses the *Static Case* option and opens the selection box with the same name (Figure 4.368).

For example, to keep as a reference the restraints established for the previous analysis, concluded with the results shown earlier, the user checks the first option *Reference,* then selects *Restraints.1* in the specification tree (Figure 4.368). These restraints are added in the new static analysis exactly as they were set in the previous static analysis, such as positioning, values, names, etc. The user also has the possibility to edit them and/or add other restraints, but any change propagates and becomes valid for both *Static Case* analyses.

In the *Static Case* selection box in Figure 4.368, the option *Hide existing analysis cases* was checked so that the program only displays the new results; the old ones are hidden, but remain within the FEM analysis for comparison. Of course, any *New* option causes the user to add restraints and/ or loads unrelated to the previous ones.

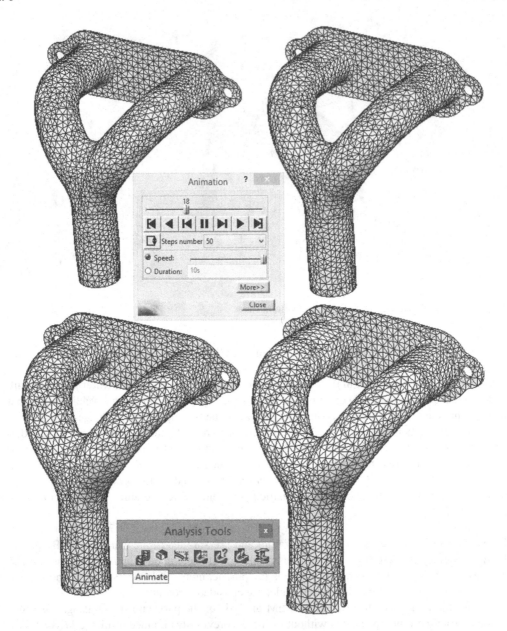

FIGURE 4.367   Four sequences of the part mesh during the animation.

Thus, the new static analysis will contain other load values, so *Loads* is checked together with its *New* option.

In Figure 4.369, the specification tree contains two *Static Case* features. The previously presented analysis in the application is hidden (name: *Static Case*), the recently added one is visible (name: *Static Case2*). Its name has been changed by the user (*Properties* option from the context menu).

First, it is observed that feature *Restraints.1* is identical to the one defined and used in the *Static Case* analysis, so the old restraints are kept. Then, the user adds a new *Temperature Field.2* = 70° C and a new *Pressure.2* = 2.5 bar = 2.5×10^5 N/m^2 = 0.25 MPa, having the names and values shown in Figure 4.369. The surfaces on which these loads are applied are the same as in the case of the

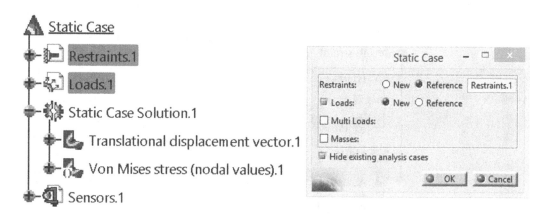

FIGURE 4.368 Inserting a new static analysis.

previous analysis. Any other loads added in the *Loads.2* feature remain valid only for the current FEM analysis.

Also, the geometry of the part remains the same, as does the discretisation of the model. Before launching the FEM analysis computations the user must specify what results are sought. The *Image* toolbar icons are inactive, however. Thus, according to Figure 4.370, the user updates *Sensors.2* in the specification tree using the *Update All Sensors* option from the context menu. This update is fast and the *Global Error Rate (%)* value is similar to that of the first FEM analysis at 14.74%.

Figure 4.371 shows the results of the second FEM analysis. The location of the maximum stress and displacement obtained by the current analysis are similar to those determined in the first analysis. The value of the maximum displacement of a network node is 0.0478 mm, and

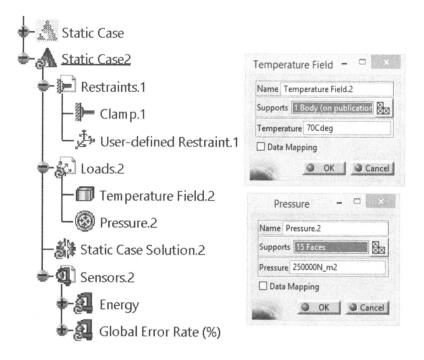

FIGURE 4.369 Adding the two new loads: temperature and pressure.

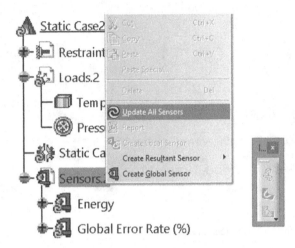

FIGURE 4.370 Updating the sensors.

the maximum stress is 2.32×10^8 N/m². Of course, these values are lower than the previous ones (Figures 4.362 and 4.365) because the values of the loads are also lower, thus, the part is less stressed.

The user should be able to continue the application by imposing other values on the two loads, by changing the thickness of the pipes wall, by removing the cut out from the free end of the part (12 × 1 mm), etc. Interesting results are obtained by changing the initial temperature of the part (double-click on the feature *Environment.1* in the specification tree).

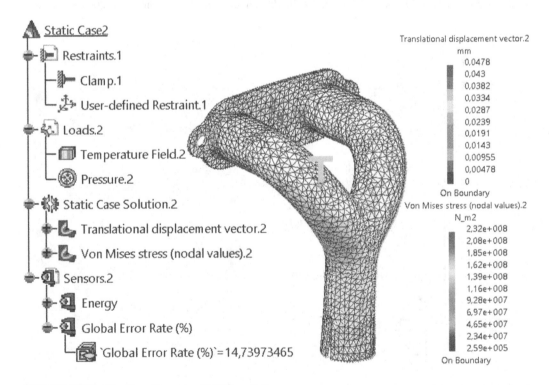

FIGURE 4.371 Results of the second FEM analysis.

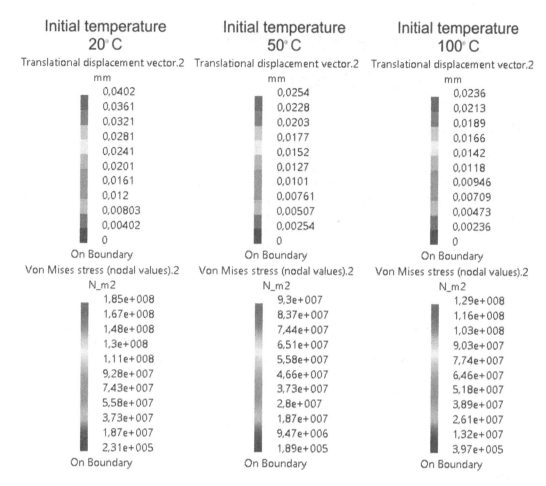

Initial temperature 20° C	Initial temperature 50° C	Initial temperature 100° C
Translational displacement vector.2 mm	Translational displacement vector.2 mm	Translational displacement vector.2 mm
0,0402	0,0254	0,0236
0,0361	0,0228	0,0213
0,0321	0,0203	0,0189
0,0281	0,0177	0,0166
0,0241	0,0152	0,0142
0,0201	0,0127	0,0118
0,0161	0,0101	0,00946
0,012	0,00761	0,00709
0,00803	0,00507	0,00473
0,00402	0,00254	0,00236
0	0	0
On Boundary	On Boundary	On Boundary
Von Mises stress (nodal values).2 N_m2	Von Mises stress (nodal values).2 N_m2	Von Mises stress (nodal values).2 N_m2
1,85e+008	9,3e+007	1,29e+008
1,67e+008	8,37e+007	1,16e+008
1,48e+008	7,44e+007	1,03e+008
1,3e+008	6,51e+007	9,03e+007
1,11e+008	5,58e+007	7,74e+007
9,28e+007	4,66e+007	6,46e+007
7,43e+007	3,73e+007	5,18e+007
5,58e+007	2,8e+007	3,89e+007
3,73e+007	1,87e+007	2,61e+007
1,87e+007	9,47e+006	1,32e+007
2,31e+005	1,89e+005	3,97e+005
On Boundary	On Boundary	On Boundary

FIGURE 4.372   Evolution of the second FEM analysis results for different initial temperatures.

Figure 4.372 contains three sets of results obtained for the temperature and pressure loads established in the second analysis (Figure 4.369) when the exhaust manifold part has the initial temperature of 20°C, 50°C, and 100°C, respectively.

The user notices that the displacements and stresses decrease with the decrease of the temperature difference between the initial one and the one imposed by the *Temperature Field.2* loading of 70°C.

5 Knowledge Assessment Applications

INTRODUCTION

This chapter contains some proposed applications. Their purpose is to test the user's knowledge gained after reading this book, but also to raise some interesting ideas about how to use the *CATIA v5* program in finite element analysis. The proposed applications contain two-dimensional (2D) drawings for the solid modelling of the analysed parts, as well as solution tips. They contain, in short, the discretization modes of each part, the applied loads and restrictions are presented, as well as some obtained results. The degree of difficulty of these applications is medium to high.

5.1   ANALYSIS OF A TWISTED PART

The part considered in the application is represented in Figure 5.1. Its modelling begins in the *Generative Shape Design* workbench by using *Multi-Sections Surface*, *Fill Surface* and *Join* tools. The result is a single surface. Then, the user applies the *Close Surface* tool from the *Part Design* workbench. At both ends of the twisted part are two parallelepiped bodies (using the *Pad* tool). A simple modelling solution of the part can be followed in this video tutorial:

https://youtu.be/NaIFWExslhU

The part can also be downloaded and used for finite element method (FEM) analysis. The volume of the solid part is 855588 mm^3.

The material added to the part is considered to be *Steel,* used several times in the applications of Chapter 4. Its characteristics are thus known by the user.

In the *Generative Structural Analysis* workbench, the user applies a *Clamp.1* restraint on the surface at the base of the part and on the side flat surfaces, according to Figure 5.2.

The solid model of the part is discretized using parabolic finite elements, having the maximum size *Size* = 2 mm and *Absolute sag* = 0.5 mm (Figure 5.3). These values ensure a good discretization of the part. The *Parabolic* type was chosen due to its twisted shape.

Back in the *Part Design* workbench, the user defines a new axis system (*Insert → Axis System* menu or the *Axis System* icon on the *Tools* toolbar; Figure 5.4). The user chooses the origin in a corner of the parallelepiped at the top of the part, then the X and Y axes along its edges. Depending on the choice of X and Y, results the orientation of the Z axis. The system of axes is important in the next step, in the *Generative Structural Analysis* workbench, for the correct setting of the load with a distributed force.

On one of the side faces located in the ZX plane, a *Distributed Force* load of 220 N is applied to the part in the Y field (Figure 5.5). The coordinate system is selected as the *User* type, *Local orientation =* *Cartesian,* and the axis system is chosen as the one previously defined, *Axis System.1.*

After completing the FEM analysis, the user notices that the maximum stress is located at the base of the part (Figure 5.6) and has a value of 4.45×10^8 N/m^2. The maximum deformation is placed in a node in the upper area of the part, it has a displacement of 1.87 mm. Sensor *Global Error Rate (%)* = 5.55%, so it has a very convenient value, it results that the FEM analysis is accurate.

The maximum stress exceeds the *Yield Strength* of the selected steel (2.5×10^8 N/m^2), so the part will deform plastically and cracks may appear in the area at its base. Therefore, a new force value for which the deformation of the part remains in the elastic domain must be identified.

 DOI: 10.1201/9781003426813-5

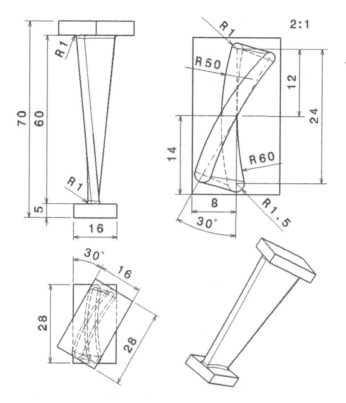

FIGURE 5.1   Two-dimensional drawing of the twisted part.

Solving this problem can be done through successive attempts/iterations, to reduce the *Distributed Force* value, but the user can also apply the tools of the *Product Engineering Optimizer* workbench.

Maximum Von Mises is added to the list of sensors because it will be used as a constraint. Thus, in the *Optimization* dialog box (Figure 5.7), in the *Problem* tab in the *Free Parameters* list, the user chooses the force on the *Y* axis (Figure 5.5), having the current value of 220 N.

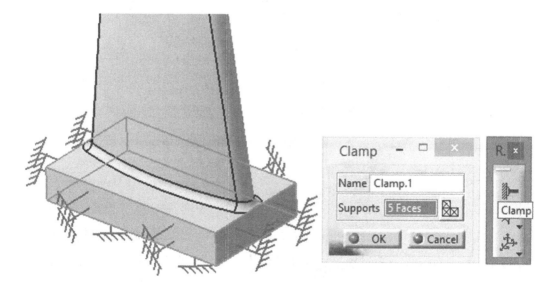

FIGURE 5.2   Applying the *Clamp.1* restraint to the surfaces at the base of the part.

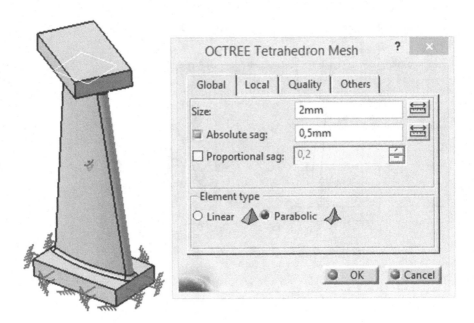

FIGURE 5.3 Discretization of the 3D part model.

The parameter to be optimized is the maximum stress resulting from the FEM analysis. Its value is kept in the sensor *Finite Element Model.1\Maximum Von Mises.3\Maximum Von Mises*. By pressing the *Select* button in the *Optimization* dialog box, a new one opens and the sensor is chosen from the list of available sensors (Figure 5.8) to be added to the *Optimized parameter* field along with its value of 4.446e + 008N_m2 (4.446 × 10^8 N/m^2).

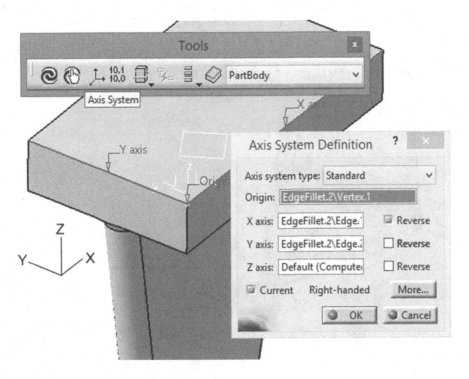

FIGURE 5.4 Inserting a new axis system.

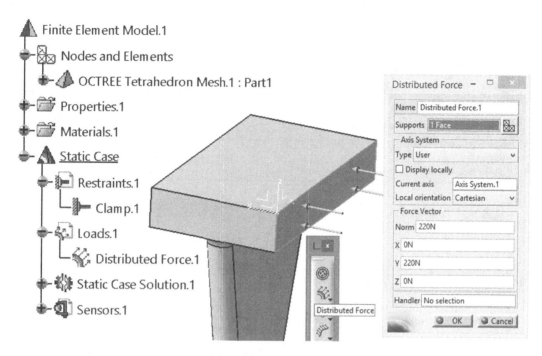

FIGURE 5.5 Applying the *Distributed Force* load.

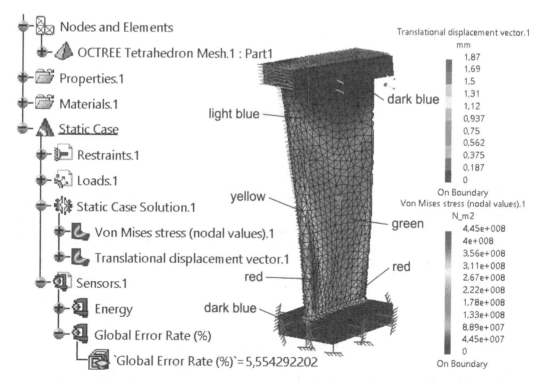

FIGURE 5.6 The results of the FEM analysis.

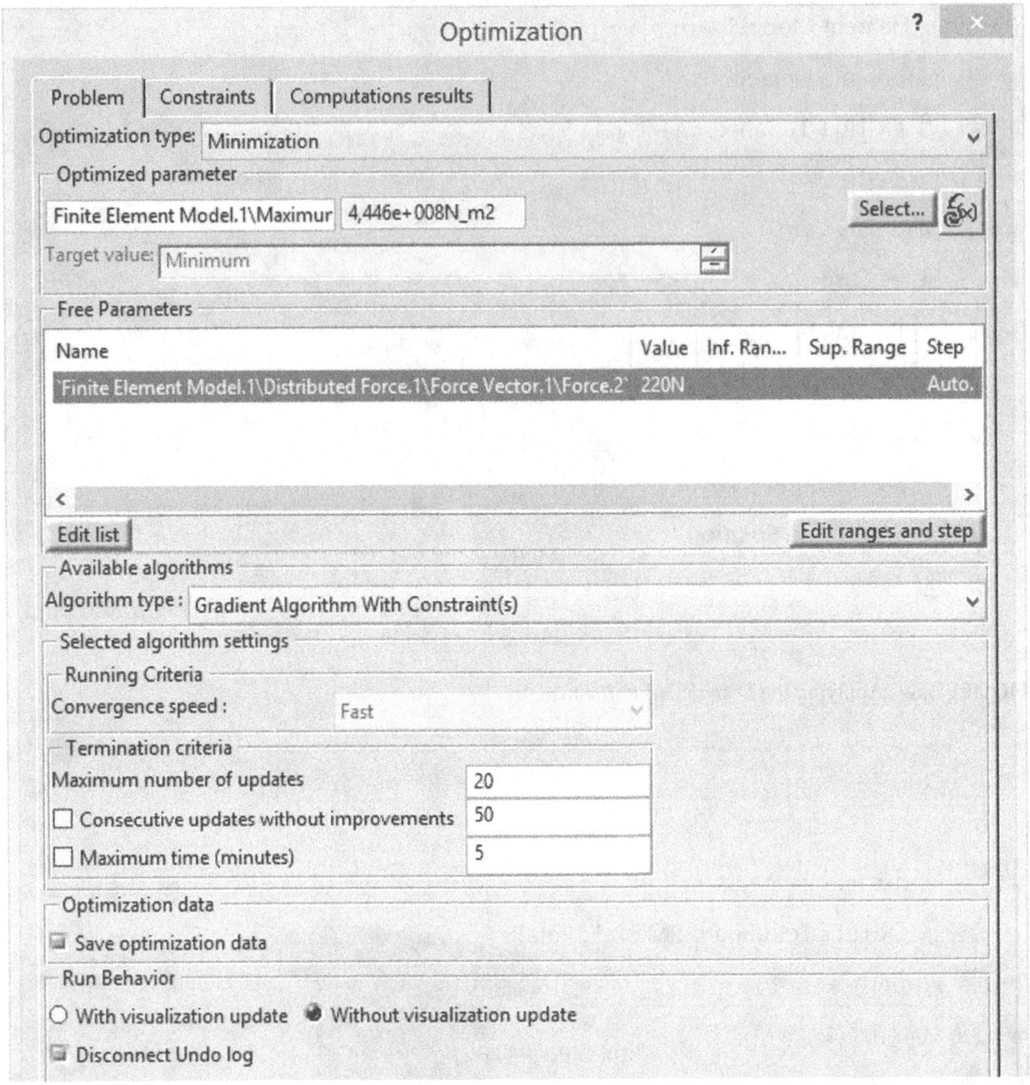

FIGURE 5.7   Optimizing the value of the loading force.

The user chooses *Gradient Algorithm With Constraint(s)* because two constraints will be defined in the *Constraints* tab (Figure 5.9), and the optimization calculation must take them into account.

Obviously, one constraint is related to the ratio between the *Yield Strength* of the considered steel material and the value of the maximum stress *(Safety factor)*. The second constraint maintains the value of the distributed force above 100 N. Thus, in the *Constraints* tab, by pressing the *New* button, both constraints are added:

`PartBody\Steel\Steel.1\Yield Strength`/`Finite Element Model.1\Maximum Von Mises.3\Maximum Von Mises`>1.2

`Finite Element Model.1\Distributed Force.1\Force Vector.1\Force.2`>100N

Identifying and retrieving the parameters to be used in the relations that define the constraints are relatively simple, from the specification tree of the FEM analysis.

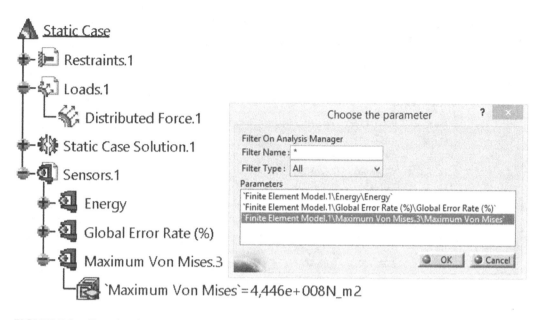

FIGURE 5.8   Choosing the parameter whose value needs to be optimized.

Each constraint automatically receives a name, the user may change it, they can be active or not, and their simultaneous fulfilment provides the desired results, in the *Computations results* tab.

Following the optimization, some useful results were obtained, their list being presented in Figure 5.10. Respecting the imposed condition of minimizing the maximum stress and the two constraints, *CATIA v5* identifies a value of 100 N (with small variations) for which the maximum stress is 2.02×10^8 N/m^2 and *Safety factor* > 1.2.

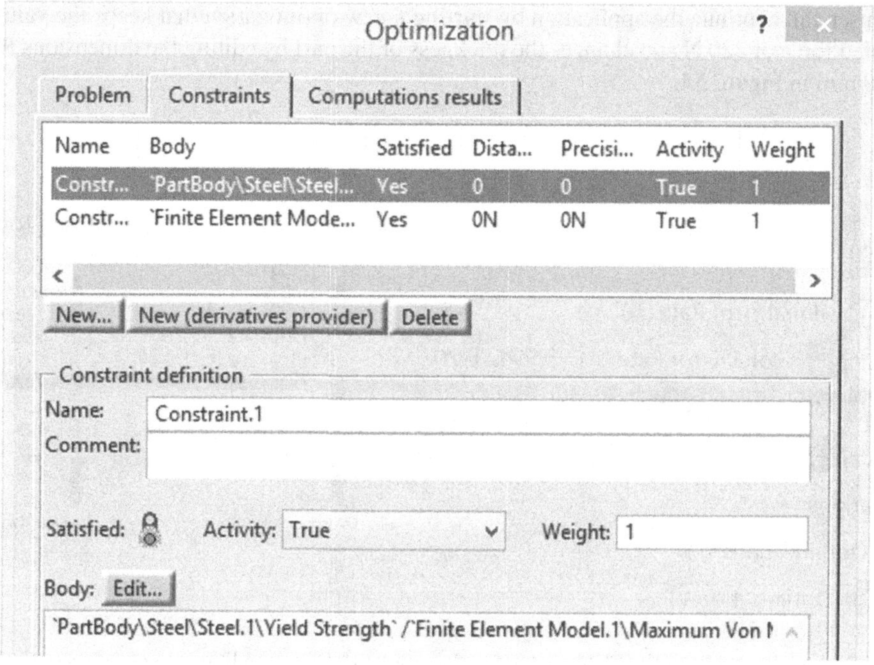

FIGURE 5.9   Setting the constraints.

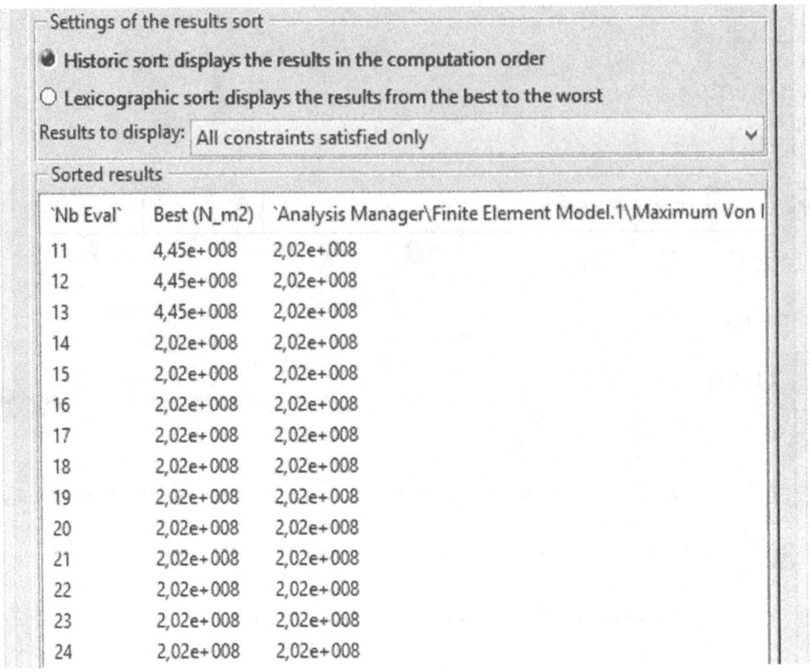

Settings of the results sort

● Historic sort: displays the results in the computation order

○ Lexicographic sort: displays the results from the best to the worst

Results to display: All constraints satisfied only   ∨

Sorted results

`Nb Eval`	Best (N_m2)	`Analysis Manager\Finite Element Model.1\Maximum Von I
11	4,45e+008	2,02e+008
12	4,45e+008	2,02e+008
13	4,45e+008	2,02e+008
14	2,02e+008	2,02e+008
15	2,02e+008	2,02e+008
16	2,02e+008	2,02e+008
17	2,02e+008	2,02e+008
18	2,02e+008	2,02e+008
19	2,02e+008	2,02e+008
20	2,02e+008	2,02e+008
21	2,02e+008	2,02e+008
22	2,02e+008	2,02e+008
23	2,02e+008	2,02e+008
24	2,02e+008	2,02e+008

FIGURE 5.10   The results obtained by minimizing the maximum stress.

In Figure 5.11 the user can see the results of the FEM analysis after completing the optimization stage, the specification tree contains the sensors created together with their values, feature *Optimization.1* and *Global Error Rate (%) = 5.55%*.

The user can continue the application by starting a new optimization that keeps the value of the distributed force of 220 N and changes the thickness of the part by editing the dimensions R50 mm and R60 mm in Figure 5.1.

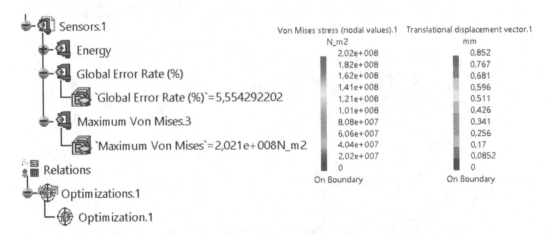

FIGURE 5.11   The results of the FEM analysis in the case of the new conditions imposed by optimization.

5.2   FREQUENCY ANALYSIS OF A TUNING FORK

A tuning fork is a fork-shaped acoustic resonator formed from a U-shaped bar of steel used in many applications to produce a fixed tone. The fork-shaped resonator produces a very pure tone, with most of the vibrational energy at the fundamental frequency.

When the tuning fork is struck, little of the energy goes into the overtone modes; they also die out correspondingly faster, leaving a pure sine wave at the fundamental frequency. It is easier to tune other instruments with this pure tone.

Another reason for using the fork U-shape is that it can then be held at the base without damping the oscillation. That is because its principal mode of vibration is symmetric, with the two prongs always moving in opposite directions, so that at the base where the two prongs meet there is a point of no vibratory motion which can therefore be handled without removing energy from the oscillation.

The tuning fork was invented in 1711 by British musician John Shore, sergeant trumpeter and lutenist to the royal court.

The application presents the two-dimensional (2D) drawing of the analysed part (Figure 5.12) and the analysis with finite elements to determine its natural frequencies. It can be seen that the part has a spherical surface, the *Clamp* restraint will be applied to it. Modelling the part is very simple, only the *Part Design* workbench with the *Rib* (for the body of the part) and *Shaft* (for the end sphere) tools are used.

The part can also be downloaded and used for FEM analysis. Its solid volume is 31000.068 mm³.

The *Steel* material from the program's standard library is applied to the part, then in the *Generative Structural Analysis* workbench the user creates a *Clamp.1* restraint on the spherical surface R7.5, according to Figure 5.13. The figure also shows the discretization of the part model: *Size* = 2 mm, *Absolute sag* = 1 mm, *Element type* = *Parabolic*. These values ensure a good discretization of the part. The part is represented as a mesh (feature *Mesh.1*) using the *Mesh Visualization* option from the context menu of *Nodes and Elements*.

By default, when accessing the *Generative Structural Analysis* workbench, the program has inserted a static analysis, but this is removed or disabled by the user because *Frequency Case* must be added from the *Insert* menu. All the icons on the *Loads* toolbar are disabled because loads (forces, moments, pressure, etc.) are not required in any modal analysis.

The user adds a result image *Translational displacement vector.1* to the specification tree to observe how the part deforms. After running the FEM analysis, the user notices that the feature *Sensors.1* in the specification tree contains *Frequency* along with the list of natural frequencies of the part (Figure 5.14).

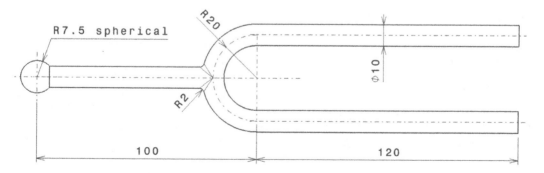

FIGURE 5.12   Two-dimensional drawing of the tuning fork.

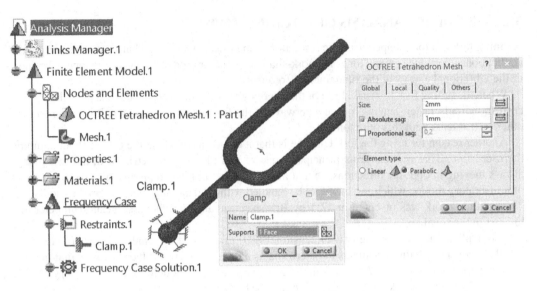

FIGURE 5.13   Discretization of the model and application of the *Clamp.1* restraint.

These frequencies that influence the part have values between 110.451 Hz and 2428.843 Hz. The deformations, presented exaggeratedly, are different and can be studied by the user by using the *Animate* tool (Figure 5.15).

Figure 5.14 contains the list of the first 10 frequencies. Extending this list by adding a larger number of frequencies to be identified by the program is possible by double-clicking on the feature *Frequency Case Solution.1* in the specification tree.

Figure 5.16 shows some exaggerated representations of the tuning fork for different values of the natural frequencies.

The user can continue the application with a study about the influence of the part dimensions on the value of the frequencies resulting from the FEM analysis. Another interesting study can highlight how two masses of 50 g placed on the free ends of the fork influence the deformations of the part and the computed frequencies.

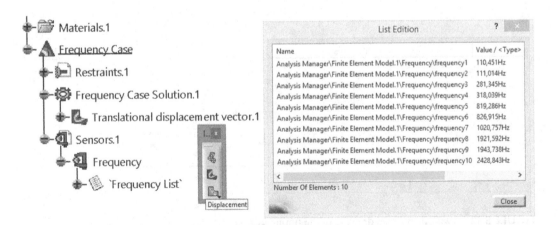

FIGURE 5.14   The list of natural frequencies of the tuning fork.

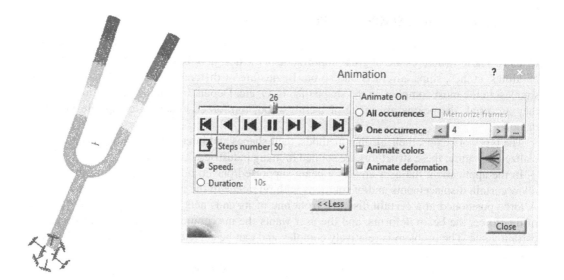

FIGURE 5.15 Animation of the part's deformations.

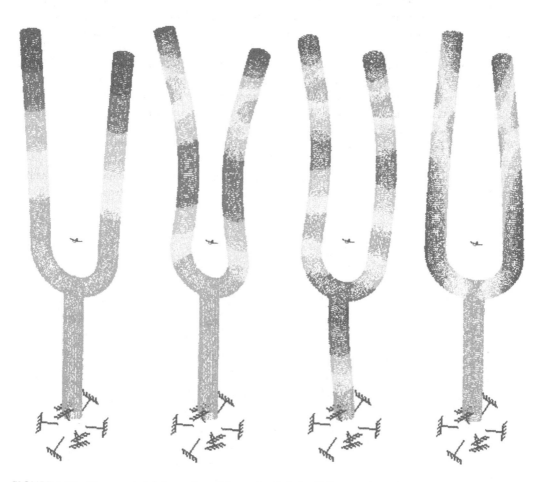

FIGURE 5.16 Exaggerated deformation of the tuning fork for different values of natural frequencies.

5.3 ANALYSIS OF A SIMPLY SUPPORTED BEAM

Beams are very important in practice because they are structures that usually carry various types of loads. These structures are of great importance in the design of bridges, beds of machine tools, buildings, cranes, ships, aircrafts and so on. Beams are of different sizes and shapes, from the simplest to the most complex, in the shape of I, U, O, etc. Depending on the destination in which they are used, the beams are made of metal, wood, reinforced concrete, composite materials or stone.

The types of beams, their loading methods, their shapes and intended uses can be found in specialized literature; these structures are studied in many industries.

In the application, the user creates a beam supported/placed on two supports. Both supports allow certain displacements and/or rotations with reference to the axes of the coordinate system. A force positioned at a certain distance from one of its ends acts on the beam. Under the load of this force, the beam deforms, and the user wants the maximum deformation not to exceed a certain value. The problem is relatively complex and can be solved by a FEM analysis with successive iterations.

The application starts in the *Generative Shape Design* workbench by drawing a 400 mm line between two points, the first is at the origin of the coordinate system, and the second along the *X* axis. The user applies the *Line* tool (Figure 5.17), chooses the *Point-Point* option from drop-down list *Line type,* in the *Point 1:* field right-clicks and opens a context menu with several options for creating points. From these options the user chooses *Create Point* and in the dialog box *Point Definition* enters 0 mm in the field of each axis. For the second point in the *Point 2:* field the point creation sequence is similar, but the coordinate along the *X* axis is 400 mm, according to Figure 5.17.

The coordinate system can also be observed in the figure, so the line is perpendicular to the *YZ* plane. The user adds two more points, as seen in Figure 5.18, as follows: *Point.3* (0, –30, 0) and *Point.4* (120, 0, 0). *Point.3* is, therefore, in the *YZ* plane, and *Point.4* belongs to *Line.1*.

Steel material is applied to the line, having the properties known from previous applications. In the *Generative Structural Analysis* workbench the user discretizes the line geometry using the *Beam Mesher* tool. In the *Beam Meshing* dialog box (Figure 5.19) the applied settings are presented (*Element size* = 2 mm, *Sag* = 0.5 mm and *Min size* = 1 mm). Thus, a good discretization of the beam is ensured, but it is not enough to be used in the FEM analysis.

So, the line is assigned an important property, which contains the type, shape and dimensions of the beam. From the *Model Manager* toolbar, the user clicks the *1D Property* icon and opens the

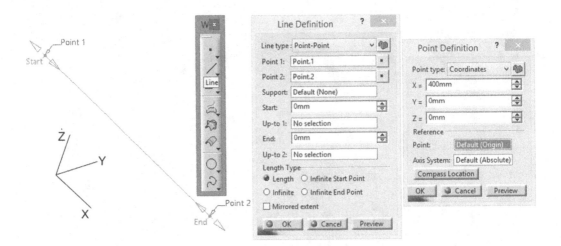

FIGURE 5.17 Drawing a line of length 400 mm along the *X* axis.

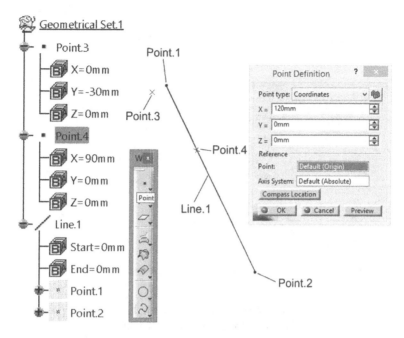

FIGURE 5.18   Insertion of *Point.3* and *Point.4*.

selection box with the same name in Figure 5.20. In the *Supports* field he selects the line, the material is added, then from the *Type* list the option *Thin box beam* is chosen. By pressing the adjacent button, *Component edition,* the user defines the inner and outer dimensions of the beam (*Beam Definition* dialog box). Its orientation, according to the detail, is possible by choosing *Point.3*.

Thus, the beam is completely defined: type, shape, dimensions, orientation in the coordinate system, discretization and material.

The user applies two complex restraints using the *User-defined Restraint* icon.

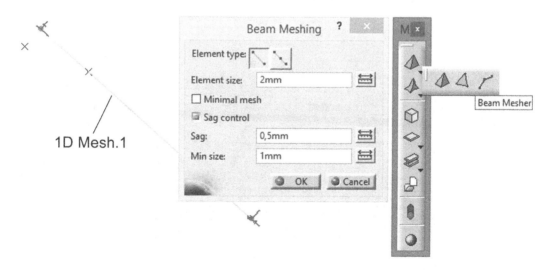

FIGURE 5.19   Beam discretization.

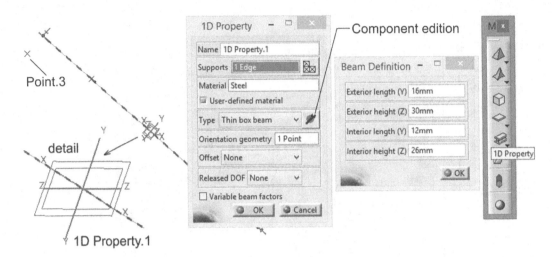

FIGURE 5.20 Defining the part as a thin box beam and its parameters.

Thus, at the right end of the beam (the end that is not in the *YZ* plane), a *User-defined Restraint.1* is inserted, whereby this point is only allowed to rotate around the *Y* axis. At the left end *User-defined Restraint.2* is added, that point has the possibility to move along the *Y* axis and rotate around the *X* and *Z* axes. Figure 5.21 shows the restrictions applied, and the user must pay attention to the axes of the coordinate system when checking the options in the two selection boxes.

To apply a distributed force load to the beam in *Point.4* the user must define a virtual part. *Smooth Virtual Part* is chosen and from the selection box with the same name (Figure 5.22). The user defines the line in the *Supports* field and the point in the *Handler* field.

Distributed Force.1 is applied along the *Z* axis, in the negative direction, and has a value of 280 N in the *Norm* field. In the *Z* field the value is –280 N. Figure 5.23 also contains all the features added in the specification tree, and the coordinate system in which the restraints and the distributed force are applied.

After computing the FEM analysis, the user obtains a maximum value of the *Translational displacement vector.1*, according to Figure 5.24. Thus, the deformation of the part is 0.0262 mm, a relatively small value in accordance with the amplitude of the force, the shape and dimensions of the beam.

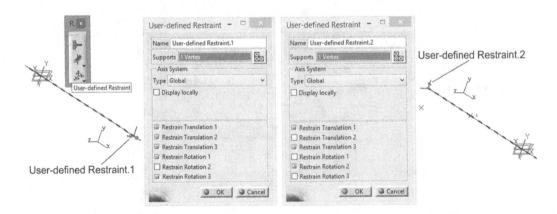

FIGURE 5.21 Inserting the two *User-defined Restraints*.

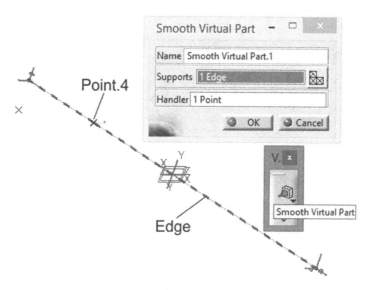

FIGURE 5.22 Defining a virtual part in *Point.4.*

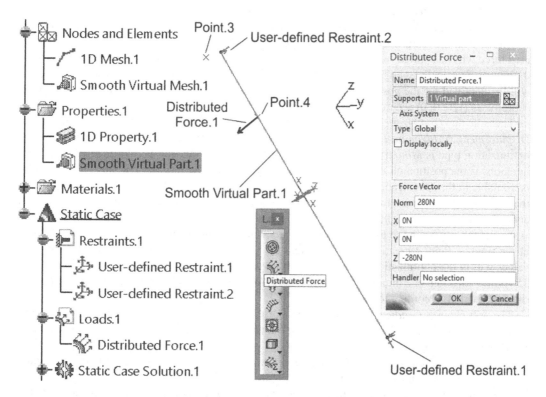

FIGURE 5.23 Applying the distributed force to the virtual part in *Point.4.*

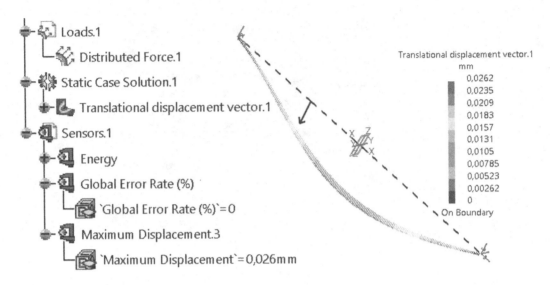

FIGURE 5.24   Maximum value of the *Translational displacement vector.1*.

The figure shows the position where the distributed force was applied (in *Point.4*), but also a new sensor created by the user, *Maximum Displacement,* having the previously presented value. This sensor will be used in an optimization iteration of the FEM analysis.

Assuming that the value of the deformation, 0.0262 mm, cannot be accepted in the structure in which the beam is integrated, the user must determine the position of *Point.4* where the distributed force will be applied (which keeps its value), so that the deformation is smaller, for example 0.025 mm. Practically, with the help of the *Product Engineering Optimizer* workbench, a new value will be identified for the position of *Point.4* on the *X* axis. Initially, the point was inserted on the coordinates (120, 0, 0) according to Figure 5.18.

Thus, in the *Optimization* dialog box (Figure 5.25), the *Finite Element Model.1\Maximum Displacement.3\Maximum Displacement* sensor is chosen in the *Optimized parameter* field, the optimization type is *Minimization,* and the parameter whose value can be modified in successive iterations is the position on the *X* axis of *Point.4* (by pressing the *Edit list* button). The parameter is available in the specification tree (Figure 5.26) next to the initial value of 120 mm. It appears in the list of available parameters (left column of the *Select the free parameters* selection box) and is added by the user to the right column (*Free parameters for optimization*).

By default, according to Figure 5.25, the *Value* column contains its initial value set by the user at the time of *Point.4* creation, then he enters the limits between which the parameter can take values within the optimization process. It is observed that the user entered the values 90 mm and 150 mm, with a step of 2 mm, between which *CATIA v5* will look for the optimal value to minimize the deformation, according to the selected sensor.

The user chooses *Gradient Algorithm With Constraint(s)* because in the *Constraints* tab he will write the condition limiting the deformation to the value of 0.025 mm. Other settings are visible in the *Problem* tab (Figure 5.25).

The constraint is entered in the *Optimization Constraints Editor* box by selecting the *Maximum Displacement* sensor in the specification tree (Figure 5.27).

The non-editable field at the bottom of the *Optimization Constraints Editor* box contains the value resulting from the first step of the FEM analysis: 0.026 mm, which means that the constraint is not met.

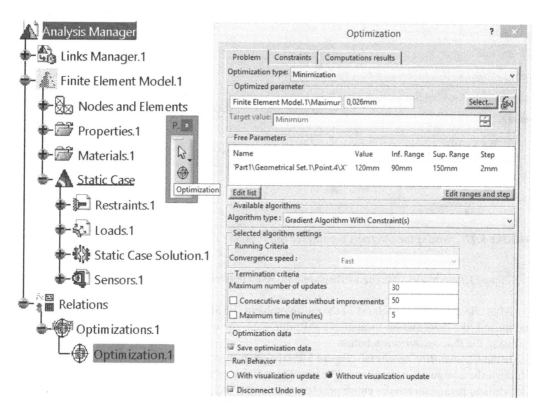

FIGURE 5.25 Establishing the parameter that needs to be optimized.

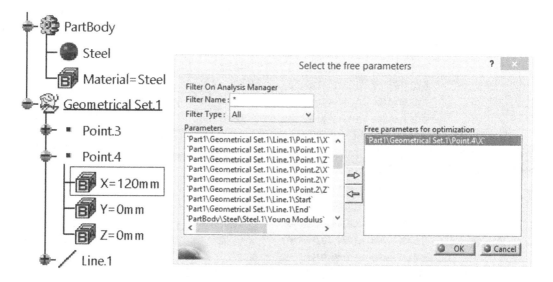

FIGURE 5.26 Identification of the parameter in the beam's parameters list.

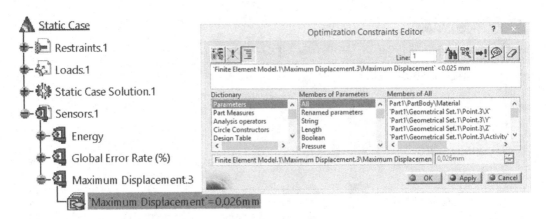

FIGURE 5.27   Writing the constraint limiting the deformation value.

`Finite Element Model.1\Maximum Displacement.3\Maximum Displacement` <0.025 mm

After checking the correctness of the entire problem (FEM analysis settings, the parameter to be optimized, the deformation limitation constraint, etc.), the user launches the optimization step by pressing the *Run optimization* button.

In the *Computation results* tab (Figure 5.28) the program displays the different solutions obtained through the computation iterations. The user chooses one of these solutions/results and applies the determined parameters to the entire problem (part, analysis).

As an example, the user chooses the solution number 12 from the list and presses the *Apply values to parameters* button. The geometry of the beam is updated, *Point.4* changes its position on the

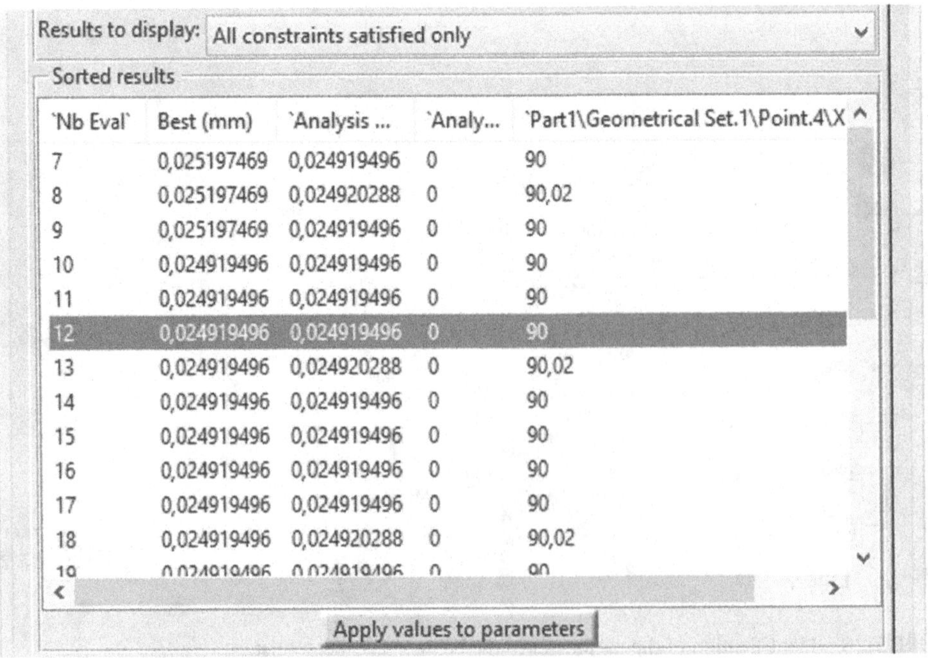

Results to display: | All constraints satisfied only

Sorted results

`Nb Eval`	Best (mm)	`Analysis ...	`Analy...	`Part1\Geometrical Set.1\Point.4\X
7	0,025197469	0,024919496	0	90
8	0,025197469	0,024920288	0	90,02
9	0,025197469	0,024919496	0	90
10	0,024919496	0,024919496	0	90
11	0,024919496	0,024919496	0	90
12	0,024919496	0,024919496	0	90
13	0,024919496	0,024920288	0	90,02
14	0,024919496	0,024919496	0	90
15	0,024919496	0,024919496	0	90
16	0,024919496	0,024919496	0	90
17	0,024919496	0,024919496	0	90
18	0,024919496	0,024920288	0	90,02
10	0,024919496	0,024919496	0	90

Apply values to parameters

FIGURE 5.28   Results obtained by optimization.

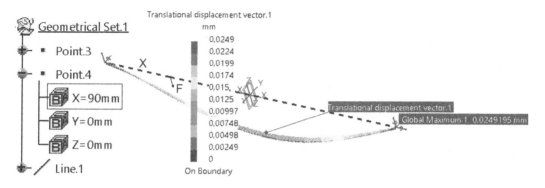

FIGURE 5.29 Updating the beam and the *Translational displacement vector.1* computation.

coordinate $X = 90$ mm, the FEM analysis results in a maximum deformation of 0.0249 mm. Figure 5.29 shows the results of applying the parameters and the location of the maximum deformation.

The user can continue the application by changing the input data of the optimization problem, as follows: delete the current optimization and start another one by proposing to minimize the value of the distributed force applied at the initial position *Point.4* (120, 0, 0) so that the value of the maximum deformation is not greater of 0.02 mm.

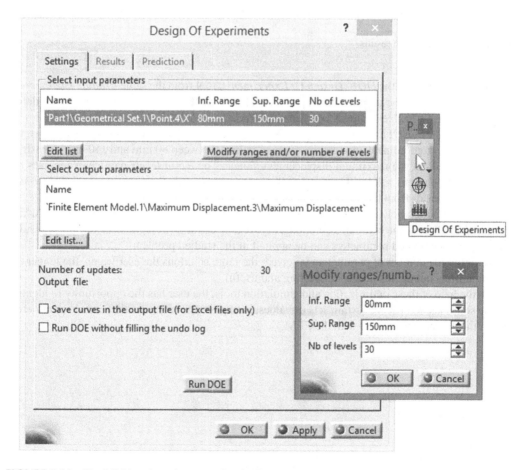

FIGURE 5.30 Establishing the value range for the X coordinate of *Point.4*.

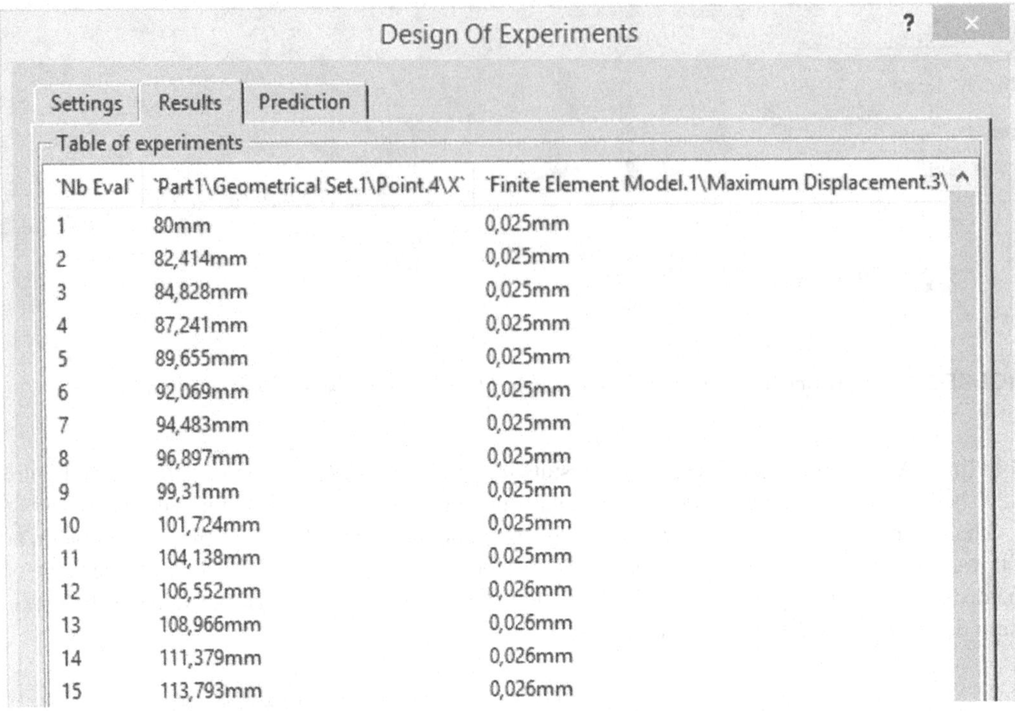

FIGURE 5.31 Simulation results.

Also, with the help of the *DOE (Design Of Experiments)* tool from the same *Product Engineering Optimizer* workbench, the user can easily simulate finding different values of the maximum deformation for a range of values of the *X* coordinate of *Point.4*.

Thus, in the *Design Of Experiments* dialog box (Figure 5.30), by clicking the *Edit list* button, the user chooses the *X* coordinate and sets the range of values (between 80 mm and 150 mm) for which he wants to find out the maximum displacement values. The *Nb of levels* parameter receives the value 30, so 30 solutions will be obtained. For the output parameters, the user chooses the *Maximum Displacement* sensor from the specification tree, according to Figure 5.30.

Running the simulation (pressing the *Run DOE* button) lets the user to save an *Excel* file in which the program will store the computed solutions. Figure 5.31 contains a fragment of the list of results. Obviously, the computed parameters can be applied in the studied problem.

The higher the number of parameters involved, the more solutions the user wants, the longer the simulation time, but the results are interesting and useful.

With the help of these optimization and simulation tools, the user has the opportunity to identify possible solutions, to observe certain sets of values, to explore the influence that certain parameters have on the geometry analysed with the FEM.

Annex

A.1 ADDITIONAL ONLINE RESOURCES: USER COMMUNITIES, FORUMS, VIDEO TUTORIALS

A.1.1 CATIA User Community

Dear *CATIA* lovers, welcome to your place to be! You are now part of the community (Figures A.1 and A.2) which shapes the world we live in.

You will discover a world of expertise, of achievements with the world leading Design & Engineering Brand: *CATIA*! Even more you will get the possibility to experience interactive services which will develop and strengthen your skills and expertise. Be social, and do not hesitate to interact with *CATIA* team and me, to grow together!

Welcome to a World of Creativity and Innovation – Welcome to the *CATIA* Cmmunity!

Olivier Sappin, *CATIA CEO Dassault Systèmes*

https://r1132100503382-eu1-3dswym.3dexperience.3ds.com/#community:6

 CATIA User Community

 Shape Healing app is very useful in...

By Dominique COZ

Some hint on healing Parts

FIGURE A.1 *CATIA* User Community.

DOI: 10.1201/9781003426813-6

FIGURE A.2   QR code to access the *CATIA* User Community, login is required.

A.1.2   *3DEXPERIENCE* Edu Hub

Discover *3DEXPERIENCE* Edu Learning Labs Network. These are learning labs created by teachers, champions of the *3DEXPERIENCE* platform (Figures A.3 and A.4) and convinced that project-based learning, that we rather prefer to call experience-based learning, is the best way for students to practice their skills and obtain new skills like project management, collaboration.

   https://r1132100503382-eu1-3dswym.3dexperience.3ds.com/#community:HNbVc2T1Qp2OBY9 BVYYngA

3DEXPERIENCE Edu | Hub
Public

3DEXPERIENCE Edu Hub: our Learning Labs network in Video!

By **Natacha BECARD**   2020-06-19

Discover our **3DEXPERIENCE Edu Learning Labs Network**, who are learning labs created by teachers, champions of the 3DEXPERIENCE platform and convinced that project based learning, that we rather prefer to call experience based learning, is the best way for student to practice their skills and get new skills like project management, collaboration.

👍 20   💬 12

FIGURE A.3   *3DEXPERIENCE* Edu Hub.

FIGURE A.4   QR code to access the *3DEXPERIENCE* Edu Hub, login is required.

A.1.3   LEARN ONLINE

From quick videos to the very complete user training for industry, hundreds of self-paced educational materials (Figures A.5 and A.6) are available for students and educators whatever their preferred learning style.

https://edu.3ds.com/en/learn-online

A.1.4   DOCUMENTATION

Discover all the documentation you need to install, get started with, and effectively use your Dassault Systèmes products.

https://www.3ds.com/support/documentation

FIGURE A.5   *3DEXPERIENCE* Edu Learn Online.

FIGURE A.6 QR code to access the *3DEXPERIENCE* Edu Learn Online.

A.1.5 CERTIFICATION FOR STUDENTS & EDUCATORS

Get certified right now and leverage your expertise on Dassault Systèmes solutions with the 3DS Certification Program (Figures A.7 and A.8). Certification brings credibility to your CV, leading to significant opportunities for career growth. By getting certified you demonstrate your expertise and prove your capability to differentiate yourself from the others in today's increasingly competitive job market.

Figures based on a survey we conducted with 7000 3DS Certified Engineers:

- Increased employability and better job opportunities: 41% said they found a better job (73% for students).
- Better paid jobs and salary increase: 10% said they received a salary increase.
- Recognition within the company and among peers: 47% said they received more recognition in their company.

https://edu.3ds.com/en/be-recognized/academic-certification-program

CERTIFICATION FOR STUDENTS & EDUCATORS

Get certified right now and leverage your expertise on Dassault Systèmes solutions with the 3DS Certification Program.

FIGURE A.7 Certification for students and educators.

FIGURE A.8   QR code to access the certification for students and educators.

A.1.6   ENG-TIPS.com

Professional forum and technical support for engineers for Dassault: *CATIA* products. Includes problem solving collaboration tools (Figures A.9 and A.10).
   https://www.eng-tips.com/threadminder.cfm?pid=560

FIGURE A.9   Professional forum and technical support for engineers.

FIGURE A.10   QR code to access the Professional forum and technical support for engineers.

A.1.7   COE

Community of Experts (COE) of Dassault Systèmes solutions help users and their company leverage their Dassault Systèmes solutions through education, training, networking opportunities, product influence and best practices available exclusively to COE members.

The primary objectives of COE (Figures A.11 and A.12) are to provide a forum for the interchange of knowledge, experiences, and technical information relating to the application of the Dassault Systèmes family of solutions and the environment in which they operate, and to communicate with Dassault Systèmes regarding the current and future capabilities and use of these products. http://www.coe.org

FIGURE A.11   Community of Experts of Dassault Systèmes solutions.

FIGURE A.12   QR code to access the Community of Experts of Dassault Systèmes solutions.

A.1.8   Complete **FREE** Video Tutorials List

This continuing growing list (Figures A.13 and A.14) contains various free video tutorials for the following:

- Sketches
- Solid parts
- Surfaces parts
- Assemblies
- 2D drawings
- Parameterization and optimization
- Tips and tricks about using the *CATIA v5* program

https://qrgo.page.link/kLqJm (Alternate link: https://www.youtube.com/playlist?list=PLdQgkscls4 OYxYeoC66cEUZRyrVKGWIfZ)

 How to create a mechanical part using Generative Shape Design and CATIA Part Design 99

workbenchstuff

 How to create a Cover model using Generative Shape Design and CATIA Part Design 98

workbenchstuff

 How to create a Knot model using Generative Shape Design and CATIA Part Design 97

workbenchstuff

 How to create a mechanical part using Generative Shape Design and CATIA Part Design 96

workbenchstuff

 How to create a mechanical part using Generative Shape Design and CATIA Part Design 95

workbenchstuff

 How to create a mechanical part using Generative Shape Design and CATIA Part Design 94

workbenchstuff

FIGURE A.13   Video tutorials list.

FIGURE A.14   QR code to access the Video tutorials list.

A.1.9   QUESTIONS TAGGED [CATIA]

Macros, scripting and programming in *CATIA v5* (Figure A.15).
  https://stackoverflow.com/questions/tagged/catia

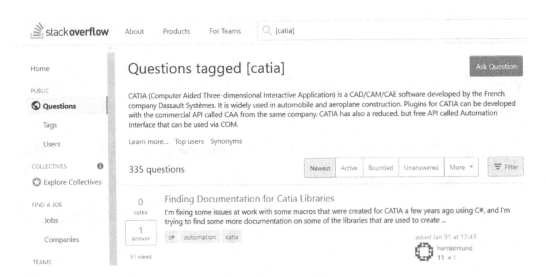

FIGURE A.15   Macros, scripting and programming in *CATIA v5*.

Bibliography

Ahmad, F., Kumar, A., Kanwar, K., Patil, P., (2016) CFX, Static structural analysis of tractor exhaust system based on FEA. CAD/CAM, Robotics and factories of the future, lecture notes in mechanical engineering, doi: 10.1007/978-81-322-2740-3_17, pp. 159–172.

Băilă, D., Doicin, C., Cotruţ, C., Ulmeanu, M., Ghionea, I., Tarbă, C., (2016) Sintering the beaks of the elevator manufactured by direct metal laser sintering (DMLS) process from Co-Cr alloy, Metalurgija, UDC – UDK 669.25.26:615.46:621.793:518.1:620.193=111, ISSN 0543-5846, eISSN: 1334-2576, vol. 55(4), pp. 663–666.

Băilă D., Ghionea I., Mocioiu O., Ćuković S., Ulmeanu M., Tarbă C., Lazăr L., (2016) Design of handle elevators and ATR spectrum of material manufactured by stereolithography, 13th IFIP WG 5.1 International Conference on Product Lifecycle Management PLM16 – In Book Product Lifecycle Management for Digital Transformation of Industries, Chapter 28, pp. 309–318, doi: 10.1007/978-3-319-54660-5_28, Springer International Publishing, IFIP Advances in Information and Communication Technology, vol. 492, ISBN 978-3-319-54659-9, ISSN 1868-4238, University of South Carolina, Columbia.

Barlier, C., Poulet, B., (1999) Mémotech. Génie mécanique, productique mécanique. Deuxième édition. Editions Casteilla, Paris, ISBN 2-7135-2063-0.

Bathe, K. J., (1976) Numerical methods in finite element analysis, Prentice-Hall.

Berta, G., Casoli, P., Canova, M., Gambarotta, A., (2002) A control-oriented model for the simulation of turbocharged Diesel engines. 2nd International Workshop on Modeling, Emissions and Control of Automotive Engines-MECA02, Salerno.

Bird, J., Ross, C., (2019) Mechanical engineering principles (4th ed.), ISBN 9780367253264, Routledge, https://doi.org/10.1201/9780429287206, 388 p., London and New York.

Bondrea, I., Frăţilă, M., (2005) Proiectarea asistată de calculator utilizând CATIA v5 (Computer aided design using CATIA v5). Editura Alma Mater, p. 247, Sibiu.

CATIA V5, (2015) Official documentation. Dassault Systèmes.

Chaskalovic, J., (2008) Finite elements methods for engineering sciences, ISBN: 978-3-540-76343-7, Springer Verlag.

Chiara, F., Canova, M., & Wang, Y., (2011) An exhaust manifold pressure estimator for a two-stage turbocharged Diesel engine. American Control Conference, vol. 11, ISBN 978-1-4577-0079-8, doi: 10.1109/ACC.2011.5991138, pp. 1549–1554.

Chivu, O., Babiş, C., Tarbă, C., (2012) The fatigue life expectancy in case of welded and non welded structures. Academic Journal of Manufacturing Engineering, vol. 10, issue 1, pp. 30–35.

Constantinescu, N. I., Sorohan, Şt., Pastramă, Şt., (2006) The practice of finite element modeling and analysis. Editura Printech, ISBN 978-973-718-511-2, p. 340, Bucureşti.

Dima, G., Velea, M., (2016) CATIA v5 Proiectare de produs (CATIA v5 Product design). Editura Universităţii Transilvania din Braşov, ISBN 978-606-19-0812-7, p. 302, Braşov.

Dubbel, H., (1998) Manualul inginerului mecanic. Fundamente (Mechanical engineer's manual. Fundamentals). Editura Tehnică, ISBN 973-31-1271-2, Bucureşti.

Enache. V., Doicin, C. V., Ulmeanu, M. E., (2022) CAD Optimization of an ankle foot orthosis using lattice structures. Macromolecular Symposia, vol. 404, Special Issue: Conference on Design and Technologies for Polymeric and Composites Products – POLCOM 2021, https://doi.org/10.1002/masy.202100393, ISSN 1022-1360, Bucharest.

Finite element method, (2022) https://en.wikipedia.org/wiki/Finite_element_method

Fish, J., Belytschko, T., (2007) A first course in finite elements. Wiley, ISBN 978-0470035801, p. 336.

Gavrilă, C., Lateş, M., (2019) Device for the study of the friction coefficient in the bolt-chain link joint. Modern Technologies in Industrial Engineering VII, (ModTech2019), IOP Conference Series: Materials Science and Engineering, vol. 591, IOP Publishing, ISSN 1757-8981, doi: 10.1088/1757-899X/591/1/012091.

Gavrilă, C., Lateş, M., (2021) 3D Modelling and FEM analysis on metal coin edge punching error. The Annual Session Of Scientific Papers - IMT Oradea, IOP Conference Series: Materials Science and Engineering, vol. 1169, IOP Publishing, ISSN 1757-8981, doi: 10.1088/1757-899X/1169/1/012006.

Ghionea, I., (2007a) A practical approach in the finite element method study of a mechanical part. Scientific Bulletin, Series C, volume XXI, Fascicle: Mechanics, Tribology, Machine Manufacturing Technology, 7th International Multidisciplinary Conference, ISBN 1224-3264, Baia Mare.

Ghionea, I., (2007b) Considerations about the methodology and results for the finite element analysis of a mechanical assembly. Proceedings of the 16th International Conference on Manufacturing Systems – ICMaS, Politehnica University of Bucharest, Editura Academiei Române, ISSN 1842-3183, Bucureşti.

Ghionea, I., (2007c) Proiectare asistată în CATIA v5. Elemente teoretice şi aplicaţii (Assisted design in CATIA v5. Theoretical elements and applications). Editura Bren, ISBN 978-973-648-654-8, doi: 10.13140/RG.2.1.1077.0642, p. 462, Bucureşti.

Ghionea, I., (2008) Study and methodology on the determination of the part's maximum stress and loading force using the Finite Element Analysis. Recent advances in visualization, imaging and simulation, ISSN 1792-4308, ISBN 978-960-474-022-2, WOS 000264044900015, Bucureşti.

Ghionea, I., (2009) CATIA v5. Aplicaţii în inginerie mecanică (CATIA v5. Applications in mechanical engineering). Editura Bren, ISBN 978-973-648-843-6, doi: 10.13140/RG.2.1.2387.7848, p. 258, Bucureşti.

Ghionea, I., (2022) Applied methodology for designing and calculating a family of spur gear pumps, MDPI Energies, vol. 15, issue 12: 4266, https://doi.org/10.3390/en15124266, WOS 000817495300001, eISSN 1996-1073, Basel.

Ghionea, I., Ghionea, A., Tănase, I., (2009) Application of CAM-FEM techniques in the establishement of milling conditions of the parts with thin walls surfaces. 4th International Conference Optimization of the Robots and Manipulators, OPTIROB, ISSN 2066-3854, pp. 27–32, Constanţa.

Ghionea, I., Munteanu, G., Beznilă, H., (2008) Von Mises stress evaluation for a mechanical part using the CATIA finite element method. Annals of DAAAM for 2008, Proceedings of the 19th International DAAAM Symposium, ISBN 978-3-901509-68-1, DAAAM International, pp. 549–550, Vienna.

Ghionea, I., Tarbă, C., Ćuković, S., (2021) CATIA v5. Aplicaţii de proiectare parametrică şi programare (CATIA v5. Parametric design and programming applications), Editura Printech, ISBN 978-606-23-1264-0, doi: 10.5281/zenodo.7016799, p. 532, Bucureşti.

Ghionea, I., Tarbă, C., Tiriplică, P., (2012) Simulation of the working conditions for a gear pump using finite element analysis method. Proceedings of the IMC 2012 International Multidisciplinary Conference, 10th edition, Bessenyei Publishing House, ISBN 978-615-5097-18-8, North University of Baia Mare – University College of Nyiregyhaza.

Ghionea, I., Vatamanu, O., Cristescu, A.M., David, M., Stancu, I.C., Butnarasu, C., Cristache, C.M., (2023) A finite element analysis of a tooth-supported 3D-printed surgical guide without metallic sleeves for dental implant insertion. Applied Sciences, vol. 13, issue 17: 9975, https://doi.org/10.3390/app13179975, WOS 001061015300001, eISSN 2076-3417, Basel.

Hughes, T., (1987) The finite element method: linear static and dynamic finite element analysis, Prentice-Hall.

Koh, J., (2012) CATIA v5 FEA Release 21. A step by step guide, ONSIA Inc., ISBN 978-1470172824, p. 386.

Liu, W., Yang, Q., Mohammadizadeh, S., Su, X., (2014) An efficient augmented finite element method for arbitrary cracking and crack interaction in solids. International Journal for Numerical Methods in Engineering, doi: 10.1002/nme.4697, vol. 99, issue 6, pp. 438–468.

Madsen, D. A., Madsen, D. P., (2012) Engineering drawing & design. 5th edition. Delmar Cengage Learning, ISBN 978-1-111-30957-2, New York, NY.

Modal Analysis – Ansys Innovation Course, (2020) Intro to Modal Analysis – Lesson 1. https://www.youtube.com/watch?v=gDtUpZm_E_Y

Ottosen, N., Petersson, H., (1992) Introduction to the finite element method, Prentice Hall, ISBN 978-0134738772, p. 432.

Popescu, D., Zapciu, A., Tarbă, C., Lăptoiu, D., (2020) Fast production of customized three-dimensional-printed hand splints. Rapid Prototyping Journal, vol. 26, issue 1, ISSN 1355-2546, https://doi.org/10.1108/RPJ-01-2019-0009, WOS 000506065900014, Basel.

Qianfan, X., (2013) Durability and reliability in Diesel engine system design. Diesel engine system design, Woodhead Publishing, ISBN 9781845697150, doi: 10.1533/9780857090836.1.113, pp. 113–202.

Ragab, S., Fayed, H., (2018) Introduction to finite element analysis for engineers, CRC Press, ISBN 978-1-1380-3017-6, p. 550, Boca Raton.

Roters, F., Eisenlohr, P., Bieler, T.R., Raabe, D., (2010) Crystal plasticity finite element methods: in materials science and engineering. Wiley, ISBN 9783527324477, doi:10.1002/9783527631483.

Shukla, A., (2019) Practical fracture mechanics in design, 2nd edition, CRC Press, ISBN 978-0824758851, p. 548, New York, NY.

SR EN ISO 4957, (2018) Oţeluri pentru scule (Steels for tools).

Süli, E., Mayers, D., (2003) An introduction to numerical analysis, Cambridge University Press, ISBN 0521810264, p. 630, Cambridge.

Song, C., Wolf, J.P., (1997) The scaled boundary finite-element method – alias consistent infinitesimal finite-element cell method – for elastodynamics. Computer Methods in Applied Mechanics and Engineering, ISSN 0045-7825, doi: 10.1016/S0045-7825(97)00021-2, vol. 147, issues 3–4, pp. 329–355

Thomas, D.J., (2018) Analyzing the failure of welded steel components in construction systems. Journal of Failure Analysis and Prevention, vol. 18, ISSN 1547-7029, pp. 304–314, Springer, https://doi.org/10.1007/s11668-018-0392-x.

Woyand, H.B., (2009) FEM mit CATIA V5: Berechnungen mit der Finite-Elemente-Methode (FEM with CATIA V5: Calculations using the finite element method), Schlembach Fachverlag, ISBN 978-3935340649, p. 260.

Zamani, N., (2017) Finite element essentials in 3DEXPERIENCE 2017x using SIMULIA and CATIA applications. SDC Publications, ISBN 978-1630571009, p. 402.

Zienkiewicz, O. C., Taylor, R. L., Zhu, J. Z., (2013) The finite element method: Its basis and fundamentals. 7th edition, Butterworth-Heinemann, Elsevier, ISBN 978-1856176330, p. 756.

Zohdi, T. I., (2017) A finite element primer for beginners. 2nd edition, Springer, ISBN 978-3-319-70427-2, https://doi.org/10.1007/978-3-319-70428-9, p. 135.

https://enterfea.com/what-are-nodes-and-elements-in-finite-element-analysis (Accessed: November 25, 2023).

https://www.nylaplas.com/plastics/pds/Ertalon_66_SA_PDS_E_17102013.pdf (Accessed: November 25, 2023).

https://onscale.com/blog/strain-energy-density-what-why-and-how (Accessed: November 25, 2023).

https://patents.google.com/patent/EP1978121A1/en (Accessed: November 25, 2023).

http://what-when-how.com/the-finite-element-method/fem-for-two-dimensional-solids-finite-element-method-part-1 (Accessed: November 25, 2023).

https://www.crystalinstruments.com/basics-of-modal-testing-and-analysis (Accessed: November 25, 2023).

https://www.simscale.com/blog/what-is-modal-analysis (Accessed: November 25, 2023).

https://skyciv.com/docs/tutorials/beam-tutorials/types-of-beams (Accessed: November 25, 2023).

https://www.materialgrades.com/aisi-1020-steel-low-tensile-carbon-steel-1401.html (Accessed: November 25, 2023).

https://en.wikipedia.org/wiki/Exhaust_manifold (Accessed: November 25, 2023).

https://rolledmetalproducts.com/stainless-steel-type-321 (Accessed:November 25, 2023).

https://www.makeitfrom.com/material-properties/AISI-321-S32100-Stainless-Steel (Accessed: November 25, 2023).

https://www.youtube.com/watch?v=ZKojtbcxBQE (Accessed: November 25, 2023).

https://en.wikipedia.org/wiki/Tuning_fork (Accessed: November 25, 2023).

https://encyclopediaofmath.org/wiki/Galerkin_method (Accessed: November 25, 2023).

https://www.sciencedirect.com/topics/mathematics/runge-kutta-method (Accessed: November 25, 2023).

Index